# VOICE MACHINES

# VOICE MACHINES

*The Castrato, the Cat Piano, and Other Strange Sounds*

BONNIE GORDON

The University of Chicago Press CHICAGO AND LONDON

The University of Chicago Press, Chicago 60637
The University of Chicago Press, Ltd., London

Published 2023
Printed in the United States of America

32 31 30 29 28 27 26 25 24 23    1 2 3 4 5

ISBN-13: 978-0-226-82514-4 (cloth)
ISBN-13: 978-0-226-82515-1 (e-book)
DOI: https://doi.org/10.7208/chicago/9780226825151.001.0001

Library of Congress Cataloging-in-Publication Data

Names: Gordon, Bonnie, 1968– author.
Title: Voice machines : the castrato, the cat piano, and other strange sounds / Bonnie Gordon.
Description: Chicago : The University of Chicago Press, 2023. | Includes bibliographical references and index.
Identifiers: LCCN 2022039099 | ISBN 9780226825144 (cloth) | ISBN 9780226825151 (ebook)
Subjects: LCSH: Castrati—Italy—History. | Castrati—History. | Singing—Italy—History—16th century. | Singing—Italy—History—17th century. | Singing—Italy—History—18th century. | Music and technology—Italy—History—17th century. | Rome (Italy)—Civilization—17th century.
Classification: LCC ML1433 .G67 2023 | DDC 782.8/6—dc23/eng/20220824
LC record available at https://lccn.loc.gov/2022039099

♾ This paper meets the requirements of ANSI/NISO Z39.48-1992 (Permanence of Paper).

# CONTENTS

# ILLUSTRATIONS

FIGURES

## MUSICAL EXAMPLES

# NOTE TO THE READER

I have used published transcriptions and translations whenever possible. I modified some either to make a precise point or because the translations themselves reflect the values of another time and place. Sometimes the gaps between languages and places are the point. Unless otherwise noted, translations are my own. In cases where the originals are hard to locate even during a massive digitization of sources, I have provided original languages.

Early modern authors were and are often known by many different names, often including a Latin and vernacular name with multiple spellings as well as an anglicized name. The poet Marino is Giambattista Marino and Giovan Battista Marini and many things in between, depending on which source you cite (I settled on Giovanni Battista Marino). The satirist Gian Vittorio Rossi also goes by the Latin moniker Iani Nicii Erythraei. For ease of reading, I have tried to use the names that are most recognizable to modern readers. Similarly, early modern books have very long titles. In cases where the entire title is not relevant, I use the conventional shorter title. Musical examples are likewise based on modern editions when they are available.

# PROLOGUE

Seventeen years ago I went to Rome with two-and-half-year-old twins. I thought I was starting a book about castrati but my miniature research assistants changed my mind. They loved the pomp and circumstance of the city and a tiny bit of pizza or gelato kept them happy in a double stroller for hours. I entertained them by chasing modern Roman spectacles, looking for *il Papa* and his entourage, visiting the Swiss guards, watching fireworks over Castel Sant'Angelo, and splashing in fountains. My mom, who was the babysitter, regularly wheeled them over the Tiber to meet me at the Vatican library, where they encountered the popemobile, feast day parades, and visiting dignitaries. With their eyes only three feet off the ground they saw things adults didn't. They fell in love with what they called "spitting statues"; an apparent taste for Bernini that led to traipsing all over Rome in search of animal fountains. It took them two months to find the courage to stick their tiny hands in the *Bocca della Verità* on Via Giulia, the first-century stone-mouth lie detector. Supposedly, if you stick your hand in its mouth and tell a lie, it bites your hand. My post library ritual became taking them to sites featured in this book with a stop at Pasquino, the talking statue who lives near the entrance to the Museo di Roma and Piazza Navona (Fig. 0.1). The sound of jackhammers crashed through Bernini's famous *Fontana dei quattro fiumi*, artists made caricatures of tourists, people dressed as iconic statues stood alarmingly still: some really looked more like stone than humans. The *gemelli* called them statues who move. Every day at five, a guy showed up with music blasted through a decrepit boom box full of reverb and did a kinky finger puppet show. One particularly bizarre afternoon the Fresno State marching band pushed him out.

On these walks, I told stories about the characters in this book, the puking Frenchman, Pasqualini, Monteverdi, and others. Piazza Navona proved especially fertile ground; adding dragons, magic, and soldiers let me review what I had read that day in the library. I knew that four hundred years ago we could have witnessed not a puppet show but a castrato singing on a boat drawn by twenty oarsmen powered by invisible wheels and water hurling through man-made rocks and out of the mouths of stone animals. But if I found it tempting even for a minute to imagine I could feel the sensation of

FIG. 0.1 Research assistants Rebecca and Jonathan Lerdau in Rome, fall 2005. Photograph by author.

prowling the streets of seventeenth-century Rome, the dying wheels of the spiffy double stroller reminded me otherwise. It worked delightfully in American cities and sometimes hardly moved on the cobblestones. As the toddlers watched workmen at the Spanish Steps using chisels to knock the rocks into shape, I cursed their quaintness.

It was not efficient to work on a book with little twins in tow, but through their eyes and ears the castrato became a point of departure and not the end game. When I returned a few years later I had two five-year-olds who could walk and an eighteen-month-old whose fifth word was *macchina*. The baby couldn't get enough of Roman vehicles. And thanks to the organic cotton backpack, he had a prime seat for exploring Roman gardens, often squealing with delight at the sometimes hilarious waterworks (Fig. 0.2). The castrato was almost a tour guide for early modern soundscapes and for a kind of sensory history. And by the time I took my last research trip to Italy in the winter of 2019, all the babies were teenagers who could read Latin inscriptions, and castrati were an inscrutable ghostly presence and an invitation to think about history as errant. We had no idea that the world would be turned upside down and that the ensuing pandemic would make such a mess of time that we could only live syncopated lives. And we had no idea that we were

FIG. 0.2 Research Assistant Eli Lerdau in backpack, spring 2008. Photograph by Joyce Gordon.

just months from an interregnum that felt alarmingly like Barberini Rome; plague, sealed off churches, falling statues, violent insurrections.

The castrato is a critical provocation for asking several questions about the interrelated histories of music, technology, sound, the limits of the human body, and what counts as human. I position castrati as sonic figures in an era that is so often understood as pretechnological. In this book I insist that boundaries between humans and machines have always been permeable in historically specific ways. These boundaries, at the turn of the seventeenth century, manifested themselves within a ripe climate for manufactured singers that has something to teach us today about both the dangers of theoretical presentism and the productively defamiliarizing possibilities of reading the castrato as an early hallmark of our current experience of (post)modernity. The castrato highlights the porous relationship of voices and instruments and the inherent materiality of sound. The connection between castrati and machines suggests that the story of mechanism and materiality, usually associated with the eighteenth century, starts much earlier and that the castrato is a central character in that story.

I approach the collection of frequencies, timbres, and sounds that constitute the castrato as an assemblage that could be heard through greater vocal power, an arresting rhetorical palette, extra flexibility, and an effective high range. This Deleuzian assemblage of material, technological and organic parts is defined not by a unified whole but by relations and aggregations. The castrato indexes historical shifts from ancient and medieval times to early modern and modern. This means that the castrato who was a product of the early modern, medieval, and ancient worlds—and exists now in the historical imaginary between fiction and reality—is a ghostly reminder that the roots of posthumanism lie within humanism itself. So, the rethinking of what it means to be human that has emerged from twentieth- and twenty-first-century inventions and philosophical interventions is a reharmonization of the past, not a radical departure. Castrati open up the possibility of premodern, novel reconfigurations of the human.

Connecting bizarre but resonant snippets and stories, this is a jagged history that moves backward and forward and makes a mess of concepts like technology, machine, and instrument. It's a story to get lost in and one with refrains that sound across time and space. And it moves to the beat of queer time, a temporal location that lives in a space where past, present, and future are often out of joint.[1] The castrato calls out the relationship between technologies of inscription and memory and issues an invitation to listen for the emotional debris of the past and the uncanny timbre of disrupted time, and some judicious anachronism may in fact be quite productive.

Even with a vested interest in embracing the circuitous imaginary, the book bears witness to the castrato phenomenon that stretched from the late sixteenth century, when castrati first appeared in Italian courts and churches, through the eighteenth century, when they occupied a celebrity status on the operatic stage, greatly expanding musical vocabularies as part of their efforts to embody poetic texts. The castrato persisted through the age of Mozart and Rousseau. Countless participants in the Grand Tour—as well as better-known and more modern writers like Giacomo Casanova, William Wordsworth, and Stendhal—recorded their impressions of castrato voices.[2] In the language of standard Western music history, the book is framed on one end by Monteverdi and on the other by Handel and thus juxtaposes already well-studied composers and works with ephemeral spectacles and festivals. And it covers the era in which female singers moved from the secret world of private concerts and cloisters to the public sphere, in which female creativity and women's voices also radically transformed musical practice: the singing ladies of Ferrara in the 1580s through the divas of the eighteenth century.

The timeline of the castrato phenomenon centers on the period that Anglophone historians think of as the early modern and asks how the early modern period, uniquely, both *produced* a caste of technologically altered male singers and *reinflected* the nature/art, techne/nature differences that it inherited from ancient and medieval times. And in doing so, it attends to the ways that settler colonialism, emergent racialized worldviews, the printing press, gunpowder, and the telescope participated in making castrati. The castrato was continuous and contiguous with a broad range of wonders including sonification instruments like pipe organs, hydraulic organs, ear trumpets, and gut strings.[3] Castrati were in other words less anomalous and less opaque than usually thought, but also more amazing for having been part of a realm of wondrous objects and phenomena.

Geographically, the book moves in and out of baroque Rome with the castrato as a poster boy for seventeenth-century Rome's theatrical, musical, and mechanical culture. In this "theater of the world," castrati participated in almost every aspect of sacred and secular ritual. Rome was the capital of a powerful state, a hub for daring and lucrative business ventures, and a site of numerous magnificent courts. It was a focal point for various early modern global networks: people, things, ideas that trafficked and were trafficked in world-changing ways. As headquarters of the Catholic drive to regain religious and cultural control over Europe, it was a point of departure for imperialist missionaries and a focal point for information captured by those missionaries. Even as Rome's intellectual world reached toward baroque ingenuity, the capital was still dominated by Renaissance Neoplatonism and the classical thought generally associated with the previous century.

Lived sonic borderlands, castrati sang on the cusp of the West and the rest, the Enlightenment and the unenlightened. In the Western European and Anglo-American imaginary, the castrato, who migrated north and west from southern Italy, was always the sound of somewhere else. Over the course of the castrato's ascendance, a global sea change shifted the access of power from the Mediterranean to the Atlantic. And by the middle of the eighteenth century, when the virtuosic castrato dominated the stage, southern Italy was an exotic Grand Tour destination residing perilously on the outskirts of Europe, a modern Enlightenment incarnation of a premodern gateway to the torrid zone.[4]

## INTERVENTIONS

Put perhaps too simplistically, I posit castrati as early modern cyborgs. In Donna Haraway's now much-quoted and critiqued words, "A cyborg is a cybernetic organism, a hybrid of machine and organism, a creature of social reality as well as a creature of fiction."[5] Cyborgs these days connote creatures like Anne McCaffrey's Helva, a human brain that inhabits the body of a spaceship in *The Ship Who Sang*. Born severely disabled, the brain gets the complex machinery of a spaceship as a shell. Haraway writes of Helva,

> Why should our bodies end at the skin, or include at best other beings encapsulated by skin? From the seventeenth century till now, machines could be animated—given ghostly souls to make them speak or move or to account for their orderly development and mental capacities. Or organisms could be mechanized—reduced to body understood as resource of mind. These machine/organism relationships are obsolete, unnecessary.[6]

Despite having read and taught Haraway's text multiple times, I didn't think carefully about the prospect of a singing spaceship until I read McCaffrey's story. In it, Helva learns to sing even without having a mouth to open. She learned by experimentation that "by relaxing the throat muscles and expanding the oral cavity well into the frontal sinuses, she could direct the vowel sounds into the most felicitous position for proper reproduction through her throat microphone."[7] After listening to her sing, one of the characters, Jennan, explains that "All [Robert] Tanner asked for was one roaring good lead tenor . . . and our sweet mistress supplied us an entire repertory company. The boy who gets this ship will go far, far, far."[8]

Jennan's comments about Helva, the ship who sang, and her mysteriously delicious voice sound like some extravagant accounts of seventeenth-century castrati. And like Helva, the castrato could sing male and female roles. We can no more hear a castrato than we can hear Helva. There are no living castrati, and the only existing recordings made by Alessandro Moreschi between 1902 and 1904 don't recover the sound or the sensation. The so-called last castrato comes out of the speakers through acoustic recording, a technique that could capture only part of the voice's resonance. That the voice exists only through texts and through recordings of such poor quality is part of the scholarly problem of hearing the castrato voice.[9]

The counterpoint between the castrato and Helva the singing ship introduces the four major interventions in this book, interventions that are unified in performing a sound-centered historiography and in asking questions about what it means to be human. They dig into the ever-tense relationship between nature and techne, not to correct the separation brought on by the nineteenth-century fantasy of transcendence but to show the ways that attention to bodies inevitably leads to awareness of bodies' literal obstructedness, which in turn leads to a renunciation of the "natural" as a truth. Bodies then and now stand in for material realities and not transcendent abstractions.

First, I challenge the presentism of current understandings of technology by inviting readers to listen to pre-electric technologies grounded in hydraulics, pumping fluids, and enhanced physicality. While modern, predominantly electronic machines tend to be understood as a product of technology, their forebearers emerged from the ancient notion of *techne*, a concept that involves the making of something and the thing that is made.

In response to this intervention, I have been cautioned by helpful interlocutors against positing the castrati as robots or machines. Critics remind me that castrati were men with lives, loves, selves. For sure, castrati and machines did not register as identical; everyone knew the difference between a mechanical singing bird and an altered man singing an aria. Likewise, we in the twenty-first century can look at Arnold Schwarzenegger and distinguish between the Terminator and the former governor of California. But the distinction between constructed singers and machines was by no means absolute in the seventeenth century. The castrato, in other words, highlights the differences between early modern and contemporary understandings of the relationships between humans and machines and figures the voice as a sounding instrument. The assumption that matter is feelable and seeable and that machines did not have hearts and souls reflects a post-Marxist fascination with the commodity that tends to equate the material with the physical.[10] But this in effect inverts the traditional Aristotelian opposition between form and matter in which form is actuality and matter is potentiality.

Castrati functioned as instruments in the maintenance of narratives that used geography to inscribe temperamental differences that, in turn, justified the subjugation and enslavement of so-called southern peoples. Castrati were human beings turned into voice machines whose sounds were consumed. This highlights the simultaneity of the castrato practice with the African slave trade and settler colonialism, both of which violently denied humanity to living humans. As Sylvia Wynter has argued, when Columbus landed in what he called an *otro mundo* (other world), he played a leading role in a system of knowledge that put man rather than God at the helm and that marked women, Black people, Brown people, children, disabled people, and queer people as inferior.[11] Man—as in white, able-bodied, Christian men—meant "us," and "them" was everyone else. Enrique Dussell has emphasized that the birth of modernity in 1492 brought "with it the myth of a special kind of sacrificial violence which eventually eclipsed whatever was non-European."[12]

The second intervention deliberately pushes against the often disembodied, presentist, and antihermeneutic nature of today's sound studies with its extreme focus on new sounds like ring tones and MP3s. Sound in this book is not just acoustical resonance, much less is it just musical or vocal. Rather it constitutes an interface between musical and nonmusical, human and animal, human and mechanical, natural and technological. Although this book emphasizes technology and materiality, it resists what Carolyn Abbate has called the material turn: the fascination with technological mediation that tends to silence human bodies and sense.[13]

Sound has a history. In their introduction to *The Oxford Handbook of Sound Studies*, Trevor Pinch and Karin Bijsterveld claim that in the modern age, "sound is no longer just sound; it has become technologically produced and mediated sound."[14] The point is significant on one level but misconceived on another, for sound has never been just sound. The phonograph hardly started the process of mediation; speaking statues and self-powered hydraulic organs existed as far back as the ancient world. Composers, meanwhile, have capitalized on echo effects, a natural sonic reproduction, for centuries. And music still matters, even as the focus on nonmusical sound by scholars like Jonathan Sterne and Emily Thompson pushes music scholars to hear outside the traditional aural boundaries of our discipline. In this my work follows and builds on Rebecca Cypess, who also addresses a paradox of instrumentality at the turn of the seventeenth century and who hears Italian instrumental practice as an artisanal form of knowledge making.[15]

The third intervention emphasizes the materiality of the human voice and song. The almost complete lack of tangible material evidence of the castrato voice, as it was thus artificially constructed, paradoxically highlights that voice's materiality. Because the castrato voice was so explicitly made by surgeons, teachers, and the singers themselves, it foregrounds the technical practices and materials that animate song. This particular voice highlights the ability of technology of all kinds, including vocal training, to separate the voice from the presence of an essential speaker, and it debunks the Western premise of the human voice as necessarily invested with unmediated presence and agency. The castrato's now totally absent sounds remind would-be listeners that voices are material—produced through bodily actions and bodily training—and that they are always produced in ways that disengage them from an essential body or subjectivity.

This reframes understandings of the human voice that are rooted in metaphysical philosophical concepts and uncritically modern assumptions about the ways the human body is made, felt, and understood. It offers a musical and sonic answer to the Western philosophical tradition that has, for centuries, been invested in the return of an Arcadian ideal: an originary fusing of word and song torn apart by modernity.[16] An alternative reading to this philosophical understanding of the voice—which runs through Plato and Aristotle to Augustine and early modern anatomists and finally to Freud and Derrida—arises when the voice exists as a sounding instrument alongside other materialities such as organs, scientific instruments, and body parts.

Plato himself insisted that there had once been a time of purity in which song and poetry coalesced through the shared ingredients of harmony, melody, and rhythm.[17] Modern philosophers who have greatly influenced scholars across the humanities and social sciences have followed suit. Rousseau, in his *Essay on the Origin of Language*, famously pitted European language and song against the speech-song that he saw as the originary form of human communication. Martin Heidegger drew on ancient Greek and Roman writers to construe poetry and song as belonging to each other. For him, song is language and "singing is the gathering of saying in song."[18] Jacques Derrida's much-cited reading of Rousseau stressed the ubiquitous subservience of song to speech and music to language across the eighteenth century.[19] In demystifying the mythos of primordial unity, he positioned writing and song as speech's alterity. The focus on what we now call music and language leaves out the crucial element of the human voice. As Gary Tomlinson has persuasively argued, even most scholars who build on Derrida do not hear that Rousseau's central text, *Essay on the Origin of Language*, attends as much to singing as it does to speech and that Derrida's reading of Rousseau opens up ways that song might resist the subordination to language.[20]

In other words, even when fortified by a Derridian critique of presence, discussions of song tend to ascribe authenticity to the voice as something that speaks from both immaterial subjectivities and the material body.[21] It is associated with an experience that is often represented as utterly unmediated and as consumed by impossible desire. And because Derrida insists on the centrality of *writing*, it is all too easy for discussions of song to persist in privileging the words, the language. To consider the castrato's voice as made implies not only that the voice is mediated but that it is materially constructed or, in my terms, that it is instrumentalized matter. The bodily singing voice is a material stream of sound, but its materiality has been obscured by discourses that posit the impalpable individuality of the soul.

The physical basis for voice and song has informed a modern theoretical tradition of hearing and feeling song as material practice but has remained still relatively unrealized. In this tradition, exemplified by French thinkers of the 1970s and 1980s, the voice becomes most obviously manifest when it intrudes on and exceeds the semantic content it transmits. Roland Barthes explained that "the singing voice, that very specific space in which a tongue encounters a voice and permits those who know how to listen to it to hear what we can call its 'grain'—the singing voice is not the breath but indeed that materiality of the body emerging from the throat, a site where the phonic metal takes shape."[22] For Barthes, nevertheless, music was chiefly a signifying practice. Adriana Cavarero—presenting song as relational and as an antidote to metaphysics and the voice as a sonorous remnant of speech—has pointed out that Barthes's vocality relates always to speech.[23] Interpreting predominantly ancient and modern texts (eliding, i.e., the early modern period), she explores the pleasurable tension between the body and language. And because she finds most interesting the voice as a force that through pleasure destabilizes masculine linguistic systems and political power structures, it's still fundamentally the semantic function of words that she is after. To this countermetaphysical tradition the castrato adds a historically specific sound, a potent reminder that the grain of the voice, the materiality of song, the physics of the metaphysical, are not new ideas but instead are a rediscovery and reapplication of early modern concepts.

The fourth and final intervention is about telling stories, about the vexed relationship between truth and language and about the ways that feeling the sensory nature of doing history changes the timbre. In more scholarly terms I use the figure of the castrato and the technologies he encompasses to break apart Enlightenment epistemologies and historiographies. Challenging received cartographic and temporal narratives that tend to read from the Enlightenment backward stresses continuities with medieval practices. Music scholars, especially those who inherit a German and Anglo-American intellectual tradition, tend to think of the Italian Renaissance predominantly through Florentine humanism, Roman papal courts, and Venetian nascent capitalism. But there is another side to the story. Walter Mignolo in *The Darker Side of the Renaissance* posits the sixteenth century as a primary justification for colonial expansion and the early stages of modernity. And I reread eighteenth-century accounts of the castrato that have been already overread by Anglo-American scholars for their role not as "evidence" but as part of an emergent idea about the castrato singer and his meaning. Attending to the ways texts have been read and heard in eighteenth-century stories about the castrato also means acknowledging a larger historiography in which, via Johannes Fabian, modernity and timeless cultural value depended on proximity to northern Europe. The point is to avoid reinscribing Renaissance Italy as the seat of modern science, modern music, and the modern body. Instead, the sixteenth and seventeenth centuries sound a radical global shift of power and a codification of dehumanizing practices empowered by colonialism and the global trade.

## THE CASTRATO: SOME BASICS

*For there are eunuchs who were born so from their mother's womb. And there are eunuchs who were made so by men: and there are eunuchs who have made themselves eunuchs for the kingdom of heaven. He that can take, let him take it.*

MATTHEW 19:12

The castrato did not just pop up in the sixteenth century. But they were special. I will argue that the connections between castrati and machines reveal some of the reasons that, though Byzantine, Greek, Roman, Chinese, and Islamic traditions have employed men with altered genitals in their sacred and secular rituals, early modern Europeans between the late sixteenth and the late eighteenth centuries capitalized, in particular, on their voices and placed them at the center of their spectacular entertainments. Ancient law acknowledged those who were made eunuchs and those who were natural eunuchs.[24] The Old Testament already debates the humanity of the eunuch. Deuteronomy (23:1) says, "No man with crushed or severed genitals may enter the kingdom of the LORD." Leviticus (22:24) not only excludes men with crushed testicles but insists, "You are not to present to the LORD an animal whose testicles are bruised, crushed, torn, or cut; you are not to sacrifice them in your land." Historians have records of men castrated for their voices in Egyptian Coptic monasteries, where in the fourth century boys got their education through polyphonic sacred music and were later sold in Turkish slave markets.[25] As Neil Moran argues, by the tenth century eunuchs appear frequently in accounts of music making.[26]

Scholars of castration in the Middle Ages stress continuities between late Roman antiquity and the Byzantine Empire and connections with the Muslim world. Eunuchs, who were admired and feared, played important roles in literature, the military, and religious cults, and they served stabilizing roles in many societies. In the Byzantine world almost every object an emperor touched came from the hand of a eunuch.[27] Their voices were cultivated from the fourth to the ninth century. One of many histories that should be written is a Mediterranean and global history of the castrato focusing especially on encounters between the Italians and the Ottoman empire.[28] This would focus on figures like Antonio Bobovi, who was captured and sold to the Ottomans as a musical page and who reported on castrati as present throughout the household.[29]

What made the Italian castrati different was that they were made for their voices, and those voices became instruments of the singers themselves, the church, and the court. In 1506 the Cathedral at Burgos in Spain recorded a castrato singer who had a good voice.[30] The papal chapel recruited the first castrati from Spain in the early 1560s, though it was by no means the only institution doing the recruiting. From their earliest appearances in European musical culture in the 1560s, castrati were actively trafficked—like other valuable objects—both among Italian city-states and across locations as far away as France, Spain, and England.[31] The novelty of the late sixteenth century rested not on the presence of castrated men but on the use of castration to manufacture high male voices for a musical purpose.

The term *eunuco* probably appears in the Sistine Chapel diaries for the first time in 1588. Since Alessandro Ademollo published his seminal history of Italian theater in the late nineteenth century, 1588 has been a magic date because of a supposed link to prohibitions of women participating in performance. Ademollo wrote,

> In what year did women for the first time appear in the theaters of the eternal city? One does not know. One knows, however, when the papal court forbade their appearance. It occurred in 1588 during the reign of Sixtus V, who, conceding to the company of the Desiosi to give public performances, ordered that they were given in the daytime and also that men play female parts.[32]

Actually, the edict forbade women not from performing but from sitting in the audience. Giulia De Dominicis, writing about Roman theater all told, cites the edict itself, which in fact says that "women should not go there."[33] Nevertheless, in 1599 the pope officially authorized the employment of castrati.

These sixteenth-century castrati were neither any more virtuosic than other singers nor predisposed to execute vocal pyrotechnics that exceeded the glottal effects used by all singers. Just as collectors sent disciples out into the world to locate artifacts and machines, so, too, did dukes and princes send their underlings searching for male sopranos to fill church choirs and take part in music dramas.[34] Unlike the late eighteenth-century singers who functioned in a quasi-star market and often made musical demands on composers who wrote for them and seemed to have had some autonomy, early castrati remained strikingly quiet during negotiations and had very little choice about where they went and what they sang.

While castrati were not owned per se, they were acquired by princes and cardinals for the perpetuation of dynastic glory. In general, compensation was paid to parents in exchange for their sons. Their bodies and voices functioned as objects subject to market forces. The notion of the human body, made or not, as a form of capital was not shocking in the sixteenth and seventeenth centuries, when bodies of all sorts underwent trafficking and mutilation. Fathers regularly gave away daughters in marriage to men older than themselves. The Catholic Church and Italian courts were active participants in the Mediterranean slave trade. Emily Wilbourne's forthcoming book will offer deep context for the relationship between voice, the castrato, and enslavement.[35]

By the middle of the seventeenth century singers were constructed, acquired, and trained for the purpose of doing special performances. Listeners increasingly heard and understood castrati as exceptional. In Rome, by 1625, Pope Urban VIII, born Maffeo Barberini, aimed for a Cappella Pontificia with nine male sopranos and nine male altos. Castrati affiliated with the Barberini family also participated in a secular ensemble that included harp, lute, harpsichord, percussion, winds, and violins. And when he vacationed at the Castel Gandolfo, he never traveled without two trumpet players and a group of castrato virtuosi.[36] Cardinal Francesco's household was composed of a vast cabinet of instruments and musicians, including several castrati. Francesco paid for the training of several castrati and eventually farmed some of them out to other courts.[37] Cardinal Alessandro Peretti Montalto and Cardinal Francesco Maria del Monte, among others, also retained castrati in their households and paid for their training.[38] Thanks in part to these active cardinals, Rome constituted a focal point for the castrato trade. Young castrati often went to Rome for training in written and improvised music, while dignitaries from other Italian courts looked to Rome to fulfill their singing needs.

Marc'Antonio Pasqualini exemplifies the ways that the Barberini supported their prized singers. Famous for the unusual timbre and breath control of the castrato voice and for an ability to sight-read anything, he performed in chapel choirs, in sacred spectacles designed to enhance the Barberini image, and in operas in Rome and other cities. The date of Pasqualini's castration is unclear, but by the age of eight he was installed as a pupil of Vincenzo Ugolini, and he entered the choir of San Luigi dei Francesi in 1623. In 1631 Antonio Barberini secured a place for him in the Sistine Chapel choir and procured for him a benefice at Santa Maria Maggiore. A year later he starred in Stefano Landi's *Il Sant'Alessio*. In addition to singing, he composed, played a variety of instruments, and copied musical scores.

Pasqualini accompanied his patron on diplomatic missions. In 1641 he accompanied the cardinal's court to Paris, where he sang in Luigi Rossi's *Orfeo* as part of a mission to bring Italian opera to France. He also caused trouble. When the singer's coach was shot at in 1637, the cardinal offered six hundred scudi for information and then apparently tried to lock the singer up for safekeeping. Chapel records obliquely explain that the singer was excused from the choir because he "has been kept in his rooms by order of Sig. Cardinal Antonio for a certain tiresome incident."[39] (Pasqualini proceeded to kill a servant the following year and suffered only temporary exile.)

Pasqualini looms large in part because of Andrea Sacchi's much-discussed painting of him being crowned by Apollo. The painting is now housed in the Metropolitan Museum of Art (Fig. 0.3). He is also featured prominently in one of the most commonly read historical sources on the castrato, Charles d'Ancillon's *Eunuchism Display'd*. Appearing in French in 1707, *Traité des eunuques dans lequel on explique toutes les différentes sortes d'eunuques, quel rang ils ont tenu, & quel cas on en a fait, &c. . . . Par M*** D*** Ancillon* celebrated more than Pasqualini's voice; it made his musicianship into extreme virtuosity. For d'Ancillon, Pasqualini's ability to sight-read was itself a mechanistic feat. He told a story of the composers Arcangelo Corelli and Alessandro Scarlatti writing a piece that looked so strange on the paper that it ought to have been unreadable. D'Ancillon stressed the composers' facility as instrumentalists. It was, according to d'Ancillon, "wonderfully shocking," and Pasqualini did indeed execute it with precision.

> And never was any one in so much Confusion as Pasqualini, who reddened and grew pale 3 or 4 times successively, for then he began to find the Design; but when all the Audience thought he must have failed, he performed that disagreeable Part with all the exactness and promptitude in the World, insomuch that the Composers themselves owned their Astonishment and Wonder; and the Audience, if they were not delighted with the Musick, were certainly very well pleased that he acquitted himself with so much Honour. I must own I cannot but think this to be a Piece of Vanity in Pasqualini to trust so much to his Knowledge, but if it discovers his Vanity, it shows at the same time the great Perfection of his Science, and I believe no one in the World can do this but himself.[40]

Castrati had higher voices and extra brilliance that came from the short, thin vocal cords and the close proximity of the larynx to the head.[41] Martha

FIG. 0.3 Andrea Sacchi, *Pasqualini Crowned by Apollo*, 1641. Oil on Canvas. New York City, The Metropolitan Museum of Art.

Feldman and others have shown that their remarkably flexible instrument could generate a huge sound with great resonance and power. They could project text unusually clearly, had exceptional breath control, and excelled at sliding easily from one note to the next. By the eighteenth century they were famous for vocal pyrotechnics, control, and brilliance of tone. They also had bodies that stood out as unusual. The alterity of every part of their physicality highlighted the made-ness of the voice and the similitude between the stuff of the body and the stuff of the voice. Boys who experienced the procedure lost the major source of testosterone before puberty. This affected singing because their larynges stayed small and did not descend into the throat. They had the voice box of prepubescent boys. But they also often developed extralarge chests, heads, jaws, and noses, which meant that their resonating chambers were much larger in proportion to their vocal materials than in unaltered bodies. Lack of testosterone meant that growth plates in the joints did not fuse at the same age as in unaltered boys, and often their limbs, jaws, facial bones, and ribs grew to extraordinary length, giving them a strange appearance. They had flat feet, never grew beards, had luxuriant hair on their scalps, developed extra fat deposits on their chests and hips, and tended toward obesity later in life.

Castrati were made by training and by surgery. Paolo Faccone, a Mantuan emissary to Rome and broker of singers, alluded to this process in a letter to his patron Ferdinando Gonzaga, sent from Rome in February of 1613. Discussing a boy who wanted to have himself castrated—with veiled references to the actual procedure, as was customary at the time—Faccone made it clear that surgery alone was not sufficient to create the prized voice of the castrato. Once an assembled team of experts had declared a boy's voice good, they prescribed daily practice and training to help ensure that it would be worthwhile to castrate him.

> And not trusting my own judgment I had him heard and auditioned by three virtuosi who think that for two or three months the boy should be sent to San Apolinaris to find out what he will be able to do with a little instruction and daily practice. The boy has a good and ripe voice, and such a good disposition; and he has a great longing to have himself castrated, and has been continually urging me in regard to this. All that is necessary is for Your Highness to order what I should do and I will carry it out. It will be necessary to have him clothed because he has nothing, then to give something to the teacher who taught and took care of him in this period, and then to pay the new teacher and whomever will provide for him.[42]

Faccone could turn this boy into a castrato via surgery *and* training. Faccone also brokered the deal for the then thirteen-year-old Caterina Martinelli, who would die before she got to sing the title role in *L'Arianna*. For her he insisted that she "be seen and tested by two or three women," by which he meant a virginity test.[43]

Many of the training techniques were used for female singers, tenors, and basses, but the implications were different for singers whose bodies had been changed to make a particular kind of voice. And though the eighteenth century represented the peak of castrato training, earlier singers were intensely trained. Castrati almost always participated in extensive training while either boarding with their teachers or residing at conservatories. This allowed for constant surveillance as they practiced for up to eight hours a day. Since they lost no time to changing voices, they could train continuously from as early as the age of six or seven. Cardinal Francesco Barberini's household included a stable of boy castrati—*putti musici* or *castratini*—trained in his household at his expense, with compensation paid to their parents. When he arranged for Girolamo Zampetti to receive instruction from the most notable musicians of his court, at the age of thirteen, Zampetti moved in with Johannes Hieronymus Kapsberger, where he received a bed, a table, and chairs. The next year saw the addition of grammar and dancing teachers as well as musical instruction, clothes washing, and money for other expenses. In 1629, the training intensified with a teacher for singing and a harpsichord instructor as well as lessons from Stefano Landi. His patrons purchased him a large Spanish guitar and a used harpsichord, also ordering a new one. He resided with Landi in 1631 and 1632 before finally joining the Sistine Chapel choir.[44]

The surgical procedure was known as a bilateral orchiectomy. The removal of both testes was generally performed between the ages of six and twelve. Musicologists have tended to focus on seventeenth- and eighteenth-century descriptions of the procedure. But the earliest description of bilateral orchiectomy comes from the influential seventh-century Byzantine medical encyclopedia of Paul of Aegina, a surgeon-physician who drew heavily on Galen.[45] He described the procedure in the sixth volume, which was devoted to surgery.

> Compression is performed thus: children, still of a tender age, are placed in a vessel of hot water, and then when the bodily parts are softened in the bath the testicles are to be squeezed with the fingers until they disappear and, being dissolved, can no longer be felt. The method by excision is as follows: let the person to be made a eunuch be placed upon a bench and the scrotum with the testicles grasped by the fingers of the left hand and stretched; two straight incisions are then to be made with a scalpel, one in each testicle, and when the testicles start up they are to be dissected around and cut out, having merely left the very thin bond of connection between the vessels in their natural state.[46]

In the sixteenth century the Venetian surgeon Giovanni Andrea della Croce included an instrument called a castrator at the very end of the Italian version of his *Universal Surgery*.[47] Croce was part of a surgical tradition in which learned artists worked with scholar humanists to solve technical problems. Croce explains that to perform a castration the artisan needed a castrator, a razor, thread, and cauterizer.[48]

This surgery stands out because it interfered with nature's perfection. Premodern surgical practices aimed to correct and restore nature to its perfect state. Surgery on hermaphrodites, for instance, in theory corrected a flaw. Prosthetics and surgical procedures repaired congenital defects or damage done by war. Surgical manuals mostly explained ways to cover up the emasculation of disease and damage with devices like "un cannellino di ottone" (a brass tube). Ambroise Paré illustrated a variety of pipes that aided injured men in the process of urination. He showed how to insert a pipe into a "urinary passage" after restoring a gland and how to use silver pipes to drain men's urine after moving what we would call kidney stones.[49]

Modern castrato scholars don't usually historicize the procedure itself. The most often quoted description of the process that made castrati comes from Charles d'Ancillon's 1707 graphic narrative. But d'Ancillon's writing is not what we consider scientific fact and was not written as a medical text. Hovering between satire and pornography, it came with all kinds of agendas. Nevertheless, according to d'Ancillon, the testicles could either be removed through an incision or withered through a crushing process that severed the ductus deferens:

> Another Method was, to take the Testicles quite away at once, and this Operation was commonly effected, by putting the Patient into a Bath of warm Water, to soften and supple the Parts, and make them more tractable; some small time after, they pressed the Jugular Veins, which made the Party so stupid, and insensible, that he fell into a kind of Apoplexy, and then the Action could be performed with scarce any Pain at all to the Patient; and this was generally done by the Mother or Nurse in the most tender Infancy. Sometimes they used to give a certain quantity of *Opium* to the Persons designed for Castration, whom they cut while they were in their dead Sleep, and took from them those Parts which Nature took so great Care to form; but as it was observed, that most of those that had been cut after this manner, died by this Narcotick; It was thought more adviseable to practice the Method I just before mentioned: However, it was by this Means, that Miracle of a Voice *Pauluccio*, and the real Wonder of the present World, a *Roman* Eunuch, was against his Will made so, by his own Uncle, (also an Eunuch) as I shall speak of more at large in its proper Place.[50]

And herein lies the rub. We moderns rightly cannot get past the violence of the cut and crush. Thanks to a preoccupation with the mutilation of young boys and the sex lives of men with mutilated genitals, scholarly discussions often define the castrato by the cut. The mutilated castrato is read as neither male nor female and thus as a nebulous creature who does not have access to personhood and is plagued by a constant longing for sexual pleasure and love.

To modern ears the castrato voice and body fell outside heteronormative desire and gender binaries. Castrati rose to prominence in Italy as pragmatic solutions to the problem of high voices. Boy sopranos had a limited shelf life, and women were considered dangerous and excessive at best. From a modern vantage point, this seems like an aggressive response to the practical problem of getting the high pitches required by new music while keeping to St. Paul's dictum that women should be silent in the church. Sixteenth-century castrati were virtuosic singers, like any other professionals (unaltered males and females), and as their popularity rose, so, too, did the demands on their voices. Female singers, not castrati, were the exotic others who provoked anxiety, desire, and ultimately containment. So in 1630 when Pasqualini entered the papal chapel, the singer-composer Barbara Strozzi could only sing at her adoptive fathers' *accademia*, a performance that signaled erotic availability. In Rome, as Amy Brosius has calculated, nearly half of the courtesans described by the governor of Rome were singers.[51]

The Freudian baggage came later. Freud, in "On the Sexual Theories of Children," wrote that all children assume that all people have a penis; male children are afraid of losing theirs, and female children want one. This leads to penis envy, castration anxiety, and of course heterosexuality.[52] Barthes's reading of the Balzac story *Sarrasine* in *S/Z*, a key text in theoretical discussions, portrays the castrato as female or sexually neutered, reinforcing a modern construction of rigid gender polarities.[53] Through this assumption Barthes started a scholarly trend of thinking about castration through a psychoanalytic lens and rendering the castrated body as, in reductive effect, symbolically dead.[54] The castrato in this line of thinking lacks sexual productivity, and the phallus moves symbolically from the sexual organs to the vocal organs.

It is true that castrati had their testicles crushed long before any reasonable age of consent—a practice unthinkable today—and it is true that castrati had secondary sexual characteristics that in modern terms are read as sexual ambiguities. But focusing on those two facts divorces the castrato from his historical moment. Though canon law had long forbidden bodily mutilation, castration was thought to cure a variety of ailments including epilepsy, gout, and hernia. It also was a form of corporal punishment. The heteronormative and post-Freudian tendency to define sexuality by penetrative sex ignores the early modern sexual imaginary. Historicizing the castrato, as Roger Freitas has explained, suggests that the erotic appeal of the castrato was effeminate but at the same time hypersexual.[55]

It's hard to find the castrato's body and desire in the tangled web of discursive and empirical histories. My interest lies not in untangling that web but in listening to it. Since the ancient world, castrato erotics and ecstasy have been the exotic, erotic other of the hegemonic European Enlightenment, which makes the truth especially elusive. The ancient Roman author Ammianus Marcellinus wrote, "When seeing the line of mutilated human beings, one would curse the memory of Samiramis, queen of that ancient time, who, first before all others, castrated young males, as if hurling violence at Nature, twisting her away from her intended course."[56] Sixteenth-century debates about eunuchs in the papal choir highlighted the fact that these singers hailed from Spanish Naples or Spain. Early work on castrati narrates Charles Burney gallivanting around Europe looking for the person who did it.

The blame game was almost a reverse colonial map. The English blamed it on the French, who blamed it on the Italians, who blamed it on the Spanish, and then it always gets to the Moors and the Turks. The above-mentioned *Eunuchism Display'd* by Charles d'Ancillon, published in 1708 in France, became famous largely through the 1718 English edition. Robert Samber, who translated it, said he was compelled to present this document when a castrato tried to run off with a young English noblewoman. Edmond Curll, the publisher, was infamous for pornography and scandal. In other words, it's a purposefully salacious document written by a Frenchman about an Italian practice. These manufactured singers always sing from a sensory place far away, and their real and imagined incarnations formed part of larger inscriptions of alterity and processes of globalization.

Modern opera singers like the alterity of the castrato, particularly the gendered otherness. In 2009 singer Cecilia Bartoli released the album *Sacrificium*, a sleek collection of music that castrati sang in the eighteenth century. She characterized *Sacrificium* as "a work dedicated to 'castratos' and the music that their magic voices inspired. . . . The title of the album refers to the sacrifice castratos did for the sake of music." Bartoli's crisp, ear-shocking coloratura, the lickety-split instrumental playing of Il Giardino Armonico, and the engineered sound give the record a sheen that exudes technological sophistication. The accompanying glossy book, titled *Evviva il coltellino* (Long live the little knife), features pictures of Bartoli's head photoshopped onto sculpted androgynous figures.[57] The figures make her look like a cross between Michael Jackson and Michelangelo's David (Fig. 0.4), while the arias hail not from the usual famous London tunes of Handel but from Naples. In her liner notes she describes Naples—often constructed as the southern, othered edge of Europe—as the late seventeenth-century capital of Western music. Despite the coloratura on this album, Bartoli said that the pyrotechnics were not the hardest sounds to make: "The beautiful sad arias are the hardest to sing, because I am moved almost to tears. I know they were singing those arias out of their own sorrow."[58]

## STUDYING THE CASTRATO

When I taught the history of Western music for the first time, in 1992, we used a set of recordings that accompanied the *Norton Anthology of Western Music*. (They were available on cassette.) The survey for nonmajors included

FIG. 0.4 Cecilia Bartoli as Michael Jackson. From Cecilia Bartoli, *Sacrificium*, Decca 460502670325, 2009, compact disc. Courtesy of Uli Weber and Decca Classics.

the concluding scene of Claudio Monteverdi's 1643 *L'incoronazione di Poppea,* in which the role of the cynical, mother-kicking emperor Nero was written for a castrato. The final scene may or may not actually have been written by Monteverdi, but it capitalizes on the conventionalized musico-erotic vocabularies, especially close imitations, luscious suspensions, and dragged-out dissonances. The sensuality sounds in the similarity of the voices. Nero and Poppea are so close in timbre and range that it's almost hard to hear who is who. But the Norton recording took the role of Nero down an octave and put it in the mouth of a tenor.

Paul Henry Lang had argued for this transposition, positing a male-female drama as the central tension of baroque opera.[59] He said that the castrato had neither sex nor personality, and he advocated for taking their roles down an octave to preserve the proper gendered relationship between high and low voices. Lang described the castrato as

> an instrument of prodigious versatility and perfection, but still a musical instrument and not a living human character. He destroyed the efficacy of the female

> voice by duplicating its register without the passion and the expressivity of the woman incarnate in her voice. Correspondingly, he could not truly express the instincts and desires of a man in the vocal range of a woman.[60]

This is an important historiographical reharmonization, one that reflects the lack of sounding of castrati in music performance and in music scholarship. The sense of modern castrato scholars has been that the castrato wrought a kind of embarrassment. The historiography, or lack thereof, until the 1980s is part of the story. German-influenced Anglophone musicology came late to Italian music of all sorts. And a focus on sources, works, and scores tended to eschew works that were in part driven by performers and were about performers. At the most visceral level musicians and music scholars saw the high-voiced male roles as a problem on the stage. Countertenors often don't have the power for which the roles seem to beg. Cecilia Bartoli's album presents the obvious pragmatic solution—a female voice—but probably the idea of a powerful Roman emperor like Nero or Julius Caesar singing through the body of a woman was unthinkable. Moreover, the spectacle of sonic erotic pleasure enacted by two female bodies was even less palatable for 1950s Anglo-American audiences and was far less acceptable than it is today.

To take an extreme example, Angus Heriot's fabulously undocumented 1956 *The Castrati in Opera* presented the story of a castrato called Balami who sang such a vigorous aria on the stage in Naples that his balls popped out. As I will elaborate in chapter 7's discussion of the castrato in the South, the story is part of a travel narrative; Heriot presented it as "fact."[61] One of the first modern treatments of the castrato, Heriot's prose bespeak a moment when the music world was, from a twenty-first-century perspective, comically squeamish about the castrato. Identity politics were also still a few decades away. Heriot wore his castration anxiety on his sleeve: "Why was so strange and cruel a practice thought worthwhile, and why should audiences of succeeding generations have preferred these half-men with voices as high as those of women, both to women themselves and to natural men?"[62] His readers were also anxious about what they thought of as an unseemly business. Winton Dean, most well known for his work on Handel's operas and oratorios, wrote in his 1957 review of the book that the castrato was "the most recondite contribution to art by the church" and went on to explain that "Mr. Heriot is able to remove the manhole covers sealed up by the intervening age of respectability without leaving a distasteful smell in the reader's nose."[63]

Beginning in the late 1980s, scholars particularly interested in gender focused on the castrati's ambiguities, erotic appeal, and supposed Lacanian "lack." At the same time, musicologists began working on the early history of the papal choir, on Handel's operas, and on individual castrati. The perceived dissonances in a high voice coming out of a male body led scholars to think in terms of gender bending. Initial studies focused on the late eighteenth century turned castrati into fantasies. This resonated with larger critical discussions in the academy and with popular culture. In 1982 Anne Rice, of vampire fame, wrote the castrato-centered *Cry to Heaven*, a novel that Alice Hoffman, a writer who embraces magical realism and strange eroticism, described in the *New York Times* as "bold and erotic, laced with luxury, sexual tension, music. Here passion is all, desires are overwhelming, gender is blurred."[64] Rice fills in the blanks, capturing in an imaginative way the desire of the castrato and the fictional experience of inhabiting that body.

The 1990s saw a renewed fascination with the castrato as gender bending on and off the stage. These stories resonated with 1980s and 1990s popular culture—a Collingwoodesque moment if ever there was one—a tendency perhaps to read the past through the seemingly immutable categories of the present.[65] Think Prince in his 1987 song "If I Was Your Girlfriend," convinced in all of his high-pitched purple glory that what he needed to connect with the girl was to be a girl. In and out of the voice of his sometimes-female alter ego Camille, he promises to wash her hair.[66] Stratospheric vocals over a sparse rhythm section, with Mendelssohn's Wedding March piped in for heteronormative effect, sang against gender norms:

> If I was your girlfriend, would you let me dress you
> I mean, help you pick out your clothes before we go out?
> Not that you're helpless
> But sometime, sometime those are the things that bein' in love's about.

The 1990s was also the decade of *The Crying Game*, whose theme song performed gender bending by melding the original version of the song by Dave Berry and the Boy George cover. And 1994 the film *Farinelli*, directed by Gérard Corbiau, cast the dreamboat-looking Stefano Dionisi as Farinelli. The singer seduces women with his voice, but his older brother finishes his sexual acts. The not-castrated composer brother slides in just before the moment of climax, and the "wounded" castrato watches silently. The movie also features scenes of Handel crushing snails with his cane as the singer's voice soars in the background.[67]

The film had a budget of nine million dollars. The movie's most luscious musical moment might be Farinelli singing Almirena's "Lascia ch'io pianga," an aria that figuratively translates to something like weeping for a wretched fate, yearning for freedom and breaking bonds of torment. As Ellen Harris pointed out, the soundtrack drives home the message of musical instrumentality at the expense of sex by putting in the mouth of the castrato the 1717 ornamented version composed by William Babel.[68]

In the twenty-first century, monographs and articles that think historically about the castrato singers as real people, not fantasies, have changed the possibilities of inquiry drastically. *Voice Machines* builds on the excellent monographs of Roger Freitas and Martha Feldman, two scholars who have made castrato scholarship central to thinking about seventeenth- and eighteenth-century musical practice and who have answered crucial questions about why so many boys were subjected to a practice that, from a modern vantage point, seems barbaric at best. Freitas's *Portrait of a Castrato* focuses in depth on one castrato, Atto Melani, by detailing the fascinating political and social interactions in the making of one singer.[69] Feldman's *The Castrato: Reflections on Natures and Kinds* relates castrati to constructions of the natural and positions them as key instruments in European cultural exchanges. Beyond comprehensive archival and theoretical work, Feldman asks, "the central unanswerable question of what would happen were we to try with all the faculties and resources we can muster to imagine the very thing that we can never locate, much less experience. What spectrum of possibilities might we hear in our mind's ear?"[70]

It is important to acknowledge here in the prologue what this book fails to do. The book implicitly and explicitly critiques the Western European and Anglo-American exceptionalism that dominates music history. It considers the castrato in the context of empire and gestures toward the long history of enslavement. It acknowledges that the practice of making eunuchs was replicated simultaneously in other parts of the world. But it does not do those histories justice. Howard Chian has used the history of castration and eunuchs in China to present a genealogy of sex and gender.[71] And I have not incorporated the explosion of literature on the history of transsexuality and on the biomechanics of trans singers, histories that will without a doubt shed light on the castrato.

The book is organized in four sections. The first two chapters set the theoretical and sound world. The first is a thought experiment called "Orfobot: Automated Orpheus." The chapter hears Monteverdi's *L'Orfeo* as a hymn to technology. I move away from the title character and attend instead to Monteverdi's use of sound effects created by musical instruments and his use of castrati as technologies. Drawing on Giambattista della Porta's and Athanasius Kircher's understandings of echo and other sound effects, this chapter ties the castrato directly to artificial magic and particularly to devices designed to enhance both sonic effects and the experience of those effects. The second chapter, "The Death of a Cicada," puts into practice the idea that the castrato is as much a provocation as a subject. I read Galileo's fable of sound as a cautionary note for modern scholars of sound and put the castrato and the cicada in counterpoint as a way to index what early modern Italians thought they knew about sound.[72]

The second section includes three chapters that get lost in the mess of changing notions of technology and nature, beginning with a short intermedio on the cat piano. A lengthy exploration of the notion of organ is framed by two shorter chapters centered on devices: the telescope on one side and the hydraulic organ on the other. The third chapter, "Organoscope: Telescoping Sound," riffs on the fact that Galileo made his first telescope out of an organ pipe and takes seriously the idea that the castrato extended the capabilities of the human voice in ways that ontologically parallel the extension of the sensorium performed by the telescope. It ends with Galileo's finger in the Florence History of Science Museum. The fourth chapter, "Organs and Organs," takes up the castrato vocal organ as the intertwining of organs and instruments. It is a theoretical exercise that allows for the castrato to be heard as a body remade and as part of a collection of organs: vocal organs, keyboard organs, body parts, and the word itself. This chapter delves deeply into early modern understandings of vocal anatomy and musical sound. The fifth chapter, "Into the Garden," is a short excursus on the fantastic hydraulic organ. These ghostly machines inhabited the gardens of villas and palaces. Like castrati, the organs played with the already murky relationship between art and nature, and like castrati, they were sites of virtuosic technical display embodied in natural organic material.

The third section begins with a deep dive into the castrato's presence in seventeenth-century Rome and moves into explorations of the ways that stories about the castrato's voice and body intersect with inscriptions of desire and otherness. The sixth chapter, "When in Rome: The Castrato as Special Effect," has fire, water, and explosions. It considers the castrato as a special effect: a body remade as a theatrical instrument. The spectacle of baroque Rome presented a variety of ephemeral events, from fireworks to operas with fountains, creating a bizarre mix of human and inanimate objects that linked illusion, machinery, magic, and music. Chapter 7, "On the Cusp," hears the castrato's song in the context of eighteenth-century European fascination, fears, and desires, with song as key in ideologies of speaking and writing, reason, and civilization. Chapter 8, "More Than One Sex," makes a series of interconnected arguments about the body of the castrato, intervening particularly in musicological engagements with the history of science and medicine. My approach to the castrato body works with the vibrations of desire resounding through those bodies, an overtone of what Carla Freccero describes as a haunting effect of the past; the past is not quite in the present, but the trace resounds.[73]

The final section of the book includes two theoretical chapters and an epilogue that critique posthumanism. Chapter 9, "Time Travel / Liquid Ecstatics," takes as a point of departure a fabulously enumerated performance of Domenico Mazzocchi's *Lagrime amare all'anima che langue* that puts on display the pleasure of the castrato voice. The last chapter, "Cyborg Echoes," critiques posthumanism and argues that the world has always had a place for creatures and objects produced by technology, but the techniques and the materials have varied according to time and place. It shows the ways that the making of the castrato challenged the boundaries of the human. The book cadences with an epilogue that embraces the relationships between inscription and memory. It ends in the gardens of the eighteenth-century castrato Gaspare Pacchierotti.

# SOUNDS LIKE . . . ASSEMBLING THE SOUND WORLD

CHAPTER 1

# ORFOBOT

## *Automated Orpheus*

In January of 1607 the young Duke Francesco Gonzaga began hunting for a castrato for the performance of Monteverdi and Striggio's *L'Orfeo* planned for February 24 in the ducal palace of Mantua.[1] Though the Mantuan court had female singers, none took the stage in that production; Euridice's few lines came out of the mouth of a castrato that one observer called "that little priest" (*quel Pretino*). The Gonzagas had been on a quest to find castrati since the 1560s, and none of the three in residence were up to the task, so they borrowed some.[2]

On December 9, 1616, almost a decade later, Monteverdi wrote to Alessandro Striggio from his post in Venice and famously invoked the Mantuan *Orfeo* in order to explain that he could not imitate winds because they do not speak. "And how can I, by such means, move the passions? Ariadne moved us because she was a woman and similarly Orpheus because he was a man." *Le nozze di Tetide*, a maritime fable, perturbed him for many reasons, including that setting a drama on an ocean would require too many instruments.[3] Generations of scholars, including me, have used this remark that a composer—prodigiously cranky, recalcitrant, and plagued with financial problems and a troublesome son—made about Ariadne's and Orpheus's potential for affective sung speech to make sweeping statements about musical speech, the passions, and human expression.[4] This makes sense given the leading role Orpheus plays in scripting the imagined powers of music.

But what if Orpheus in his many iterations is not just a figure who personifies the divine powers of music but is one who manipulates nature and matter? What if he were an automaton? This thought experiment began as a paper for an invited conference that brought together early music and robotics.[5] Imagine Orfobot as a little musical windup toy hopping along with a tinny version of Monteverdi's music spewing out of the little plastic body. The energizer bunny Orpheus is of course absurd. But maybe not totally. Orpheus is the character that musicians summon, and Orfobot is the one who summons. Offspring of the mortal muse Calliope and the god Apollo, the hybrid bard is already an assemblage of technological, material, and organic forces. He plays a lyre that Mercury made by gutting a tortoise, covering it with cowhide, and stringing it (Fig. 1.1). He transformed an animal into a seven stringed instrument.

FIG. 1.1 Wooden lyre with a tortoise shell voice box restored from shards, fifth century BC–fourth century BC. Accession Number 256226001. © The Trustees of the British Museum.

The automaton is as much a mode of thought as an automated object, and it is a concept that perpetually troubles the boundaries between human and not human.[6] Both the automaton and the castrato ooze what Jane Bennet calls "vibrant materiality," one that "runs alongside and inside humans."[7] This concept is especially applicable to the premodern world where as Carolyn Walker Bynum writes, all matter was understood as malleable and "pregnant with creative potential."[8] Addressing a slightly later time period, Pamela Smith says that artisans in the European Renaissance engaged in an ongoing bodily struggle against matter and that they had to come to know and master it though experience.[9]

The admittedly tongue-in-cheek Orfobot begins to knit together the castrato with the story of objects that pushed the limits of the human, and it deliberately draws on artisanal and philosophical writings. It is a way to write about the body as convoluted time and the castrato as untimely matter reverberating with traces of the past and projecting into the future. Orpheus famously looked back, and with that backward glance he killed his young bride. He was always already a figure who existed in a hybrid present, a circle of transmission that Elizabeth Freeman calls temporal drag, by which she means "a kind of historicist *jouissance*, a friction of dead bodies upon live ones, obsolete constructions upon emergent ones."[10]

Modern readers usually know Orpheus from Ovid and Virgil. Ovid's *Metamorphoses* is a series of stories in which nature constantly transforms. Bodies are defined by transgression and instability and vital matter, and the human is constantly reconfigured. Ovid's Orpheus wants to animate nature. He softens stones and makes them move, and he tells the stories of Echo and Pygmalion, figures who are and were tied to animation and sound production. Echo is a sound phenomenon and a technology, a coalescence that early modern authors loved to play with. Pygmalion has come to embody a vitalist notion of music where sound breathes new life into characters, figures, and spaces. He abhorred women so much that he made one out of stone. But then he begged for a living likeness, a prayer Venus answered by turning stone to flesh. Ovid's words invite the reader into Pygmalion's fantasy: to feel her flesh soften and turn red, sense her movement. The literary fantasy of animating matter made it possible for the artist to imbue his statue with life itself.

There are prequels and postludes to Ovid and Virgil. Despite no evidence that a "real" Orpheus existed, the name appears in ancient sources; lyric fragments of the sixth century BCE make him a signature for mystical ideas and the power of lyric.[11] In the Roman empire, condemned prisoners acted out the Orpheus myth on their way to death.[12] In 1600 the French king Henry IV hired an Italian engineer to build, among other things, an Orpheus grotto full of magical hydraulic waterworks (Fig. 1.2).

Shakespeare's *Henry VIII* includes one of Queen Catherine of Aragon's women singing these words:

> Orpheus with his lute made trees,
> And the mountain tops that freeze,
> Bow themselves when he did sing:
> To his music plants and flowers
> Ever sprung; as sun and showers
> There had made a lasting spring.[13]

John Quincy Adams proposed the lyre of Orpheus as a federal seal because it could lead men and beasts to harmony. In 1820, when he was secretary of state, he used the symbol of the lyre on the passport (Fig. 1.3).[14] In 1866 Arthur Sullivan made "Orpheus with His Lute" the first of his five *Shakespeare Songs*. In

FIG. 1.2 Grotto of Orpheus at Saint Germain-en-Laye, Château Neuf, by Thomas and Alexandre Francine. Engraving by Abraham Bosse ca. 1650. Paris, Bibliothèque Nationale de France.

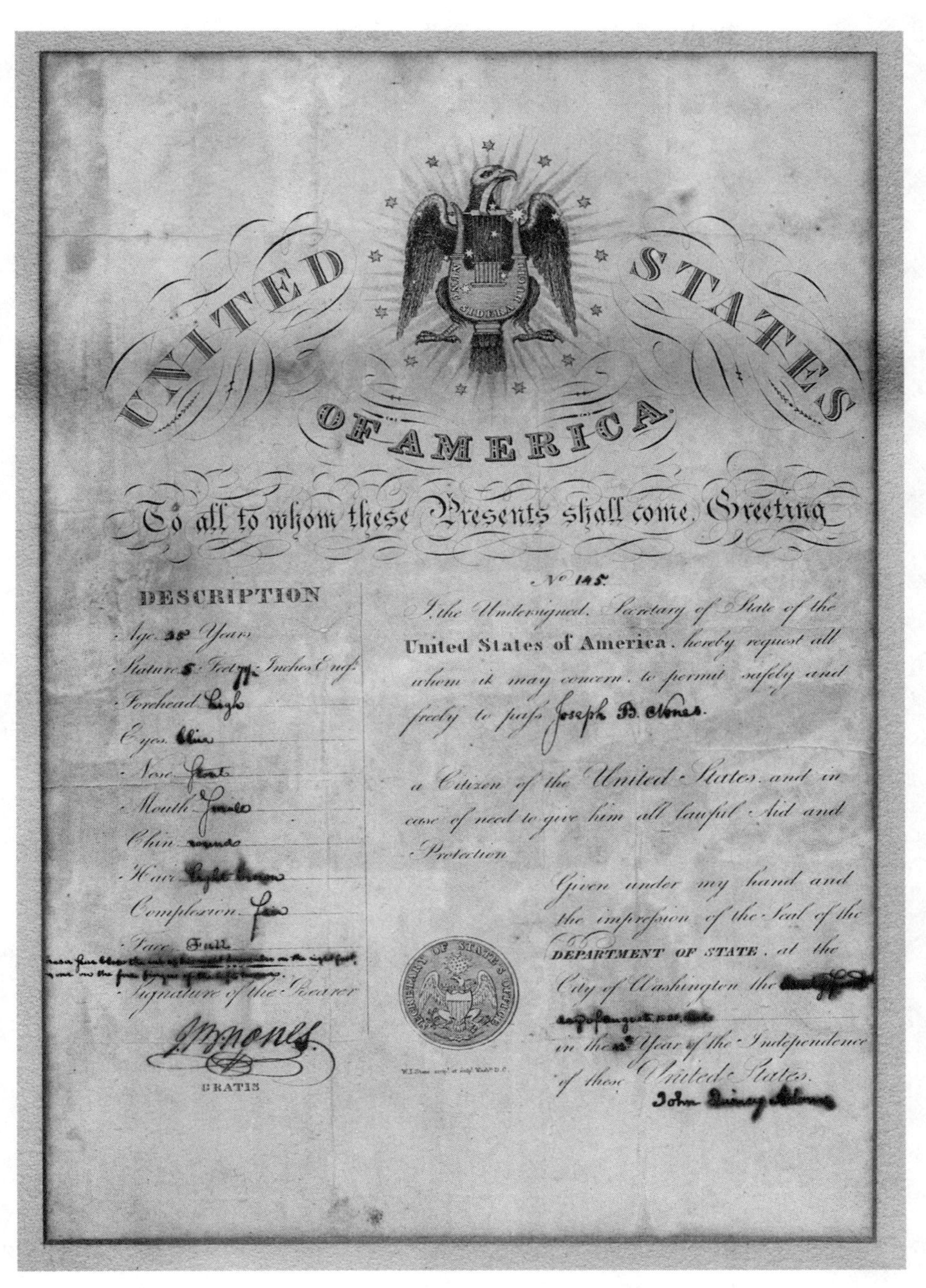
UNITED STATES OF AMERICA

To all to whom these Presents shall come, Greeting

No 145.

DESCRIPTION

Age 28 Years
Stature 5 Feet 7 Inches Eng.
Forehead high
Eyes blue
Nose straight
Mouth small
Chin round
Hair light brown
Complexion fair
Face Full
[illegible]

Signature of the Bearer
J B Nones

GRATIS

I, the Undersigned, Secretary of State of the United States of America, hereby request all whom it may concern, to permit safely and freely to pass Joseph B. Nones, a Citizen of the United States, and in case of need to give him all lawful Aid and Protection.

Given under my hand and the impression of the Seal of the DEPARTMENT OF STATE, at the City of Washington the [illegible] day of August 1821, in the [illegible] Year of the Independence of these United States.

SECRETARY OF STATE'S OFFICE

John Quincy Adams

FIG. 1.3 Passport issued to Joseph B. Nones, August 22, 1821, signed by John Quincy Adams, with eagle and lyre emblem. Courtesy of Basil Williams.

the twenty-first century, Anais Mitchell's Broadway production *Hadestown*, set in a New Orleans jazz club, staged a critique of climate change and Donald Trump.[15] Alternative rock band Nick Cave and the Bad Seeds tells Orpheus's story as a lugubrious blues in which the music god has to make his own lyre out of wood and wire, and—oops—he accidentally kills her.

> Look what I've made, cried Orpheus
> And he plucked a gentle note
> Eurydice's eyes popped from their sockets
> And her tongue burst through her throat

Chronologically this chapter pivots around the turn of the seventeenth century. In 1600 Jacopo Peri and Giulio Caccini each wrote a *L'Euridice*. The two of them and Emilio de' Cavalieri claimed to invent a new style of music called *stile rappresentativo*. Monteverdi entered the fray a few years later. Frederick Sternfeld charted twenty works featuring Orpheus that hit the stage between 1599 and 1620.[16] The novel genre that the Orpheuses populated came with increased virtuosity and complex recitatives, both of which required professional performers, and, as Andrew Dell'Antonio has argued, new kinds of listeners.[17] To be sure, the investment in the human voice as able to excite and entice passions ran deep, but that voice worked with mechanical instruments of all kinds. The sixteenth and early seventeenth centuries witnessed new instruments and a parallel flowering of manuals devoted to describing those instruments as well as to developing skills as both a soloist and an accompanist.

The year 1607 saw multiple milestones. Vittorio Zonca's *Novo teatro di machine et edifice* was published posthumously in Padua. Zonca set the trend for using images and descriptions of machines in virtuosic printed books called "theaters." Galileo purchased his first telescope in Venice. The English settled Jamestown, and the English, like Orpheus, tried to "tame" what they imagined as wild beasts. And the crew of the *Red Dragon*, a British trading ship, performed Shakespeare's *Hamlet* on the ship somewhere off the coast of Sierra Leone.[18] Francis Bacon penned an essay called "Orpheus; or Philosophy" that positioned Orpheus as the founder of civilization because he could tame the wild beasts, as in the Savage.[19] He gave conquest a sung body.

This game of historical concordance is not a temporal coincidence. Scholars date the years around 1600 as the beginning of the Anthropocene—the age of us—and the age that marks the period when Earth's geology, writ broadly, reflects the signature of human activities. Whether or not this ought to mark the beginning of the Anthropocene matters less than the fact that trade in people and things across the Atlantic Ocean changed the earth.[20]

Katharine Park and Lorraine Daston have described this as the emergence of new domains of inquiry.[21]

When this chapter turns to Monteverdi's and Striggio's *L'Orfeo*, the reader should ask why this much-discussed canonic piece of music warrants yet another discussion. Unlike much of the material in this book, Monteverdi's *L'Orfeo* appears in textbooks, and any discussion provides opportunities for even the most convinced reader to object, to be bitten by the snake that killed Eurydice. That's part of the point: to engage familiar tunes and their even more familiar stories to think about why stories are told again and again. My hearing lingers on "Possente spirto" and the Echo finale because they bring to the fore the complicated relationship between physical phenomena, literary histories, natural philosophy, and stories. Pushing together Echo, the mythical figure, and echo, the sound object, amplifies instrumentality and automated sound. Athanasius Kircher, who plays a supporting role in this book, exemplifies the ways that the mythical echo and the sound object coalesced in early modern thought, and his writings reveal a fascination with music's animating potential.

## THE NOT-AUTOMATED CASTRATO

Most discussions of castrati and automatons veer toward eighteenth-century and early nineteenth-century descriptions of the virtuoso divo as mechanical and machinelike. Scholars frequently cite the 1825 *London Magazine and Review*'s description of Giovanni Battista Velluti, one of the last operatic castrati. Velluti attracted attention as a divo and as something of a playboy whose sexual exploits mystified husbands all over Europe.[22] Will Crutchfield hears Velluti's ornamentation as the basis for the pianistic ornamentation of Chopin and others.[23] James Davies highlights the anxiety Velluti incited in early romantics, epitomized in responses by Mendelssohn, Balzac, and the British press. The *London Magazine*'s negative comments were part of a press campaign against Velluti and picked up on the association with vapid mechanism:

> That he is a perfect master of the science it is easy to perceive, and his execution is wonderful, and not to be resembled to any thing we have ever heard; but if we could imagine an automaton as skilled in singing, as Roger Bacon's fabled clockwork head was in speaking, we can fancy that the effect would be similar; for the precision with which Velluti executes the most difficult passages, can only be compared with that of a piece of machinery, and the likeness would hold good also in respect of an occasional want of modulation in his highest tones, and a certain grating sharpness of finish. Some pieces of music he performs exactly as

> a steam-engine would perform them, if a steam-engine could be made to sing, taking each note with unerring accuracy, and taking each by a separate impulse, instead of floating on the gamut as less perfect singers commonly do.[24]

Quotations like this preoccupy modern scholars' reading of both castrati and automata.[25]

Similarly, for the most part, music scholars take up the story of automata at this late eighteenth-century moment when they did, in effect, become androids who could make music and imitate the actions of humans.[26] The Swiss watchmaker Pierre Jaquet-Droz made the lady keyboard player "La Musicienne" out of 2,500 pieces, and she could play five tunes. Jacques de Vaucanson famously made a life-sized automated flute player in 1737, which could play twelve different tunes. Emily Dolan writes elegantly on E. T. A. Hoffmann's story "Automata," which features a concert produced by all mechanical instruments and players.[27] Hoffmann's 1817 "The Sandman" is now almost required reading in some musicological circles and has inspired fantastic scholarship in part because of Offenbach's operatic version, *Les contes d'Hoffmann*, where Hoffmann loves a mechanical doll.[28]

Much musicological writing approaches the mechanical doll through a Benjaminian anxiety about automated music in which the doll is oracle of modern machines, industrialization, and electronic music. These interpretations correctly posit automated music as uncanny, in the Freudian sense. Olympia is uncanny because it's impossible to distinguish her from a real live singing woman. But perceiving automation through a nineteenth-century lens leaves out quite a bit. Before the eighteenth century, automata answered the desire to control and manipulate others as well as to animate, and they were a zone of inquiry. The ancient world buzzed with automata: the glass spheres of Archimedes, described by Cicero, which depicted motion of the planets; the flying wooden dove of Archytas; and the hissing serpents of Boethius. Ancient automata centered on the root words *self* and *action*.[29]

Take the quotation about Velluti above. It certainly bespeaks nineteenth-century negative fantasies of mechanization. But name-checking Roger Bacon also signals that the tradition of automata ran continuously from ancient times to the present moment. The learned Latin Middle Ages abounded with automata that inhabited a space between natural philosophy and technology. By the twelfth century, legends of oracular heads attributed to natural philosophers had begun to appear in Latin historiography, and they appeared in literature as marvelous objects. For instance, Mariano Taccola published a book on devices in 1433 and one on machines in 1450. He painted a world with underwater breathing equipment and boats that paddled themselves.[30]

The Roger Bacon legend is famous in part because of its appearance in Robert Greene's sixteenth-century *Friar Bacon and Friar Bungay*.[31]

Associating the castrato figure with Bacon's head doesn't just liken him to the automaton and the steam engine. Roger Bacon was also known as Doctor Mirabilis, or Doctor of Wonder, and was equally famous for his ability to incite wonder and to animate the motions of emotions. Bacon's experimental science harnessed nature's hidden forces to make wondrous effects.[32] In the premodern period, especially the Middle Ages and the Renaissance, thinkers like Heinrich Agrippa embraced a mechanical branch of natural magic that involved setting things in motion.[33] These thinkers continued the artificial magic of Daedalus and Archimedes but stopped just short of using the language of automation to describe the world. Artisans and engineers identified with magicians in part because of the belief that art could in effect make nature alter its course.

In the 1490s Leonardo da Vinci attempted to prove that nature was mechanically uniform and that the same mechanical principles governed humans and artificially created moving things. "A bird is an instrument working according to mathematical law," he wrote, "which instrument it is in the capacity of man to reproduce with all its movements but not with as much strength."[34] He invented several mechanical devices that through force produced a birdlike flapping.

> Force is nothing but a spiritual power, an invisible power, which is created and imparted, through violence from without, by animated bodies to inanimate bodies, giving to these the similarity of life, and this life works in a marvelous way, constraining and transforming in place and shape all created things.[35]

When he turned to the human body after 1500, his work depended on his mechanical work and focused on motion.

Giambattista della Porta, a Neapolitan nobleman who wrote extensively about natural magic at the end of the sixteenth century, described speaking tubes, trick mirrors, and automata. Della Porta's treatise on natural magic comprises twenty books of spells, recipes, experiments, and observations that include such activities as the generation of new plants, the conversion of metals, recipes for hair dye, and the construction of speaking tubes and magic mirrors.[36] Della Porta never left nature to her own devices. Instead, he prescribed manipulation: ripening strawberries out of season, making lettuce grow, and so on. Building on ancient practitioners, della Porta defined the magician as "a skillful workman, both by natural gifts, and also by practices of his own hands."[37] The art of magic alters nature and paves the way for her:

> If she find any want in the affinity of Nature, that it is not strong enough, she doth supply such defects at convenient seasons, by the help of vapors, and by observing due measures and proportions; as in Husbandry, it is Nature that brings forth corn and herbs, but it is Art that prepares and makes way for them.[38]

The magician is an artisan who first understands the natural form of a thing and then learns how to make nature do his bidding.

At the same time, the materials of sound had spiritual properties, vibrant materiality. Musical instruments, made of natural materials, carried the spiritual properties of the animals from which they were made. Della Porta claimed that music could lead men into battle and calm beasts because of the materials of the instruments.

> Whence I think it is not against reason that the same may be done by the Lute or Harp alone, but what is done by art or cunning, is more to be wondered at, which none can deny. But if we seek out the causes of this, we shall not ascribe it to the Music, but to the Instrument, and the wood they are made of, and to the skins; since the properties of dead beasts are preserved in these parts, and of Trees cut up in their wood.[39]

He explained that because bears hate horses, a drum made with horse skin will scare away a bear:

> A Horse, that is a creature made obedient to man, has capital hatred with a Bear, that is a Beast hurtful to man; he will know his enemy that he never saw before, and presently provide himself to fight with him, and he uses art rather than strength for it; and I have heard that Bears have been driven away in the Wilderness by the sound of a Drum, when it was made of a Horse skin.[40]

Because of the danger of snakes and the like, fiddle strings made of serpents can wreak all kinds of havoc:

> If Fiddle strings be made of Serpents, especially of Vipers, for being put on a Harp and played on, if women with child be present, they suffer Abortion, and Vipers are wont to do as much by meeting them, as many write.[41]

For della Porta, the power of sound came from the cunning mobilization and modification of a natural material.

## ORFOBOT AND THE SEVERED HEAD

Thanks in part to Carolyn Abbate, I can't forget the head floating down the river still singing. She hears Orpheus as reflecting modern anxieties about the self as mechanical, and his drowned out and decapitated head is an image that haunts the history of opera. Orpheus's lament in Gluck's opera, with its organ-grinder-coded tune, sounds like a music box, and Orphic prowess in *Die Zauberflöte* (*The Magic Flute*), sounds in Tamino's flute, not anyone's voice. These moments reverberate with dread of the mechanical human and the uncanny power of something that is strange and recognizable all at once.

Abbate understands the Orpheus story as a tale of dismemberment and disembodiment: "Orpheus's head is a musical instrument, an object given life as long as a master plays it."[42] It "represents singing that travels far from the body in which it originated, as a physical object that is cousin to the classic poetic image, the echo."[43] The head defies the notion of the opera singer as an inert vessel operated by a force outside itself. It "represents the uncanny aspects of musical performance in particular, precisely because one cannot say how it sings, who is in charge, who is the source of the utterance, and what is the nature of the medium through which musical ideas become physically present as sound."[44] Abbate's language about the head and lyre resemble the *Oxford English Dictionary*'s definition of a robot: a machine capable of automatically carrying out a series of programmable movements.

Abbate notes that the head doesn't really appear on the stage regularly until the nineteenth century when it turns uncanny, and Apollo plays it like a distorted instrument. Think Pierrot Lunaire grotesquely playing a song on Cassander's bald head: "He grasps Cassander by the collar. / Dreaming, he plays on the man's bald spot / With a bow monstrously gross."[45] Or think of Rainer Maria Rilke's 1919 "Primal Sound," which makes a literary fantasy of recording skulls by running gramophone needles across the cranium.[46]

But the head and the violent end were always already there. In Ovid's version the Bacchantes drown him out, decapitate him, and leave his head murmuring and floating down the river still singing, already a self-propelling sonic mouth in creepy counterpoint with the lyre also floating down the river. In Ovid's rendition, "The poet's limbs lay scattered all around; but his head and lyre, O Hebrus, thou didst receive, and (a marvel!) while they floated in mid-stream, the lyre gave forth some mournful notes, mournfully the lifeless tongue murmured, mournfully the banks replied."[47] The lyre continues to play without human assistance, and the tongue could merely murmur; it couldn't make words. The duet between lyre and tongue reverberates in nature until the head eventually lands on the island of Lesbos.

Judging from the 1607 published libretto for Monteverdi and Striggio's production, the first performance did stage the Bacchantes and allude to Orpheus's fate: "He shall not escape, for the later Celestial anger descends on guilty heads, the harsher it is."[48] The published score replaces the severed head with a Christianized deus ex machina in which Apollo comes down on a cloud and takes his son off to heaven with him. The circa 1600 Orpheus dramas didn't need the actual decapitation because it was a palimpsestic presence: the ghostly head complicates the assumed relationship between sound and human, and it existed in a hybrid presence with past iterations of the story.[49]

Striggio and Monteverdi were deeply influenced by Angelo Poliziano's 1480 *Favola d'Orfeo*, performed in Mantua, and like him based their story on Virgil and Ovid, texts in which the severed head makes disembodied sound, thrown voices, and sound effects. Virgil gives the head words.

> Even then as his head, torn from his marble neck,
> was rolled along by the Oeagrian Hebrus, carried on the
> gurgling rapids, "Eurydice," cries his very voice and cold tongue,
> "Ah! poor Eurydice!" as his breath of life flees.
> "Eurydice" the banks, along the whole course, echo back.[50]

Poliziano's staging concluded with an offstage decapitation followed by a gruesome John the Baptist–style parade of the head on stage.[51] Orpheus circulated in Giovanni Andrea dell'Anguillara's popular and widely disseminated reworking of the *Metamorphoses*, first published in Venice in 1561 and reprinted frequently for the next half century, a source likely known to most literati.[52]

> Another threw a stone, which, even as it flew through the air, was overcome by the sweet sound of voice and lyre and fell at his feet as if 'twould ask forgiveness for its mad attempt. But still the assault waxed reckless: their passion knew no bounds; mad fury reigned. And all their weapons would have been harmless under the spell of song; but the huge uproar of the Berecyntian flutes, mixed with discordant horns, the drums, and the breast-beatings and howlings of the Bacchanals, drowned the lyre's sound; and then at last the stones were reddened with the blood of the bard whose voice they could not hear.[53]

These fifteenth- and sixteenth-century heads sound a lot like the oracular talking heads that populate ancient and medieval literature.[54] And they sound like the talking puppets Francis Bacon wrote about in his *Sylva Sylvarum, or, a Natural History in Ten Centuries*, first published in 1627, just after he died.

> If a Man (for curiosity or strangeness sake) would make a Puppet, or other dead Body, to pronounce a word: Let him consider on the one part, the motion of the *Instruments of Voice*; and on the other part, the like Sounds made in *Inanimate Bodies*; and what Conformity there is, that causes the Similitude of *Sounds*; and by that he may minister light to that effect.[55]

Like automata more generally such talking heads lived in liminal spaces: fairs, gardens, labs, and labyrinths. They lived on the margins of life and death, built and natural worlds, and they occupied a space between spirit and matter. Orpheus, in all his incarnations, lived in exactly these places. He appeared in gardens and sang his song at the gates of hell.

Descartes's somewhat traumatic encounter with an Orpheus automaton sits at the center of his famous understanding of the human body as a machine. His 1644 treatise on man describes, in detail, an Orpheus automaton who lived in the royal gardens of Saint-Germain-en-Laye and was built by the Italian hydraulic engineers Tommaso and Alessandro Francini.[56] The gardens featured a hydraulic music-playing automaton that sprayed visitors with water, giving them an early modern 3D experience. The automated Orpheus made music on his lyre as birds sang and animals danced around him. "Animals, birds and trees are seen coming toward Orpheus as he plucks the strings of his lyre, the animals lowering their heads and stretching themselves, the birds flapping their wings, the trees moving on to hear the music of the divine singer."[57] The mechanical Orpheus spewed artificial music out of his lyre as animals danced mechanically.

Descartes used this encounter as a platform for explaining the nerves of the human body and the power they sent coursing through the muscles, which resembled the pipes and water contained in the statues. He had already used mechanism to assert that animals and humans were automata. But machines and automata that could have vitality and divinity meant something very different to Descartes and to his predecessors than they did to his Cartesian followers, who would empty mechanical creations of soul and, eventually, life. The distinction between Descartes and the Cartesians is as important as the distinction between Marx and Marxists.

## INSTRUMENTS AND THE ARTISANAL TURN

The ancient and fifteenth-century prequel helps to modulate standard musicological narratives into modes that consider instruments as tools and that embrace artisanal practices as well as Neoplatonic ideals.[58] The Homeric ode written down around 520 BCE linked Orpheus to the lute and described Mercury the artisan making the instrument out of vibrant flesh and guts.

> There, he found a tortoise, and gained immeasurable
> prosperity: yes, Hermes first engineered a tortoise to be a
> singer—the one who crossed his path at the courtyard gates,
> browsing on the flourishing vegetation in front of the house,
> stepping with a swagger.[59]

The process was gruesome, with the young god boring out "the life-giving marrow" of the tortoise. In Isidore of Seville's encyclopedic *Etymologiae,* written in the seventh century, the word *string* (chorda) comes from *heart,* because the throbbing of the strings in the cithara is like the throbbing of the heart in the chest. He writes that Mercury "made them first, and was the first to pluck sound from musical strings."[60]

In the performance of Poliziano's Orpheus story, the lyre played a key role. Baccio Ugolini, who sang the title role, was a celebrated Florentine improvisatory singer and ambassador who seemed never to travel without his lute and was well schooled in Latin and Greek. Blake Wilson's work paints a picture of a virtuosic singer who, on a 1473 mission to Rome, sang unseemly drunken odes in the streets of Rome and accompanied himself singing the role of Orpheus at banquets.[61]

When in the late sixteenth century there was an uptick in vocal and instrumental treatises, the singer was not yet totally severed from the instrumentalist. In practical terms the vocal and instrumental treatises shared vocabulary and technique, documenting an artisanal and technical practice. Sixteenth-century singers were trained predominantly in churches, where they also learned to read and write. Instrumentalists learned their trade in guilds through a largely oral tradition. In his 1544 *Dialogo della musica,* Antonfrancesco Doni prescriptively described a violin player in ways that mirror vocal treatises, including strict instructions on bodily comportment:

> Signor Giovaniacopo Buzzino was playing the soprano *violone,* as he does so marvelously, when some nobody came up to him in the midst of his performance to say "O Signore, move your fingers a bit more slowly; it looks so ugly to see you move your fingers so much on the neck of the instrument." Buzzino, accommodating himself to this bit of insolence, began to play without diminutions, whereupon the poor dolt, hearing the melody so lacking in grace, shamefacedly asked him to start moving his fingers again.[62]

Silvestro Ganassi's *Opera intitulata Fontegaro*, first published in Venice in 1535, gives formulas for mastering ornamentation and reminds instrumentalists to imitate the voice. The virtuoso cornet player Zenobi made the connection more explicitly in a letter about choosing musicians that he wrote from Naples, probably around 1600:

> Now all the requirements described above, or the greater part of them, are sought in an instrumentalist, whether he plays the cornett, the viola da gamba, the violin, the flute, the shawm, or similar melody instruments. As concerns those who play instruments that produce all the parts I shall tell Your Highness later briefly and truthfully what I think and know about that, without prejudice to anyone. Indeed this was and is my intention, as in everything else already said, since it seems to me that this is what the command of Your Highness and modesty and virtuous sincerity on my part demand.[63]

By the end of the sixteenth century, these instrumental and vocal practices began to coalesce in print.

Musicians frequently performed polyphonic music by singing one part and accompanying themselves on an instrument, usually a lute. Ercole Bottrigari, a professional singer and lute player from Florence, published, for example, a book essentially of madrigals arranged for singer and lute in 1574.[64] It contains arrangements of madrigals for lute and voice made for his own use.[65] Antonio Gardano's 1577 edition of *Tutti i madrigali di Cipriano di Rore a quattro voci, spartiti et accomodati per sonar d'ogni sorte d'istrumento perfetto, et per qualunque studioso di contrapunti* (All the madrigals by Cipriano de Rore for four voices, scored up and adapted for playing on all sorts of perfect [i.e., chordal] instruments, and for any student of counterpoint) comprises a collection of thirty-five pieces scored for study; it made sound an object and captured the sounds of singing—without the words—in print for instruments.[66]

Most composers began careers as players, and most castrati played musical instruments. Monteverdi was, as Matteo Caberloti remarked, hired first in Cremona as a string player and was described as "novello Orfeo col suono della sua viola, di cui non ebbe pari" (novel Orpheus, who with the sound of his viol [viola bastarda] had no rivals).[67] A well-known letter sent on May 4, 1583, by the Duke of Mantua, Guglielmo Gonzaga, to his agent in Paris, explaining the attributes of the kind of castrato he was seeking for his court, described his preferred castrato as a good Catholic of even temperament, of good voice, and stated that he would be more highly prized if he were also to know counterpoint and were able to accompany himself on the lute in chamber music.[68]

Around the turn of the seventeenth century, while composers were increasingly writing idiomatically for individual instruments, there was a surge in treatises on how to sing and how to play virtuosically but with taste. Scholars tend to think of the seconda prattica as predominantly vocal music and its consequents and instrumental music as *stile moderno*. Andrew Dell'Antonio has used collections of instrumental pieces that were published in the 1620s and 1630s to demonstrate shared compositional and stylistic tendencies across instrumental genres.[69]

Giovanni Battista Doni, in his *Trattato della musica scenica*, written in about 1633–1635, described the acoustical problems of making instrumental music audible to the audience. When poorly done it

> renders so little sound that it can scarcely be heard by those nearest the stage.... The result is, at the very most, that their sound reaches the ears of those who are in the middle of the hall. But if it is so powerful that it can reach the ends, no doubt it will excessively cover the voices (which are usually heard but little); and those who are in the foremost places will be unable to endure it.[70]

He went on to describe the many practical difficulties:

> The pains, the disgusts, the anxieties, and the griefs that the poor musicians feel in arranging together so many players and sounds in so narrow a place, would scarcely be believed. For, with much loss of time and [much] confusion, they must arrange the instruments, distribute the lamps, order the seats, erect the music stands, and tune the instruments. And God knows if, after tuning them well, they don't often have to do the whole thing again from the beginning, because of the multiplicity of the strings and their slackening on account of [the heat of] the lamps, and as well as they can be readjusted while the others are playing. To say nothing of the trouble and time it takes to make so many copies of the tablature of the bass, and of other disorders which result from this miscellany, introduced without any ground.[71]

Despite the long-standing privileging of the voice over any "artificial" instrument, such as a viol or keyboard, by the turn of the seventeenth century composers had begun to recognize the potential of musical instruments to create new soundscapes. Vincenzo Galilei, following his teacher Zarlino, derogatively referred to instrumental music as an artificial manipulation of nature, even as he based his theories and music on those same instruments. For Galilei, recovery of the lost Orphic music demanded more than the voice alone; it also required the use of musical instruments.

Amid his discussion of instruments, Galilei invoked Orpheus in an unusual way. He compared the harsh tones of "modern instruments" to the sounds that Neantius made when he used the cithara of Orpheus. Neantius failed at playing not because he lacked skill.

> When this Neantius played upon the cithara in question, it was revealed by his lack of skill that the strings were partly of wolf-gut and partly of lamb-gut, and because of this imperfection—or because of the transgression he had committed in taking the sacred cithara from the temple by deceiving, believing that the virtue of playing it well resided in it by magic, as in Bradamante's lance that of throwing to the ground whomever she touched with it—he received, when he played it, condign punishment, being devoured by dogs.[72]

Orpheus derives his magic from his ability to play his instrument, to manipulate nature with his hands. And what makes instruments fail, as in tickle the senses rather than nobly and respectfully imitate the ancients, is the materials themselves:

> If this property of tickling (for it cannot with truth be called delight in any other sense) resides in a simple piece of hollow wood over which are stretched four, six, or more strings of the gut of a dumb beast or of some other material, disposed according to the nature of the harmonic numbers, or in a given number of natural reeds or of artificial ones made of wood, metal or some other material, divided by proportioned and suitable measures, with a little air blowing inside them while they are touched or struck by the clumsy and untutored hand of some base idiot or other, then let this object of delighting with the variety of their harmonies be abandoned to these instruments, for being without sense, movement, intellect, speech, discourse, reason or soul, they're capable of nothing else.[73]

In each of these cases, the human hand molds a natural material into something that produces sound. Without the proper intelligence and spirit, the instruments do nothing more than create base sounds, but with the proper play they accomplish something special:

> But let men, who have been endowed by nature with all these noble and excellent parts, endeavor to use them not merely to delight, but as imitators of the good ancients to improve at the same time, for they have the capacity to do this and in doing otherwise they are acting contrary to nature, which is the handmaiden of God.[74]

## REHEARING MONTEVERDI

What does this have to do with Monteverdi's *L'Orfeo?* In the wood, throats, and bodies, the intimate Mantuan performance becomes a key moment in the history of music technologies. Casting this as a production consumed with echoes, sounds bringing back the dead, and objects that make people feel things dethrones the lamenting Orpheus. It turns the drama from a Neoplatonic re-creation of the human voice as Orphic magic to one of mechanical manipulation in which Orpheus is hurled through his own production by musical instruments and castrati. As I have argued elsewhere, in *L'Orfeo* song as a powerful incantation depends not just on its human animator but on its mechanical extension in the form of instrumental effects and castrati singers.[75]

I want to avoid reading Orfeo as just a character drawn by Monteverdi with the singer as human playback machine, the familiar composer-as-genius-driving-performer model. Instead, Monteverdi's *L'Orfeo* and the figure of Orpheus are connected points of departure for asking questions about the instrumental potential of humans, the enlivening potential of music, the affective potential of instruments, and the stories musicians and music scholars tell.

La Musica's prologue doubles as a manifesto for the seconda prattica. In the first performance Giovanni Gualberto Magli, a Medici castrato borrowed by the Gonzagas for the production, delivered the "I am music, hear me roar" moment, or in Renaissance terms, "Now while I alternate my songs, now happy, now sad, let no small bird stir among these trees, no noisy wave be heard on these river-banks, and let each little breeze halt in its course."[76] The body of a castrato with a calculated ritornello sets the stage for music and the title character to do their affective work. She trumpets her own powers to move, allure, and enchant and in doing so reinforces what the audience already knew: that Orpheus's song could stop nature in its very tracks.

The prologue's words explain the intellectual commitment behind the multiple stagings of the Orpheus story that occurred around 1600. To be sure, Orfeo, envoiced by the famous tenor Francesco Rasi, was a musical magician whose musico-poetic utterances carried a long-heralded occult power. But he's rather impotent and is set in motion by La Musica. This was of course the dramatological setup in production in the seventeenth century, but here the effect highlights musical automation. She spins him into affective action. La Musica describes and demonstrates the power of recitative to incite and mimic the invisible psychological motions that make up that magic. She has descended from on high to praise the Gonzagas and call the audience to attention, but she has also come to celebrate music itself, championing her own abilities to calm every troubled heart and inflame frozen minds in the second and third stanzas:

> I am Music, who in sweet accents
> can calm each troubled heart,
> and now with noble anger, now with love,
> can kindle the most frigid minds.
>
> I, with my lyre of gold and with my singing, am used
> to sometimes charming mortal ears,
> and in this way inspire souls with a longing
> for the sonorous harmony of the singing lyre.

The third stanza, like the others, begins with a stable harmony that emphasizes the declamatory style of the opening and then is interrupted by the ornamented word *singing*. Both harmony and melody set off the opening, which sounds completely different from the rest of the strophe, with its easygoing tune.[77] La Musica sings in ritornello form, her voice punctuated after each stanza by identical instrumental interludes. Not only is her voice instrumental but her execution of affect is a kind of instrumentally driven form.

The ritornello animates the action and affect. La Musica's ritornello comes back without La Musica herself at two crucial moments in the opera, in the process becoming the musical sign created by musical instruments for the loss of Euridice. In act 2 the ritornello sounds just after the messenger announces Euridice's death, and the chorus reiterates her distress with the words "Alas, bitter chance! Alas, wicked cruel Fate! Alas, injurious stars! Alas, you acquisitive heavens" (Ex. 1.1). The refrain marks what sounds like a one-line aria; a descending diminished fifth filled in and elongated.

The melody imitates the messenger, but it's less dissonant, a smoothing cadence. And it comes again between Euridice's second death at the end of act 4 when Orfeo's fatal glance banishes her back to the underworld. After a loud offstage noise Euridice is commanded back to hell and sings her own shock in a series of musical sighs that move down a tritone: D to G♯ (Ex. 1.2). La Musica's accompaniment precedes Orfeo's final act 5 lament, "These are the fields of Thrace, and this is the place where my heart was pierced with grief at the bitter news" (Questi i campi di Tracia, e quest'è il loco Dove passommi il core Per l'amara novella il mio dolore).[78] Orfeo's perhaps overzealous lament sounds an accelerated grief.

"Possente spirto," Monteverdi and Striggio's version of Orfeo convincing

*Un organo di legno & un chit [arone].*

**Messagiera (C1)**

no Ahi ca-so a-cerbo ahi fa-t'em-pio e cru de le Ahi

Stel-le in-giu-ri - o - se, ahi Ciel' - a-va - ro.

EX. 1.1 Monteverdi, *L'Orfeo*, act 2, "Ahi, caso acerbo"

EX. 1.2 Monteverdi, *L'Orfeo*, Euridice's lament

Caronte to let him bring Euridice back to life, is a hymn to technology and the phonographic potential of the voice.[79] The song occurs in the soundscape of hell, low brass and music mostly more speechlike than songlike, except this one moment. For the 1609 score Monteverdi notated two versions of the aria, a scaffolding-like declamatory and syllabic melody printed above an embellished line (Fig. 1.4). The first elaboration occurs on the word *spirto* and adds visceral harmonic tension to the sound. Orfeo musters up all the musico-rhetorical prowess he can to woo everyone who hears his powerful plaints. The published score documents the tricks of a well-trained seventeenth-century virtuoso. However, the scene does not just create a climax of vocality, it makes one of instrumentality as well. A pair of violins, followed by a pair

FIG. 1.4 "Possente spirto," Claudio Monteverdi, *L'Orfeo favola in musica* (Venice: Amadino, 1609). Biblioteca Estense, Modena.

of cornetts, and finally double harps stand in for his lyre and answer the voice *passaggi* with figures that are specific enough to each instrument that they can show off the unique timbres and sounds. The instruments sing.

With each phrase Orfeo sings, two instruments echo, one after the other, neither of them exactly completing a thought or exactly playing the same thing. The wood of the instruments tosses the sounds back. The piece escalates in virtuosity as Orfeo's pleas are increasingly ornamented, up until the fourth strophe when it climaxes in an aria-like segment. Orfeo's thoughts turn to Euridice, and indeed toward a perfect Neoplatonic consummation through meaningful glances: "O serene lights of my own eyes" (O de le luci mie luci serene). Seeing that he has still not succeeded, the singer, abandoning virtuosity and song, hurls his final plea in an emotional recitative: "You alone, noble God, can help me" (Sol tuo, nobile Dio, puoi darmi aita).

The bard fails.[80] Orfeo didn't exactly convince the guard of anything; he merely sang a lullaby: "He sleeps, and though my lyre cannot wrest pity from the hardest heart, at least his eyes cannot escape from sleep at my singing. Let us go then. Why wait longer?"[81] The song works on Plutone and Proserpina, but no one knows exactly what they hear. The couple appears at the beginning of the next act, with Proserpina telling Plutone that Orfeo has wrested from her heart such pity that she can do nothing but beg her husband to yield to Orfeo's desires.[82] She responds to something offstage—maybe the severed automated head. More specifically, Orfeo doesn't move Plutone himself but appeals to the prospect of pleasing his young wife.[83] Once Plutone agrees to Euridice's release, Proserpina sings her love of him, rewarding him for acquiescing to her plea.[84] This climactic, or maybe nonclimactic, song demonstrates that Orfeo needs his lyre, and music needs its mechanical extensions, its artificial sounds. Here, the instruments sound their most vocal.[85]

While others, including me, have read the aria-like segment at the end as the defeat of virtuosity, it might also be a triumph of technology, a moment when voice and instrument try hardest to imitate one another, effectively fusing Orfeo with his lyre. The instruments supplement his voice, and the echo technology emerges as a compositional tactic to merge instruments and humans. This represents the height of musical ingenuity, of orchestral tricks. And perhaps the ghostly echo serves to usher in the Echo scene that comes next. It might also be a thorn in the side of the entire myth: music's superpower is not persuasion but the complete erasure of reason. And Orpheus can only manipulate those who do not have a seat at the table of the academies; Proserpina is a woman, and Charon is a creepy ferryman.

As if to emphasize the role of instruments, the chorus of spirits that follows Euridice's second death, after "Possente spirto," also sounds a brief hymn to the lyre, celebrating the artisan's ability to alter nature:

No undertaking of man is tried in vain,
Nor can nature arm herself against him.
He has ploughed the uneven fields
And scattered the seeds of his labor

The chorus sings accompanied by regal and wooden pipes, five sackbuts, two bass viols, and a violone—making it one of the densest sonorities in the drama. Their music moves directly into the ritornello that rings in the entry of Plutone and Proserpina.

In the original libretto (the version with the severed head) but not the published score (the one with Apollo taking his son up into the clouds), "Possente spirto" leads into a praise of Daedalus, whose artificial wings gave humans the gift of flight:

For his journey through the airy regions
The ingenious Daedalus unfolded light wings,
And neither the hot rays of sun
Nor the swampy dampness loosened his feathers.
But seeming in his progress a new bird
He made the winged family
Stop its flight in wonder.[86]

After serenading the ancient craftsman, Striggio returns to Orpheus as if to remind listeners that Orpheus, like Daedalus, makes his music with a bodily extension.

Daedalus, whose Greek name means "cunning worker," constructed wings for himself and his son Icarus by securing feathers with thread and wax. Icarus met his demise when he flew too close to the sun, but Daedalus went on to invent the saw, the compass, and other mechanical things. And in Ovid, Daedalus "applied his thought to new invention and altered the natural order of things," much as, say, the castrato who envoiced La Musica sang in a body whose natural order was altered by intervention.[87] Ovid connected sound and flight by a similar material process: "He laid down lines of feathers, beginning with the smallest, following the shorter with longer ones, so that you might think they had grown like that, on a slant. In that way, long ago, the rustic panpipes were graduated, with lengthening reeds."[88] After all of this comes a hymn to the lyre. In act 4, when Orfeo wins Euridice back, Il pastore introduces "the gentle singer / Who leads his bride to the Heaven above" (Ecco il gentil cantore Che sua sposa conduce al Ciel superno).[89]

Orfeo then serenades, in a strophic song, his own musical appendage—the appendage that, in Ovid and Virgil, sang even after the lyrical decapitation:

What honor is worthy of you,
My irresistible lyre,
Since in the realm of Tartarus
You had the power to sway all hardened hearts?
You shall have a place
Among the fairest images of heaven,
When to your sounds, the stars
Shall dance in rounds, now slow, now fast.[90]

And he heard only himself. The sounds of the lyre make the stars dance. Here, as in La Musica's prologue, an instrumental ritornello provides structure. But unlike the declamatory sounds of the muse, he sings a ritornello song, one of the moments that sounds most songlike. The three-stanza tune comes with a bass line that never stops moving, almost a walking bass with chord changes that allow Monteverdi to make melodies for the violins, sonic stand-ins for the lyre. The accompaniment is harpsichord, viola da braccio basso, and chitarrone. Orfeo's words sound over a walking bass interspersed with a violin ritornello, perhaps in deference to the *lira da braccio* that sixteenth-century iconography always placed in the bard's hands. Orfeo's serenade to himself captures and includes the dead wood that in "Possente spirto" served as his material echo. He is underscored by his answer.

## ECHO EFFECTS

The echo effects of "Possente spirto" turn human in Orfeo's act 5 lament. In 1607 these echo effects were already a favorite trick of Francesco Rasi, the first Orfeo.[91] And almost all of the music dramas produced around 1600 had an echo scene, including the various *Orfeos*.[92] The echo scene continued to be a feature of Orpheus operas. Stefano Landi used echo effects in *La morte d'Orfeo*, and Gluck embedded his echoes in the musical structure with instrumental offstage echoes.[93]

Like Orpheus, Echo is a mythological character who represents what Lynn Enterline calls the phonographic imaginary, a sound that dismantles our understanding of the voice and removes the speaking tongue and listening ear.[94] Orfeo's offstage echo, also a tenor, oozes myth, technology, and musical practice. That she speaks through a voice that matches Orfeo's highlights her status as sound object: sound bounces back transformed from its point of origin. In addition to doing narrative work in the drama, Echo represents a sound phenomenon, a reverberation. She is the separation of voice from body that Abbate writes about, the fragmented resonance that follows disembodiment.

A bodiless ghost who faded away into nothingness, capable only of repeating the words of others, she speaks without thought, breaking language into little bits.

Before Ovid, Echo often appeared as a lover or wife of Pan. Pan, angry that she was a better pipe player than he was, had the shepherds tear her to pieces and toss her "singing limbs" all over the place. Ovid, a master of literary sound effects, made Echo such a skilled conversationalist that Jupiter asked her to amuse his wife while he cavorted with nymphs. Juno was not fooled and, in a rage, reduced Echo to what we today might call a voice-activated device, unable to make her own words. "Only her bones and the sound of her voice are left. Her voice remains, her bones, they say, were changed to shapes of stone."[95] She works like Orpheus's stone head: "Echo still had a body then and was not merely a voice. But though she was garrulous, she had no other trick of speech than she has now: she can repeat the last words out of many."[96]

In the next chapter of Echo's saga, the soon to be bodiless voice falls in love with Narcissus but cannot speak until he addresses her. "Avoid me not" he says, and, of course, "Avoid me not" she answers. Eventually she rushes from the bush, throwing her arms around him. He says, "Better death than such a one should ever caress me." She answers with "Caress me." Embracing him, she is still here an expressive body. Rejected, she hides her blushing face in the lonely caverns of the hills until she in "her miserable body wastes away, wakeful with sorrows; leanness shrivels up her skin, and all her lovely features melt, as if dissolved upon the wafting winds—nothing remains except her bones and voice—her voices continues in the wilderness; her bones have turned to stone."[97] And then, finally, when Narcissus does himself in looking for his lover in the water, it is Echo who answers his "Farewell" with "Farewell." She has lost her uniqueness—her very subjectivity. However, she, unlike Orpheus, knows how to listen, how to make a time delay.

Echo scenes were regular conventions in pastoral theater and poetry, functioning partially as experiments in special effects and what we might call phonographic or automatic sound. Composers exploited it in madrigals, operas, and antiphonal music. In Orlando di Lasso's 1581 eight-voice madrigal "O la o che buon echo," the singers play with echo effects, joking with echo. Echo in this tune becomes somewhat hostile, saying "Taci, dico! Taci tu," which loosely translates as "Shut up! No, you shut up."[98] The echo motet in Monteverdi's 1610 vespers *Audi coelum* features a six-voice ensemble that turns into a solo with echo, whose puns mirror Orfeo's echo in that the last syllable of text is repeated or truncated in such a manner that it reveals a slightly different meaning. In this case, the echo comes as a voice from heaven answers first a call and then a series of questions, like Orfeo's echo.

EX. 1.3 Monteverdi, *L'Orfeo*, act 5, mm. 19–21

In the Mantuan *L'Orfeo*, Echo appears just after Orpheus fails again. His act 5 singing feels oddly dispassionate and disengaged. During this lament to the rocks in a valley designed to bounce sound, Monteverdi prescribed the accompaniment to include "duoi Organi di legno" and "duoi Chitarroni," one played at the left-hand corner of the scene and the other at the right. Echo repeats everything Orfeo says and in so doing dissociates sound from body. "You grieved and wept," Orfeo tells the mountains, "ahi pianto," and then she answers, "Hai pianto," changing his meaning slightly by telling Orfeo that *he* has wept (Ex. 1.3). "Gentle, loving Echo, you who are disconsolate and would console me in my grief, although these my eyes are already, through weeping, made into two fountains, in such heavy and cruel misery I still have not wept tears enough." Echo spits back only the word *basti* (enough), and in an odd reversal he is to be silenced by his own echo. Monteverdi writes for Echo very clichéd cadential phrases, which do Orfeo's work for him yet again, concluding his very thoughts, but on the other hand undo what he has done.

Echo, like the sound phenomenon she shares a name with, sends reverberations back to Orfeo.[99] Where Orfeo has moved from impassioned recitative to extreme virtuosity, his own speaks only in a monosyllabic monotone. This displeases him. "If you have pity on my misfortune, I thank you for your kindness. But while I lament, say, why do you answer me only with my last words? Return my laments entire." And then, of course, Echo vanishes. If, for Ovid's Echo, the act of repetition turns rejection into desire, for Orfeo the act of Echo's repetition turns his plaints into a harsh banishment of his articulated feelings.

The rocks and trees fail to answer back, and all he hears is his own voice dumbed down to a punctuation mark. For many scholars of opera, Orpheus's echo and his severed head point toward Wagnerian metaphysics. For Daniel Chua, Echo embodies an Adornian fear of the mechanistic, for "the disembodied voice is no longer under the subject's control."[100] He goes on to say that "in its attempt to master nature through its voice, the monodic subject ends up being mastered by nature, becoming like it in death."[101] For others the head points toward the machinic. Klaus Theweleit suggests that Echo's closing reverberations replace Euridice. Her loss entices Orfeo to codify her image like a sound recording. "You might laugh but . . . Orpheus is asking for Edison."[102] Theweleit references the story about Edison designing a metal cylinder and giving it to his favorite mechanic, John Kruesi, who produced the instrument in a mere thirty hours. Edison covered the cylinder with foil, turned the crank and screamed "Mary had a little lamb" into it. In effect, he set the reproducer on go and came out with a reproduction of his own voice, proclaiming, "I was never so taken aback in my life."[103]

Orpheus may or may not be asking for Edison, but he is gesturing toward a fascination with the technological production of sound, and he represents a way in which the music drama privileges technology and automated sound, not just the Orphic voice. Repetition turns into mechanistic effect. And it reflects an inextricable intertwining with the natural phenomenon. Echoes do not, in fact, reproduce exactly what they echo.

## A PAUSE FOR ECHO

It is worth pausing for a moment to think about just how mysterious, magical, and perhaps even scary echo effects would have seemed in a world of preindustrial, unrecorded sounds—a prereverb world where the phenomenology of sound was still fundamentally tied to mythologies and inexplicable forces. Echoes were some of the earliest examples of nonhuman, nonnatural sounds and played key roles in exploring the nature of sound and sensation. And imaginings of echoes have always involved an overlap of mythology and acoustics.

Aristotle, in *De anima*—the writing known for articulating a philosophy of living things—takes the soul as more of a principle of animation than something that depends on a mind. Aristotle's discussion of echo is part of his account of hearing, in which sound is the striking of bodies and reverberations:

> An echo occurs, when, a mass of air having been unified, bounded, and prevented from dissipation by the containing walls of a vessel, the air originally struck by the impinging body and set in movement by it rebounds from this mass of air like a ball from a wall. It is probable that in all generation of sound echo takes place, though it is frequently only indistinctly heard. What happens here must be analogous to what happens in the case of light; light is always reflected—otherwise it would not be diffused and outside what was directly illuminated by the sun there would be blank darkness; but this reflected light is not always strong enough, as it is when it is reflected from water, bronze, and

> other smooth bodies, to cast a shadow, which is the distinguishing mark by which we recognize light.[104]

Lucretius discussed the echo in a long section of the fourth book of *De rerum natura* devoted to hearing, which, by extension, discussed sound effects like echoes. Echo effects were a means, like loudspeakers, to broadcast voices to a populace—corporeal voices flung into nature, voices with no bodies, and voices that sometimes copy words but sometimes alter them just a bit, seemingly of their own volition. Lucretius agreed preemptively with modern acoustics about the things that make a good echo chamber: rocks, hills, shady mountains. He explained that people in those places often imagine their world inhabited by nymphs and Pan. He thus connected myth, magic, and technology.

> Beating on solid porticoes, tossed back
> Returns a sound; and sometimes mocks the ear
> With a mere phantom of a word. When this
> Thou well hast noted, thou canst render count
> Unto thyself and others why it is
> Along the lonely places that the rocks
> Give back like shapes of words in order like,
> When search we after comrades wandering
> Among the shady mountains, and aloud
> Call unto them, the scattered. I have seen
> Spots that gave back even voices six or seven
> For one thrown forth—for so the very hills,
> Dashing them back against the hills, kept on
> With their reverberations.[105]

Virgil, too, understood the echo chamber and its reproductive potential. If for Lucretius humans learned their sounds from nature, Virgil's Tityrus reversed the process, teaching the woods to echo—to resound. As Meliboeus remarked, "You, Tityrus, lie under the canopy of a spreading beech, wooing the woodland Muse on slender reed, but we are leaving our country's bounds and sweet fields. We are outcasts from our country; you, Tityrus, at ease beneath the shade, teach the woods to re-echo 'fair Amaryllis.'"[106] The ninth eclogue, in effect, describes a pitch-perfect echo chamber:

> Your pleas merely increase my longing. Now the whole sea plain lies hushed to hear you, and lo! every breath of the murmuring breeze is dead. Just from here

> lies half our journey, for Bianor's tomb is coming into view. Here, where the farmers are lopping the thick leaves—here, Moeris, let us sing. Here put down the kids—we shall reach the town all the same. Or if we fear that night may first bring on rain, we may yet go singing on our way—it makes the road less irksome. So that we may go singing on our way, I will relieve you of this burden.[107]

In the Middle Ages, thinkers for the most part interpreted echoes and acoustics through a wave metaphor. A wave worked something like the concentric circles that a skipping rock produces. The rediscovery of Archimedes in the sixteenth century led to understanding sound through theories of light, and the rediscovery of Vitruvius-inflected theater design. Sixteenth-century natural philosophers began more aggressively to incorporate echoes into experimental practice and attempts to create speaking machines.[108] For della Porta, the echo was a trick of natural magic:

> It is wonderful, that as the light, so the voice is reverberated with equal angles. I shall show how this may be done by a glass. It is almost grown common, how to speak through right or circular walls. The voice passing from the mouth goes through the air: if it goes about a wall that is uniform, it passes uncorrupted; but if it be at liberty, it is beaten back by the wall it meets with in the way, and is heard, as we see in an echo.[109]

And later he wrote this:

> We see that the voice or a sound, will be conveyed entirely through the air, and that not in an instant, but by degrees in time. We see that brass guns, which by the force of gunpowder, make a mighty noise, if they be a mile off, yet we see the flame much before we hear the sound. So handguns make a report, that comes at a great distance to us, but some minutes of time are required for it, for that is the nature of sounds; wherefore sounds go with time, and are entire without interruption, unless they break upon some place. The Echo proves this, for it strikes whole against a wall, and so rebounds back, and is reflected as a beam of the sun.[110]

In the seventeenth century Giuseppe Biancani called the study of echoes "echometria." Marin Mersenne and Athanasius Kircher both performed a variety of echo experiments that not only continued the quest to create automatic sound but inextricably connected the myth of Echo to the sound object and to music making as experimental practice.

Kircher called his study of echoes "Phonosophia Anacamptica."[111] He wrote about echoes in the 1650 *Musurgia universalis* and the 1663 *Phonurgia nova*. The *Phonurgia* is a literary theatrical production dedicated to sonic experimentation and sound objects. The word *phonurgia* comes from the Greek for *sound* and *work*, and the full title can be translated as "New modalities of sound." He began this tome with a discussion of echo:

> The echo, that jest of Nature where she is in a playful mood, is called the "image of a voice" by the poets, in accordance with the line of Virgil: *The rocks resound and the image of the voice that has been struck bounces back*. It is called a reflected, rebounding and alternating voice by philosophers and "the daughter of voice" by the Hebrews.[112]

Kircher explained the echo as the ability to provoke the "marvelous" with sound. And he infused his experiment with mythology:

> But as I pursue her, she runs away, while I run away, she pursues me, while I speak seductively, she seductively tricks me, as I cry aloud, she redoubles her voices by taking on additional voices like attendants, not knowing how to stop.[113]

Kircher's published echo experiments focused on the production of artificial echoes and instruments that extend the range of the voice and hearing, enabling secret communications at a great distance. In 1672, for example, he demonstrated his speaking trumpet from the rock of S. Eustachio, near the church of S. Maria de Monterella in Sabina, not far from Tivoli, using a conical tube whose range extended even to certain nearby villages between two and five miles away. He supposedly summoned the peasants and succeeded in assembling a crowd of two hundred for evening vespers. He also worked on partial echo chambers that worked like portable mirrors.

Kircher's echo did not repeat exactly. He experimented even with multilingual echoes. One could speak in French and receive back an echo in Spanish. Kircher explained an experiment (Fig. 1.5) involving the Latin word *clamore* (clamor) that resulted in the tossing back of a variety of words, including *auri* (gold) and *amore* (love).

> Let us set up for example a four-syllabled echo at the reverberative objects B C D E F arranged in such a way that each of these reverberates one syllable later. The voice or the word that follows shall be the three-syllabled Latin CLAMORE. Moreover, the vocal center or out-going voice is A, because the reverberative

FIG. 1.5 Athanasius Kircher's echo formation, *Musurgia universalis* (Rome, 1650), 2:264. Courtesy of the Albert and Shirley Small Special Collections Library.

> objects are so ordered that each reflects back one syllable later, and thus it is clear that each object will reverberate back a distinct and different word. At the reverberating object B, CLAMORE; AMORE at the reverberating object C. MORE at the reverberating object D. ORE at the reverberating object E, and finally RE is reverberating at object F. In this way, constantly different words resound back from a many-voiced echo, as if someone is shouting.[114]

Each of these words that comes from the echo comes out as a resounding body. It is phonographic for sure; Kircher's drawing shows this viscerally. The effect is animated by an originary human, who seems to shoot forth sound.

## PUPPET THEATER

To conclude, Gaetano Guadagni was, in his day, most famous for singing the role of Orpheus in Gluck's 1762 *Orfeo ed Euridice*. In retirement he made his house in Padua a phantasmatic marionette theater where, among other things, he had a puppet Orpheus. The theater had all the stuff of a major theater in miniature: machines for scenery changes, stage sets for a garden with waterfalls and a sea. The tenor Michael Kelly, on his Grand Tour, described it like this:

> He had a very neat theatre, and a company of puppets, which represented *L'Orpheo e Euridice*; himself singing the part of Orpheo behind the scenes. It was in this character, and in singing Gluck's beautiful rondo in it, "Che farò senza Euridice," that he distinguished himself in every theatre in Europe and drew such immense houses in London. His puppet show was his hobby-horse, and as he received no money, he had always crowded houses. He had a good fortune, with which he was very liberal, and he was the handsomest man of his kind I ever saw.[115]

By manipulating marionettes and envoicing them with his failing voice, Guadagni turned Orpheus into an automaton, a marionette who is made vital and animate by the castrato's voice and the puppeteer's hands.

Guadagni was famous not for pyrotechnics but for a natural voice and for his acting, and thus he could sing the Orpheus role even as his voice aged. Gluck even said that "Nothing but a change in the mode of expression is needed to turn my aria 'Che farò senza Euridice,' into a dance for marionetters."[116] The aria for Gluck lies in the hands of the singer, animated by the voice. But the voice here came out of a puppet. Here, in the aging castrato's little puppet show, was a materially constructed man animating his own production by moving puppets.

Later the puppets and the castrato became ghostly figures. Vernon Lee wrote a dialogue called "Orpheus in Rome" in which Baldwin, the character who voices her alter ego, spins a deep fascination with Guadagni. For Baldwin, Guadagni the puppeteer becomes his own ghost.

> Perhaps I have a clearer impression of Signor Gaetano Guadagni—his name was Guagni, and he was a Lombard like you, Dona Maria—than of myself in the days when I made his acquaintance in old music books and memories. It's odd by what caprice one singles out some forgotten creature of the past or rather by what caprice some particular ghost chooses to manifest himself and haunt.

> Anyhow, I used to see and hear Signor Guadagni whenever I turned over the pages of Orpheus, or when I hummed over any of its airs in my memory. Do you care to hear about my friend the ghost?[117]

For Baldwin, Guadagni is the ghost in the machine, the ghostly voice leaping out of the book.[118] In Lee's writing Orpheus is a never a figure of one time but is rather a cacophonous mixture of elements from multiple temporal spaces.

There's one last point to this thought experiment. If Orpheus traffics in the same matter as the rest of us humans and machines, we might be able to move away from disembodied metaphysics to the material details of human experience. There are consequences to this kind of materialism: if everything is alive—humans, machines, ideas—then the distinctions between active subjects and passive objects start to crumble. Such questions lead quickly to questions of personhood: what or who counts as a person and what or whose voice gets heard? If Orpheus is intimately connected to the material world and not a figure apart from it, then he does not stand as an individual, transcendent creative genius. He does not speak in a universally affective human voice, and maybe no such voice exists.

CHAPTER 2

# THE DEATH OF A CICADA

In June of 2021, as the world hoped it was waking from a global pandemic, cicada Brood X was waking up from a long slumber. Listeners to NPR might have heard the poet Margaret Atwood read her erotic poem about the cicada, "the piercing one note of a jackhammer vibrating like a slow bolt of lightning, splitting the air." The male cicadas make their striking noise to attract a mate.

> And close by, a she like a withered ear, a shed leaf brown and veined shivers in sync and moves closer. This is it. Time is short. Death is near. But first, first, first, first in the hot sun, searing all day long in a month that has no name, this annoying noise of love, this maddening racket, this admitted song.

Anyone who has spent time in what used to be called the torrid zone knows the sound of the cicada. On steamy days thousands of them create a symphonic surround sound. Often invisible, the acousmatic voice, like echo, has no visible body.

The clamorous love-death has fascinated thinkers, artists, and musicians for thousands of years. According to various ancient renditions of the myth, cicadas were men so consumed by the pleasure of song that they forgot to eat and died. The muses turned them into insects who survived on song and the morning dew. Anne Carson describes the cicada as "at odds with time."[1] Galileo Galilei made the dramatic literary vivisection of a cicada the punch line of his fable of sound. He published the story on papal lockdown in 1623, and it appeared first in the twenty-third section of a gorgeous little best seller called *Il saggiatore* (The assayer), which told the story of a curious hermit who goes on a sound walk to discover the origins of multiple sounds.

The point of musings on the cicada in a book about castrati might seem connected to the violence of science and creativity—Galileo's lyric vivisection that sits at the center of this chapter reads like the animal vivisections that studies of anatomy depended on from the second through the eighteenth centuries. The cicada always sings himself to social death, so both might be interpreted as songs of addicted men for whom song catalyzes metamorphosis. And the cicada and the castrato were enmeshed in mythologies of denied desire and thwarted sexual reproduction; the cicada can only sexually reproduce by dying; the castrato can only have desire and love without sexual reproduction.

For me the connection resides in stories and in the ways that the castrato and the cicada invite attention to the historically constituted materiality of sound. Both index what early modern Italians knew, or thought they knew, about sound as vibration, motion, resonance, and political force. They sang and still sing through filters of myth, philosophy, aesthetics, and natural philosophy in part because their sounds were so little understood for so long. This excavation is a textual version of what happens when we produce a record. An audio engineer takes raw sound blown into multiple microphones and works magic: maybe they turn up the singer to out blow the brass player, maybe they use reverb to make it sound like a huge cathedral.

A Google search reveals that the song of the cicadas comes from little, tiny drums in their bellies, or tymbals, that resonate when muscles contract and release. But as late as 2013 entomologists were still claiming to discover how the cicada made its sound. In that same year archaeologists exhumed the remains of the castrato Gasparo Pacchierotti, and a team of scientists from the University of Padua Medical school determined that the castrato's skeleton changed from both vocal training and the hormonal effects of castration. The causal relationship between crushing or removing a boy's testicles and retaining a high singing register was well known in the ancient world. But, the concept of hormones, for instance, didn't emerge in scientific thinking until around 1900. The word came from the Greek term for motion, *hormone*, and the idea of actions of one organ affecting another was new and a radical change from the prevailing assumption that human bodies worked by electric messages sent along nerve fibers. That meant that in some fundamental way the castrato's high voice remained mysterious no matter how many throats were explored.

This chapter begins by retelling Galileo's story as a sound studies text and a cautionary tale. While on his quest, the hermit wandered farther and farther from the pleasure of sound and, indeed, from the sounds themselves. He found himself seduced not by the vibrations of air but by the mechanics—the tools that make the noises. He no longer cared about understanding why sounds create pleasure and moved toward separating the mechanics of sound from sensory experience. The fable's soundscape suggests ways in which the field of contemporary sound studies often evades sounds, bodies, and history. This makes for an antihermeneutic practice that embraces a new formalism, one obsessed with objects as removed from bodies and culture, much as notes and style once were in music studies.

I move on to a quick historical survey of accounts of the cicada; the little bug is a rich site for the way sound was imagined and felt at different historical moments. Think about lightning: before Benjamin Franklin flew his kite, educated people thought that the thunder and lightning were signs of God's wrath and that the damage came not from the fiery lightning but from the noise of thunder. As late as the seventeenth century, reputable thinkers thought that gunpowder and bells could serve as defenses against the wrath of thunderstorms. Finally, I conclude by listening carefully to the vivisection of the cicada for what it reveals about the anatomical exploration of the vocal cords in the early modern period and for what it says about the complicated relationship between humans and animals—for a sense of what this sonic relativism says about how humans inhabited the earth (Fig. 2.1).

## KILLING THE BUG

Not surprisingly, Galileo was a competent musician, apparently an accomplished lute player. It was sort of the family business. His professional lute-playing father Vincenzo Galilei did important musical experiments on tuning and wrote a counterpoint treatise that prefigured the seconda prattica. Vincenzo was part of the collective that wanted to recreate ancient drama and that arguably set the stage for the *Orfeo*s of 1600. It makes sense that he used a story about sound to stage a confrontation between humans and technological objects. Scholars outside of music know Galileo's fable mostly for its polemic attack on Orazio Grassi's treatise on the comets and as one of Galileo's earliest statements on the importance of reading nature through the language of mathematics as opposed to ancient texts. It is often read as delivering a message about the power of science. Science leads the hermit from an ignorant place of isolation to a place of knowledge and variety.[2] It has also been understood as demonstrative of the limits of science: you can know something is true even if you can't understand or explain it. The hermit knew the cicada made a sound, but even as it died in his hands, the mechanism remained mysterious. Crystal Hall analyzes the fable in terms of its connections to contemporaneous literature, and Mario Biagioli analyzed it as a critique of the culture of absolutism.[3] Galileo's patron and friend Pope Urban VIII liked it for its message about the impossibility of humans understanding everything God created—the buzzkill inherent in trying to understand it all.

For my purposes it's a sound studies text: acoustic investigations, explorations of sound, literary flair, contemporary philosophy, and an engagement with the larger world. Once upon a time there lived a curious hermit who raised birds. He loved to hear them sing, and he loved that "they could transform at will the air they breathed into a variety of sweet songs." One night he heard something that sounded like a bird but was actually a shepherd boy

FIG. 2.1 Branch of pomegranate with lanternfly and cicada. Maria Sibylla Merian, *Metamorphosis insectorum Surinamensium* (1705). Dumbarton Oaks Research Library and Collection, Trustees for Harvard University, Washington, DC.

> who was blowing into a kind of hollow stick while moving his fingers about on the wood, thus drawing from it a variety of notes similar to those of a bird, though by quite a different method. Puzzled, but impelled by his natural curiosity, he gave the boy a calf in exchange for this flute and returned to solitude. But realizing that if he had not chanced to meet the boy he would never have learned of the existence of a new method of forming musical notes and the sweetest songs, he decided to travel to distant places in the hope of meeting with some new adventure.[4]

With this sound the man began to perceive that technology can indeed reproduce the sounds of nature and that material objects can make sounds otherwise produced by animals. There is no objective truth to sound. The man set out initially to capture the sound of the bird, an endeavor that sounds like capturing sound, long before technological reproduction as we know it. In the early modern world, recording sound came about not through recreating an entire tune but through technologically manipulating objects and what we would call sound waves.[5]

When the hermit failed to find a bird, he embarked on a sonic quest that taught him that sounds are not necessarily what they sound like and that different materials can produce the same sounds. When he thought a sound was of nature, he discovered that it was made by human hands.

> His amazement was increased when upon entering a temple he heard a sound, and upon looking behind the gates discovered that this had come from the hinges and fastenings as he opened it. Another time, led by curiosity, he entered an inn expecting to see someone lightly bowing the strings of a violin, and instead he saw a man rubbing his fingertip around the rim of a goblet and drawing forth a pleasant tone from that.[6]

The hermit learned that similar sounds sometimes have different origins. His wonder feels like the surprise listeners felt when radio emerged in Weimar Germany and, suddenly, they could hear at their neighbor's house the very same Beethoven symphony that was playing in their very own living room.[7]

The hermit embraced sound sculpture. He heard the door just after he figured out that sounds could be produced in multiple ways. So some of the work that musique concrète and other modernist avant-garde movements did to extend the category of music had already been done. That is, the creative as well as the critical aspects of modern sound studies are recapitulating previous efforts. Galileo's door resonates with Pierre Henry's 1963 *Variations for a Door and a Sigh*, a twenty-four-movement piece that transforms sounds generated by a creaking door and a human sigh into musical movements. In Brian Kane's words, "The doors lose their characteristic 'doorness' and are metamorphosed into flatulent tubas, rumbling contrabasses, or honking baritone saxophones."[8] Henry worked to make sound plastic, to make it something to manipulate, but such a materialist and flexible sense of sound already sat at the center of early modern philosophical understandings of sound.

The curious hermit thought he had discovered every possible means of making sound, but the cicada stumped him.

> Well, after this man had come to believe that no more ways of forming tones could possibly exist—after having observed, in addition to all the things already mentioned, a variety of organs, trumpets, fifes, stringed instruments, and even that little tongue of iron which is placed between the teeth and which makes strange use of the oral cavity for sounding box and of the breath for vehicle of sound . . . he suddenly found himself once more plunged deeper into ignorance and bafflement than ever.[9]

Galileo cadenced the fable with the death of a cicada, using language that Crystal Hall characterizes as the vibrant lyricism of one of Ariosto's battle deaths.

> For having captured in his hands a cicada, he failed to diminish its strident noise either by closing its mouth or stopping its wings, yet he could not see it move the scales that covered its body, or any other thing. At last he lifted up the armor of its chest and there he saw some thin hard ligaments beneath; thinking the sound might come from their vibration, he decided to break them in order to silence it. But nothing happened until his needle drove too deep, and transfixing the creature he took away its life with its voice, so that he was still unable to determine whether the song had originated from those ligaments. And by this experience his knowledge was reduced to diffidence, so that when asked how sounds were created he used to answer tolerantly that although he knew a few ways, he was sure that many more existed which were not only unknown but unimaginable.[10]

The cicada sent the curious hermit into "ignorance and bafflement." He could not shut the cicada up even after closing its mouth and stopping its wings. The cicada's sound is not lovely like a bird but is "strepido"—noisy—and the thirst for discovery justifies the vivisection. The lyrical presentation of mechanisms for making sound suggests that sound production in the natural and material world works the same way. The story resonates with the early modern tradition of using inquiry, instruments, and creative practice to break through hard surfaces in an effort to expose the inner mechanisms; the cracked-open and murdered cicada parallels descriptions of the innards of machines that I'll discuss later in this monograph.

## CICADA HISTORIES

The cicada is a flashpoint for sound studies. It was only a few hundred years ago that the loudest repeated sounds came from musical instruments. As humans learned to harness fossil fuels and electrons to make machines and instruments that could blast repeating sounds that exceeded even the cicada, the bugs were heard as extensions of the mechanical world. This puts the cicada at the center of a process that Carolyn Abbate describes as "the ringed and bonded equivalencies we devise to connect music and sound to the culture in which they originated."[11] The cicada, in her terms, has repeatedly rearticulated the sensory apparatus. When Plato heard the repetitive noise he associated it with music and crafted a myth around it. In Plato's early dialogue *Phaedrus*, the cicadas provide a soundtrack for the conversation between Socrates and Phaedrus that connects cicadas, the birth of the muses, and the invention of singing:

> When the Muses were born and singing had been invented, the story goes that some of the men of that time were ecstatic with pleasure, and were so busy singing that they didn't bother with food and drink, so that before they knew it they were dead. They were the origin of the race of cicadas, whom the Muses granted the gift of never needing any food once they were born; all they do is sing, from the moment of their births until their deaths, without eating or drinking.[12]

The first-century Roman Christian thinker Clement of Alexandria amplified the cicada even more. His "Exhortation" featured Arion, Amphion, Eunomus, and Orpheus. He retold the story of Eunomus of Locrian participating in the Pythian games, the Olympic precursor that took place at Delphi and involved musical as well as athletic contests. While Eunomus played, the string on his cithara broke. According to Clement, the cicada heard the harmony falling apart and jumped onto the string to offer up the missing tone, singing himself to death in the process.

> A string breaks in the Locrian's hands; the grasshopper [cicada] settles upon the neck of the lyre and begins to twitter there as if upon a branch: whereupon the minstrel, by adapting his music to the grasshopper's lay, supplied the place

> of the missing string. So it was not Eunomus that drew the grasshopper by his song, as the legend would have it, when it set up the bronze figure at Pytho, showing Eunomus with his lyre, and his ally in the contest. No, the grasshopper flew of its own accord, and sang of its own accord, although the Greeks thought it to have been responsive to music.[13]

In this version of the story the cicada is not working for Eunomus. Instead Eunomus tunes his music to the cicada. A bit later the text exposes Orpheus as a trickster who uses music to deceive.

Jump ahead a millennium. In the nineteenth century, the terminology involved animalistic machines. When thousands of them hang out together, their clicking sounds amplify and can reach somewhere between one hundred and one hundred and twenty decibels, the volume of a stadium rock concert or an airplane. Before trains, car alarms, and electronic amplification, such decibels hardly ever resounded in the world, and they seemed mysterious and mythical. Charles Darwin explained the cicada as a miniature misogynist percussion instrument:

> Everyone who has wandered in a tropical forest must have been astonished at the din made by the male Cicadae. The females are mute; as the Grecian poet Xenarchus says, "Happy the Cicadas live, since they all have 'voiceless wives.'" The noise thus made could be plainly heard on board the "Beagle," when anchored at a quarter of a mile from the shore of Brazil; and Captain Hancock says it can be heard at the distance of a mile. The Greeks formerly kept, and the Chinese now keep these insects in cages for the sake of their song, so that it must be pleasing to the ears of some men.[14]

Musically, Cicadas were described in terms of instruments, perhaps reflecting that emergent concert halls featuring symphonies, pianos, and violins (all of which celebrated human-made sounds) became the locus of music. In June of 1864 the Wisconsin State Register reprinted an 1844 article that called the cicada a musical instrument:

> As instruments of music, the male Cicada is furnished with a pair of bellows, one on each side of the body, consisting of two large oval plates, formed of convex pieces of parchment, and placed just behind the wings and thorax. When this exterior membrane is raised, a second one of much greater delicacy is exposed to view, tensely stretched between the cavities of the body of the insect, having the transparency of thin mica, and exhibiting the colors of the rainbow. These drumheads are played upon by a bundle of muscular cords fastened to

> the inside of the drums, and the vibratory motion occasioned by alternately relaxing and contracting these cords, produces the buzzing sound. The male Cicada commences his drumming about sunrise and seems to consider himself privileged to keep it up unceasingly until near sunset, while his silent though industrious companion busily performs her household duties. The sound can be heard on some calm morning nearly a mile away. "This," Mr. Kirby says, "is as if a man of ordinary stature, supposing his powers of voice increased in the ratio of his size, could be heard all over the world."[15]

But as the Industrial Revolution rolled across America, descriptions of the cicada went from musical toward mechanical; the cicada was still a living tambourine, but the anthropomorphizing tone of the earlier accounts is gone. This 1843 account from the *Daily National Intelligencer* is a good example:

> In two or three days, if the weather is warm, they commence singing, at first low and feebly; but in about a week the singing becomes loud and incessant, from a little after daybreak until near night. They also sing at intervals through the night when it is warm. When the weather is cool they are scarcely heard day or night. When a large number are singing together they can be heard nearly a mile. Their instrument of music is a kind of tambouret, situated under the wings of the males, between the thorax and abdomen. Its exterior part is convex, with several bands or cords stretched across it. Their note may be represented thus: "whir-irrh, oh," or thus, "cho-o-oke," like one calling hogs at a distance, the voice being prolonged and elevated at the first syllable, and short and low at the last. These notes they repeat at intervals of a few seconds. When a large number are heard singing at a little distance, the first notes only are distinguished, and there is then an incessant "whirrh"; the concluding note is only heard when near. The body is stretched out in sounding the first note, and relaxed again in sounding the second. There is a strong cord-like muscle reaching from the sternum to the back, to which the tambouret is attached, and it is, no doubt, by the rapid contractions of this muscle that the membrane is made to vibrate.[16]

By the end of the nineteenth century, the bugs were heard entirely in machine terms. That was the Industrial Revolution talking. Factories, railroads, and other machines introduced continuous, repeating sounds that gave the bugs a whole new sonic context. The *Bismarck Daily Tribune* reported that "the sights and sounds of 1885 in Indiana were among the most remarkable I ever witnessed. There was a continued monotonous roar, or rather a rattle, like many threshing machines at a distance."[17]

## CONTRA SOUND STUDIES: AN EXCURSUS

The long multidisciplinary history of talking about the cicada suggests that the recent auditory turn in the academy is not so new and that sound is a constant in debates on matter, vitality, nature, body, and sensation. The cicada, like the castrato and like Orpheus, inhabits an interstitial space between natural and artificial sounds: a space that has much to tell modern listeners about sound in their respective eras. Sound studies tend to focus on the modern and postmodern eras because of the implicit assumptions that late nineteenth-century industrialization and electricity made sound production technological and that the instruments of mechanical reproduction such as the gramophone and phonograph put sound into commercial circulation. But sound as a technology came much earlier. Similarly, although sound studies scholars usually trace the field's mode of thinking to the mid-1940s, its origins are much earlier. This timeline may result from the received understanding of modernity as responsible for a visual turn that then inspired an auditory turn. Christoph Cox and Daniel Warner, for example, write that "a culture of musicians, composers, sound artists, scholars, and listeners attentive to sonic substance, the act of listening, and the creative possibilities of sound recording, playback, and transmission" has led to a "culture of the ear."[18] But such a culture of the ear and such creative partnerships were the norm in the early modern period; the relationship between the arts—especially music—and natural philosophy was not occasional and infrequent but was all-encompassing.

Scholars working on music of the sixteenth and seventeenth centuries have been discussing science and technology in relation to musical practice since the mid-twentieth century. Rather than discussing mechanical reproduction, as in the more recent work of Jonathan Sterne or David Suisman, musicologists and historians of science, even those steeped in positivism, looked to discussions of temperament and instrument invention. In the 1960s Claude Palisca from musicology and Stillman Drake from the history of science posited the inextricable intertwining of natural philosophy and music. Palisca argued that scientific discoveries nudged music along, and Drake argued that musicians who were accustomed to using their ears stood at the center of the move toward experimental science, where seeing and hearing with one's own organs mattered more than what the ancients said.[19] Writing about Galileo Galilei and his father, they noted that the astronomical discovery of the universe lacking any essential harmony rendered implausible the notion of music as a microcosm of the universe. And they showed that studies of the dynamics of vibration and sound displaced the Pythagorean theory of number symbolism in which consonance and dissonance reflected the spiritual power of numbers and the harmony of the spheres as the root cause of music's effect on the senses and the mind.

Musical productions have always been about sonic substance making, playback, acoustics, and effects. And those who write about music cared about technique and soundscapes long before electricity. Monteverdi, for example, perceived instruments, singers, stage machines, and sound projection as connected technical and philosophical problems. When he brought a castrato to Saint Peter's for an audition in 1610, he made sure he sang with an organ, and in 1613 he insisted on the particular positioning of two organs for the *Vespers* performance. He often acted in the capacity of what we might think of as a sound man, as in his contributions to the much-discussed marriage festivities of Odoardo Farnese and Margherita de' Medici in 1628. In these productions he worked closely with theater architect Francesco Guitti and Antonio Goretti. The creative team engineered sound through instrument placement and instrumental alteration.

In Monteverdi's musical practice, the singers, the instruments, and the sounds all functioned as both technological problems and as sonic mediations. When the production required a flood, the engineering challenge involved figuring out a way for the water to flood without drenching the instruments and the instrumentalists. To complicate matters further, the production team faced the typical seventeenth-century problem of working within a classically designed theater that left no space for an orchestra; they were still a century away from a proscenium stage or orchestra pit. About this vexing process, Goretti explained,

> We have been several times to the *salone*, to begin to organize the place for the musicians both for accompanying the singers and as decoration where it is needed, and we have repeatedly found many difficulties because of the narrowness of the place for this blessed music, which is an essential part, and they never think at the right time about how it is to be done, and yet it is so very necessary.[20]

For Monteverdi, harmony involved not just musical sonorities and dramatic order but materiality. This is another spin on the materialist underpinnings of the seconda prattica that I articulated in the last chapter. In a letter to Bentivoglio, Guitti described the harmony coming not from sound per se but from putting the instrumentalists beneath the proscenium and between two flights of movable stairs that could also be used for scenery. Making the orchestra an intricate part of the stage machinery was different from the usual model of musicians playing behind the stage on hidden balconies: "Signor

Monteverdi has finally found harmony [*armonia*], for I have provided him with a place to suit him—which much pleases him."[21]

*Harmony*, already a loaded word, pertains to a technological feat and makes use of just the kind of materials that Galileo's hermit "discovered." Monteverdi and the engineers effectively created a new instrument for the occasion. Goretti described the *claviorgano* in material and practical terms, seeming to worry as much about the pipes getting wet as about the actual sound:

> We have also adjusted the sound appropriate for the theatre by having had made some pipes to help the harpsichords, and we have also added them to the front of the stage, in the space between the arms of the stairways, where the machinery of the stairs stands, and this, too, for the music, so that it will be in the same manner as we did for Your Most Illustrious Lordship, so that the sound will be adjusted. Guitti says that all this will provide decoration for the stage-front. There remains only a small difficulty, and that is over the water, for which it will be necessary to build part of a wall in the said place of the stairways to save the instruments and the claviorgano which we propose should be placed there.[22]

Monteverdi made an especially explicit statement about creating sound and about understanding the workings of sound in the composition of *Il combattimento di Tancredi e Clorinda*, first performed in 1624 at the home of Girolamo Mocenigo in Venice, just a year after Galileo's fable of sound was published. The production used musical instruments to recreate the sounds of nature, putting into practice what Galileo's hermit learned by listening. While the hermit puzzled out what made the sounds that sounded like nature, Monteverdi used his compositional power to make them. To dramatize Tasso's impassioned description of the conflict between Tancredi and Clorinda, Monteverdi created a new sound world. A narrator told the story, and instrumental music created a soundscape. In this piece Monteverdi was the first composer to specify the two-fingered pizzicato—*si strappano le corde con duoi diti*—and the smooth bow stroke—*arcate morendo*—which he used to accompany Clorinda's dying words. Throughout the piece, musical instruments create a sound world for the text. The slow repetition of D major chords at the beginning of the piece—accompanying the narrator's description of Clorinda circling the mountain on foot while Tancredi slowly approaches on horseback—sounds like the stomping hooves of the horse (Ex. 2.1).

Such sonic gestures continue to underpin, or represent in music, the narrator's portrayal of Tancredi and Clorinda's battle in progressively faster

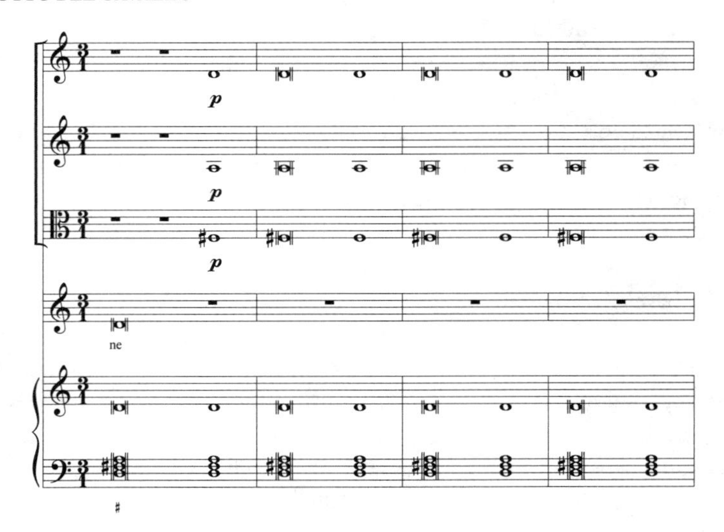


EX. 2.1 Monteverdi, *Il combattimento*, Motto del cavallo, mm. 18–26

rhythmic patterns: eighth-note and quarter-note figures turn into dotted-eighth and sixteenth-note figures. The narrator speaks of clashing swords and flailing arms, the music mimics the battle actions, and the protagonists gesture in the manner of fighting soldiers. The battle is punctuated by whole measures of instrumental sixteenth notes between vocal declamatory gestures. Instruments mimic the described effects by plucking their strings, making a percussive sound that imitates clanking metal.

Monteverdi's preface to his *Madrigali guerrieri et amorosi* explicitly articulates a theory of technological affect grounded in aural evidence in what Nino Pirrotta described as "a theorization ex post facto for manners and movements that Monteverdi must have found spontaneously and madrigalistically."[23] The composer detailed the discovery of the musical representation of contrary passions. "With no little research and effort, I set myself the task of discovering it." After "cogitating on the semibreve" and performing a series of experiments with faster and faster notes, he determined that, after dividing the semibreve into sixteen semiquavers (sixteenth notes) and combining them with a text of vexation, he "heard in this small example the similitude of the emotion [he] was seeking."[24] The three primary affections of "anger, temperance, and humility" corresponded to three vocal ranges—high, medium, and low—and three styles—*concitato* (agitated), *temperato* (temperate), and *molle* (languid). The *concitato* or *agitato* style demanded the most cognitive effort since ancient sources provided no examples of sounds that Plato described as "those that imitate the voice and accents of a man going bravely into battle."[25]

### ON SOUND

*Firstly, a sound and every voice is heard,*
*When, getting into ears, they strike the sense*
*With their own body. For confess we must*
*Even voice and sound to be corporeal,*
*Because they're able on the sense to strike.*
*Besides voice often scrapes against the throat*
*And screams in going out do make more rough*
*The wind-pipe—naturally enough, methinks,*
*When, through the narrow exit rising up,*
*In larger throng, these primal germs of voice*
*Have thus begun to issue forth.*

LUCRETIUS, *DE RERUM NATURA*, BK. 4, LINES 524–34

By the first century BCE Lucretius had imagined sound and voice as deeply intertwined and as a laboratory for the ways sound strikes the body, even when the thing that speaks is a technological proxy. The following paragraphs articulate some of the fundamental differences in the experience of sound in early modern Europe as compared with our twenty-first-century world. Briefly stated, sound for the early moderns was about motion and substance, and was transmitted predominantly through bodies, instruments, and print. There was an already long-established tradition of what we understand as acoustics and technology, but the materials were strings, weights, and pulleys. And sound was multimedia: think books, part books, dramatic productions, and more. You don't need a projector to be multimedia.

These days, while the philosophers of perception debate the true nature of sound, there is a shared basic understanding of sound as vibration that is perceived and knowable through the ears. The *Oxford English Dictionary* defines sound as "that which is or may be heard; the external object of audition, or the property of bodies by which this is produced."[26] Sound functions as a wave or vibration that moves through the air and is heard through a medium. It is a physical force whose musical iterations are experienced largely through sound waves captured and transmitted through formats like the MP3.

In the early modern period, sound stood as movement of air and as motion. This goes back to Aristotle, who gave his most complete discussion of sound in *De anima* (*On the Soul*). In his text the sounding object causes the air to move and vibrate, for "sound is a movement of what can rebound from a smooth surface when struck against it."[27] He understood sound as two objects forcibly striking each other; sound moved through the ear with air as a medium.

According to both Aristotle and Plato, perception occurred when an external thing—either immaterial or material—acted on a perceiving subject through a sense organ or through the body as a whole:

> Such disturbances would occur when the body encountered and collided with external fire (i.e., fire other than the body's own) or for that matter with a hard lump of earth or with the flow of gliding waters, or when it was caught by a surge of air-driven winds. The motions produced by all these encounters would then be conducted through the body to the soul, and strike against it.[28]

Plato went on to explain sensation as a chain reaction: parts affected other parts until they arrived at the "center of consciousness." Song, a material entity that worked with animating spirits, sounded when air from the singer's mouth met air from the listener's ear: "Sound is the percussion of air by way of the ears upon the brain and the blood which is transmitted to the soul, and hearing is the motion caused by the percussion that begins in the head and ends in the place where the liver is situated."[29] Pitch depended on the speed of percussion and volume on the force of percussion. Sound acquired meaning through its physical properties. So, the physical motion of music propelled a particular disturbance of air between one person's mouth and another person's ear, a motion that paralleled the material disturbances of anger. Similarly, Aristotle imagined sound as a collision of air moving at different speeds: "The air in the ear is built into a chamber just to prevent this dissipating movement, in order that the animal may accurately apprehend all varieties of the movements of the air outside."[30]

In the Middle Ages Boethius's *De institutione musica* served as a foundational acoustic text. And Dante's Circle of the Violent is literally a noisy hell. The suicide woods terrify with acousmatic sound, "so many voices moaned among those trunks from people who had been concealed from us."[31] The scholarly place of sound shifted over the course of the medieval and early modern periods in the same ways that other branches of natural science did. The object of study, the people who did the studying, and the practice of study remained in a state of flux. At the same time the humanistic rediscovery of ancient ideas turned them into historical texts reflective of a specific time and place, not of universal truth.

Galileo and other natural philosophers refined Aristotle's ideas into something like a wave theory. Galileo explained this at the end of *The Assayer*:

> Sounds are made and heard by us when the air—without any special property of "sonority" or "transonority"—is ruffled by a rapid tremor into very minute waves and moves certain cartilages of a tympanum in our ear. External means capable of thus ruffling the air are very numerous, but for the most part they may be reduced to the trembling of some body, which pushes the air and disturbs it.[32]

He went on:

> If ears, tongues, and noses were removed, shapes and numbers and motions would remain, but not odors or tastes or sounds. The latter, I believe, are nothing more than names when separated from living beings, just as tickling and titillation are nothing but names in the absence of such things as noses and armpits.[33]

Similarly, in his 1628 *Discorso sopra la musica*, Vincenzo Giustiniani explained, "I will say only that sound, in my opinion, is proper to inanimate things, for it proceeds from air struck or compressed or confined, which is then exhaled and diffuses with varying proportions of time and degrees of violence."[34]

Galileo's ideas about music and sound intersected with those of Giovanni Battista Benedetti, a key figure in mechanics before Galileo. Benedetti put forth his sound-wave theory in two letters to an imagined De Rore that were probably written in 1563 and published in *Diversarum speculationum mathematicarum & physicorum liber* (Turin, 1585). The letters critiqued the harmony of the spheres. As Ross Duffin notes, Benedetti probably addressed De Rore in part because of the same chromaticism that attracted the Monteverdi brothers.[35] Benedetti used numbers to explain sounds but did not insist that numbers caused sounds. And what he demonstrated was a theory of sound grounded in listening and dependent on the movement of air, not numbers.[36]

Galileo is also credited with being one of the first to understand what is now called *frequency*: the number of sound waves per unit of time and the vibrations of the sound object causing the sound waves. Typically modern scientists measure it in Hertz (Hz), the number of vibrations in one second. Humans can hear between 20 and 20,000 Hertz. Pitch indicates perception—it is how human ears interpret the frequency. Pythagoras's understanding of pitch as dependent on string length was already well known. Remember that Pythagoras heard the harmony in the blacksmith's forge and noted that hammers that produced pleasant sounds existed in whole number ratios to one another.

Galileo added to this the concept of motion and vibration as determinant of pitch. His 1638 *Two New Sciences* described scraping a hard iron chisel across a brass plate in order to remove some spots. The chisel made a screeching sound, which varied according to the speed at which it was dragged. He linked the pitch of the sound produced to the spacing of the chisel's skips. This is, in effect, frequency:

> Whenever the stroke was accompanied by hissing I felt the chisel tremble in my grasp and a sort of shiver run through my hand. In short we see and hear in the case of the chisel precisely that which is seen and heard in the case of a whisper followed by a loud voice; for, when the breath is emitted without the production of a tone, one does not feel either in the throat or mouth any motion to speak of in comparison with that which is felt in the larynx and upper part of the throat when the voice is used, especially when the tones employed are low and strong.[37]

I will come back to the specific connections to the voice later, but for now I'm interested in the embodied language and the physicality of Galileo's understanding of sound. Frequency affected the entire body. His experiment articulated a direct correlation between pitches heard by the ear and frequency, which he called the "number of pulses of air waves."[38]

The instrument of measurement was the human body: his ear and his tactile surfaces. Galileo went on to claim that he tuned two spinet strings by using the scraping tones as a kind of tuner.

> At times I have also observed among the strings of the spinet two which were in unison with two of the tones produced by the aforesaid scraping; and among those which differed most in pitch I found two which were separated by an interval of a perfect fifth. Upon measuring the distance between the markings produced by the two scrapings it was found that the space which contained 45 of one contained 30 of the other, which is precisely the ratio assigned to the fifth.[39]

## PRINT CULTURE

It has become commonplace in discussions of the castrato voice and of other early modern sounds to lament the lack of recordings and the fact that the few existing recordings hardly measure up to contemporary high-fidelity standards. Martha Feldman's fascinating excavation of Alessandro Moreschi's voice points to the almost impossibility of hearing that sound. Moreschi's recording of "Ave Maria," made in 1902 has lots of distortion and no noise reduction so that what tends to stand out to students is—by our standards—the poor quality of the recording. But perhaps in some fundamental way we are no more able to capture sound than our early modern predecessors. Think about the MP3, as brilliantly analyzed by Jonathan Sterne. This now most popular form of music dissemination is itself ephemeral, and its compressed sound brings to the fore just how partial and fragmentary the nature of sound recordings is all told. Perhaps printed and even manuscript texts were phonographic. Written words, notes, and images made sound visible, capturable, and dissectible, and print was a form of instrumental mediation and knowledge making.

Print culture operated as a mode of reproduction, one that leaned into the discrepancy between the ephemerality of performance and the permanence of the printed page. That musical practice grew in tandem with print culture does not count as news. Music scholars have written extensively on the impact of print on composition, production, and dissemination of madrigals and solo vocal music. Likewise, expansions in early modern knowledge depended highly on print culture. Galileo himself paid for the printing of *Sidereus nuncius* in 1610 and worked in all of his writings to create new readerships.

Shane Butler takes the phonographic potential of text all the way back to the ancient world, where there were radical attempts to capture the voice and the specificity of vocal performance. Sound and text are phonographically linked, "not so much as original and copy as they are by a more basic resemblance rooted in the stuff of which they are made. In this regard, phonographic inquiry goes looking for a certain strand of ancient materialist thought."[40] And the Aristotelian and Platonic notions of sense perception that held sway through the early modern period were inherently phonographic. Aristotle derived his assertion that sensation was similar to "the way that wax receives the impression of a signet ring" from Plato's materialist conception of sensory perception first articulated in *Timaeus*.[41]

Natural philosophers from Aristotle onward excelled at describing in writing what they saw. Galileo himself wrote lyrically, in the style of his favorite poet, Ariosto; *The Assayer* also told a poetic story of the discovery of the moons of Jupiter, complete with words and perspective illustrations. Galileo wanted to communicate in printed word the thrill of discovery. He invited readers to view other worlds and implied that seeing them was visiting them. Lyrical descriptions and illustrations turned reading into a visual, sensual experience of discovery.

Galileo's antischolastic rhetoric came off as critical of reading as the sole mode of discovery but committed to reading as a form of mediation and, by extension, to print as a technology for mediating discovery:

> Possibly he thinks that philosophy is a book of fiction created by some man, like the *Iliad* or *Orlando furioso*—books in which the least important thing is whether what is written in them is true. Well, Sig. Saris, that is not the way matters stand. Philosophy is written in this grand book—I mean the universe—which stands continually open to our gaze, but it cannot be understood unless one first learns to comprehend the language and interpret the characters in which it is written. It is written in the alphabet of mathematics, and its characters are triangles, circles, and other geometrical figures, without which it is humanly impossible to understand a single word of it; without these, one is wandering about in a dark labyrinth.[42]

Galileo acknowledged that fiction and fables work as didactic explanations of observable phenomena. And he articulated reading as a process of mediation: looking at a text does not give the same information as looking through a telescope or a microscope.

Scholars of printed music have long understood the complexities of print as a medium for experience and sound. Music studies by Cristal Collins Judd, Jane A. Bernstein, and Kate van Orden changed modern understandings of the ways that the circulation of music in print affected musical practice.[43] Printed music had a special materiality and ability to reproduce sounds. Printed versions of ornamented song functioned as both prescriptive texts and written records of individual performers. Part of what made print useful for composers was its ability to fix and standardize a musical text in multiple copies, to phonograph it. Most famously, Luzzasco Luzzaschi's *Madrigali per cantare et sonare a uno e doi e tre soprani* of 1601 amplified the musical singing ladies of Ferrara, making public and repeatable sounds that had been restricted.[44] As I have argued, Monteverdi's setting of Guarini's "Mentre vaga angioletta" uses descriptive rhetoric and musical gestures to capture and copy the female voice.[45] Both a printed text and a performance, the tune stages the effects of the female voice; the male composer puts into the mouths of male singers the virtuosic tricks of the enchanting female voice. Guarini described the actions of the woman's voice—its flexible sounds, twists, turns, and texture.

Technically inclined musical thinkers imagined sung music and music print as phonographic performances. They were repetitions of an original, but unlike modern recordings they required live singers to infuse them with life. To hear print culture as technological reproduction means losing the fantasy of permanence and returning briefly to sixteenth- and seventeenth-century concepts of multimedia arts, which often took the form of comparing the expressive and sensory potential of painting, poetry, and music. Perhaps we moderns have been too quick to believe a certain version of Leonardo da

Vinci, who mourned the lack of permanence in music in his *Trattato della pittura*, a volume compiled in 1542, after his death, and not published until 1651. Leonardo positioned music as inferior to painting because the latter "does not perish immediately after its creation, as happens with unfortunate music. Rather painting endures and displays as lifelike something that is in fact on a single surface."[46] Modern scholars of technology and sound often know this quotation from all three versions of Walter Benjamin's "The Work of Art in the Age of Mechanical Reproduction," first published in 1935.[47] For Benjamin, this quotation was one of many that operated pastiche-style in his argument about the failure of recordings to capture the aura. For him, mechanical reproduction destroyed something vital in art. Benjamin was obsessed with aura as a thing that withers when it is copied, and he used the idea of the aura to create a distinction between modes of art that are about feeling and those that are about finished products.

But Leonardo was not Benjamin. A competent and committed musician, he located painting's superiority in the permanence of the medium. Music was the sister to painting, and he ranked it above sculpture. Only written music achieved permanence; the score was a tangible collection:

> With painting the images of gods are made, around which religious cults arise, in the service of which music is used as an adornment; semblances of those who inspire love are provided for lovers; beauty, which time and nature renders fleeting, will be preserved; and we are able to preserve the likenesses of famous men. If you should say that music lasts forever by being written down, we are doing the same here with letters. . . . Therefore, seeing that you have placed music amongst the liberal arts, either you should place painting there or remove music. And if you say that vile men can make use of paintings, music can be similarly corrupted by those who do not understand it.[48]

Music had to be reborn again and again, so much that it wore out everyone involved. It sang itself to death like the cicada.

Leonardo, as well as generations of thinkers and creative practitioners after him, thought of print as a medium for capturing the essential motion of the human body and of forces like sound. The sixteenth-century writer and painter Giovanni Paolo Lomazzo followed Leonardo and characteristically articulated emotions as motion and the visual representations of emotions as capturing motion. The artist must

> discover all the several passions & gestures which man's bodies is able to performe: which here we tearme by the name of *motions*, for the more significant

> expressing of the inward affections of the minde, by an outward and bodily Demonstration; that so by this meanes, men's inward motions and affections, may be as well, (or rather better) signified.[49]

Music most effectively captured the motions of emotions.

Galileo followed in Leonardo's and Lomazzo's tradition by using comparisons, or *paragoni*, to make crucial points about the connections among the arts, sensation, and natural philosophy. Panofsky explored this in his 1956 essay "Galileo as a Critic of the Arts." Panofsky's essay focused on correspondence between Galileo and the Florentine painter Lodovico Cigoli. In these letters Galileo explained that the sculptor could never measure up to the painter because the three-dimensional medium of the sculptor makes imitating nature too easy:

> The farther removed the means of imitation are from the thing to be imitated, the more admirable the imitation will be. . . . Will we not admire a musician who moves us to sympathy with a lover by representing his sorrows and passions in song much more than if he were to do it by sobs? . . . And would we not admire him even more if he were to perform silently, on an instrument only, and achieve his aims solely by dissonances and passionate musical accents?[50]

Panofsky used this quotation to explain that Galileo preferred instrumental music to sung music. I will come back to the question of that relationship and argue for similitude between instrumental and vocal sound in the next chapter.

But for now, the quotation suggests that Galileo cared about sound and its capacity to move the emotions and that music as a mode of expression worked at least as well for him as painting. Galileo criticized Tasso for being a strange fellow who collected curiosities in contrast to Ariosto who created marvelous worlds. In his "Considerations on Torquato Tasso," unfinished and unpublished until after his death, Galileo wrote,

> Here I feel as if I am entering the studio of some funny little fellow, someone who took pleasure in decorating it with things that, either because of their antiquity or their rarity or some other feature, have an out of the way air about them, but which are in fact just trifles, perhaps a petrified crab, or a dried chameleon, a fly and spider frozen in a piece of amber, some of those little clay figurines that they say are found in the ancient tombs of Egypt, and then, by way of painting, some small sketches of Baccio Bandinelli or Parmigianino, or other such trinkets. When, instead, I enter [Ariosto's] *Furioso*, I see opening before me

> a treasury, a tribunal, a royal gallery, adorned with one hundred ancient statues of the most celebrated sculptors, with an infinite number of complete stories, and the very best of the most famous painters, with a great number of vases: crystal, agate, lapis lazuli, and other gems, and filled, finally, with rare, precious, marvelous, and the most exquisite things.[51]

Eileen Reeves points out that Ariosto stood in for what Galileo and modern critics imagined as the ideal space for natural philosophy: the performative and refined space of public display where a particular mechanics was practiced as opposed to the arts of alchemy and curiosity collecting, which were associated more with magic and imagined to be less legitimate.

Both passages have been read by numerous scholars for what they reveal about Galileo's humanist tendencies, his investment in the arts, and his participation in larger discourses around sensory perception. To me, the first of these passages suggests that Galileo inherited a technoculture that was deeply tied to sound. And the most important message in both is that, for Galileo, natural philosophy, the arts, and performance flowed into one another. And they suggest that natural philosophy—or, in our terms, technology—constantly drew on creative expression and was a practice of reading and storytelling.

## VIVISECTIONS AND VOCAL CORDS

As the sixteenth- and seventeenth-century investment in anatomy and dissection grew, practitioners frequently strove to bolster their cases with references to animal dissections and vivisections. Just as Galileo argued that knowledge required observation, not just reading ancient texts, and as composers used their ears for guidance, not Pythagorean numbers and old treatises, so, too, did doctors and artists set about discovering the inner workings of the human body generally, and the voice in particular, for themselves.[52] The first description of the actual sound mechanics of the cicada, as we understand them, appeared in Giulio Cesare Casserio (Julius Casserius), *De vocis auditusque organis historia anatomica* in 1600–1601. Described as the first treatise on vocal cords, the text enumerates the structure and function of the larynx in part through the sound mechanisms of the cat, ape, dog, pig, goat, ox, sheep, horse, rabbit, turkey, goose, cormorant, heron, frog, pike, cricket, and locust. While exploring these sound machines, he discovered the cicada's tymbals. About the cicada he wrote, "for in addition to the fact that a similarity which is not to be underestimated exists between these very things and the larynx, [this makes] the function of the larynx more evident."[53] Casserio presented the cicada's mechanisms in beautiful and detailed anatomical illustrations (Figs. 2.2, 2.3).

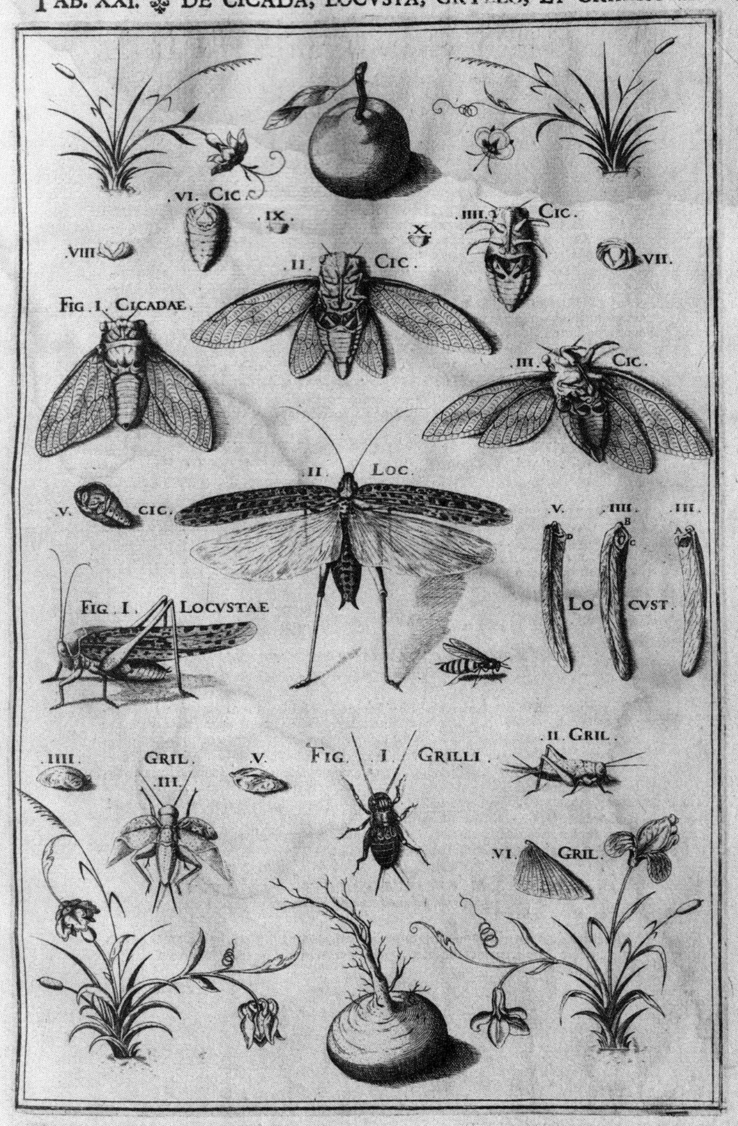


FIG. 2.2 Cicada, plate 21, Giulio Casserio, *De vocis auditusque organis* (Ferrara, 1601), 115. Courtesy of the Medical Historical Library, Harvey Cushing / John Hay Whitney Medical Library, Yale University.

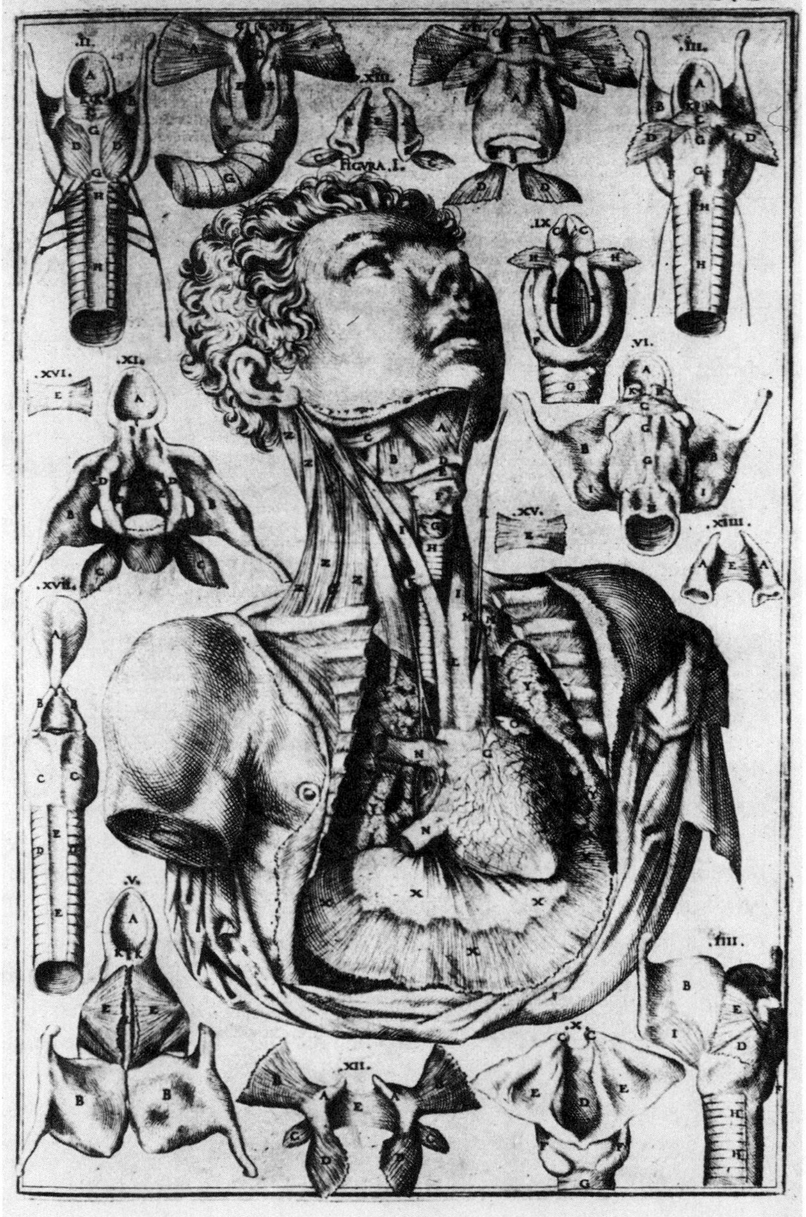


FIG. 2.3 Vocal cord, plate 13, Giulio Casserio, *De vocis auditusque organis* (Ferrara, 1601), 69. Courtesy of the National Library of Medicine.

So, Galileo's vivisection of the cicada was in keeping with attempts to discover and explain vocal noises, and his lyrical Ariosto-like language matched his scientific descriptions. The story merged technical skill, drama, ritual, and spectacle in a literary encapsulation of the practice of anatomy theaters.[54] In 1594 the anatomist Fabricius ab Aquapendente, who made important discoveries about embryos and vocal apparatuses, also oversaw the building of the first public anatomy theater in Padua, which made him an architect of the institution of ritualized anatomy demonstrations. Fabricius understood anatomical theater as a research method and used his public performances to connect philosophy and anatomy. Attempts to discover the inner workings of the human body were out vividly and publicly and noisily. Klestinec notes that in 1597 Fabricius summoned lute players to the theater to halt tumult and bring tranquility.

The vivisection also reflects the turn of the seventeenth century fascination with discovering the innards of the vocal cords.[55] One of the most common vivisections from the time of the ancients through the Renaissance involved working on the recursive nerves, or what is now called laryngeal nerves. In the late sixteenth century, Andreas Vesalius and his teaching assistant/rival Realdo Colombo both claimed to prove that the larynx was the "principal organ of the voice."[56] Vesalius seemed to relish holding pulsating organs in his hand and offered rather disgusting descriptions of the vivisections he conducted to identify recurrent laryngeal nerves.

> Therefore, next I perform a long cut in the neck with a sharper razor, in order that it divide the skin and the muscles underlying it all the way to the rough artery [*sc.* trachea]: taking care that the cut not deviate to the side, and injure the vein, which is worth observing. Then with my hands I grasp the rough artery and, exposing it from the muscles laying on it only by the force of the fingers, I search for the sleepy arteries at its side and the nerves stretched along it from the 6th pair of the cerebral nerves. Then I see the recurrent nerves indeed arising to the side of the rough artery which sometimes I block with ties, and sometimes I cut; and this I do on only one side first, in order that it can plainly be seen when I cut the nerve how half of the voice is destroyed, and when both nerves have been cut, it totally disappears; and how it comes back again if I loosen the ties. Done swiftly and without conspicuous flow of blood, one can neatly hear how powerfully the animal gasps without a voice, with the recurrent nerves divided by a knife.[57]

First he cut one side of an animal's throat and observed that it lost half of its voice, and then he cut the other to observe it lose the second half. Vesalius explained the voice through the motor nerves that make the voice and through a process that injured the animal he vivisected.

Colombo also made use of vivisections, and like Vesalius he explored the vocal organs as part of a way to understand the operation of the voice, breathing, and the flow of humors. He describes the glottis as ending

> in two processes, which nature certainly wanted to use, as though they were little tongues of a sort not only to close the width of the larynx and the opening of the rough artery [*sc.* trachea], so that nothing which might do harm, especially from vomit, could fall down into its interior and be carried down to the lungs. But also so that it might control that cleft for the sake of making various tones, not differently than in shepherd's pipes, or as little tongues are customarily placed in flutes, joined together from two slender pieces of reed. For this reason, the union of those processes making up a little tongue of this type is called the glottis.[58]

Fabricius ab Aquapendente, whose most well-known work with animals involved his discoveries about human embryos, also used animal dissection and vivisection in his work with the larynx and the voice. His 1603 *De brutorum loquela* focuses exclusively on animal speech.[59] This was in part to pursue an anti-Aristotelian project that rallied against making humans completely distinct and exceptional from animals. Fabricius distinguished humans from animals and flutes by the ability to articulate letters of the alphabet. Humans could engage animal language through empathy even if the "words" remained unintelligible.[60] He reminded readers that they might not understand the words in other languages, like Syrian, Hebrew, or Persian.

Anatomical discoveries, particularly around the vocal cords, had long been especially theatrical and performative. Galen performed his public vivisection of a squealing pig in the streets of Rome, and beginning in the 1540s, a print of this public vivisection served as the frontispiece for his collected works. In fact Galen chose the pig because of its loud voice: "animals . . . with loud voices are much better at providing a convincing demonstration to prove the point at issue than those with soft."[61] In this spectacle, Galen dramatically silenced the pig by slicing its nerve in order to show that the vagus and laryngeal nerves enabled voluntary speech. In effect, he turned the pig's voice off, like a fountain. The sixteenth century anatomists who aimed to correct Galen's errors that they blamed in part on his reliance on animal dissection still relied on vivisection especially to understand the mechanics of the voice. Vesalius made the same point about the loud pig: "A sow is better because of its voice; for after a dog has been tied down for some time it will often not bark or growl no matter what you do to it, so you cannot investigate whether its voice has been lost."[62]

Beyond the theater of anatomy, the vivisection of the cicada begs questions about the vexed borders between humans and animals. In Western thought, speech has stood as an indicator of "man's" unique ability to make rational thoughts, so questions around what made vocal noise were inevitably questions about what makes a human. And there has been a line of thought in posthumanism that embraces a nonhuman turn and emphases on human sympathies with nonhuman organisms.[63] I'll come back to this at the end of the book. But for now I'll just say that late sixteenth- and early seventeenth-century vivisections in the service of understanding the voice made understanding the apparatus of the voice depended on understanding animal functions. This in turn suggests a connection with nonhuman others. When anatomists did what Galileo did—dissecting a living animal to figure out how it made sound—they made a mess of the divide between humans and animals, finding no evidence for a "soul" or a "voice" that would mark a rigid distinction between humans and other living, breathing things.[64] The relationship between humans and animals has continued to pivot around song and sound.

While Galen, Vesalius, and Galileo killed pigs and cicadas to figure out the sound mechanisms of singing things, these days doctors—or anyone with enough nerve and either a nice dentist mirror or good web browser—can look at vocal cords. Manuel Garcia, seeking the causes of a cracked voice, plunged a dentist's mirror down his own throat in 1854, and it has been up- or downhill ever since. And then there is today. When scientists wanted to prove that the late Freddie Mercury, of Queen fame and now the subject of a mainstream movie, had an exceptional voice, they hired a professional rock singer to imitate his voice and then filmed his larynx at four thousand frames per second so that they could see the crazy growls, giant range, and gritty vibrato in action, literally. The answer to his magical voice lay in his throat and in the use of subharmonics usually associated with throat singers. It is tempting to see the scientific analysis of Freddie Mercury's voice, or the videos of vocal cords as more accurate and less mediated than Galileo's lyric description of the cicada's voice or Vesalius's description of a dying pig in the service of explaining the human speaking voice. But these digital productions are every bit as distancing and mediated as the analogue productions of centuries ago.

# FEEDBACK LOOPS

*Entanglements of Voices and Instruments*

INTERMEDIO

# THE CAT PIANO

The following three chapters play with the concept of organ and approach the residue of long vanished sounds as layers of entanglements. Moving through Galileo's telescope, musical treatises that interweave the human voice and musical instruments, and hydraulic organs, they trace a continuous feedback loop among humans making musical sounds, humans making instruments that simulate humans and nature, and instruments made from the materials of formerly living beings. These feedback loops are made even more convoluted by layers of mediation and the fluid relationship between books, theaters, gardens, and museums in which the space between print and performance was compressed into almost nothing. Though these chapters center on the vocal and instrumental sounds of the sixteenth and seventeenth centuries they listen also to their prequels and afterlives as a way to hear the resonances between premodern modes of thought and posthumanism.

The probably imaginary cat piano that Athanasius Kircher described in 1650 at the height of the Roman castrato's reign is a creepy and fantastic overture to these entanglements. The supposed keyboard had seven to nine cats penned up in cages that corresponded to their meowing. Meticulously selected for their vocal pitch and volume, the cats' tails were stretched out like strings and a mechanism sent a nail pounding through the tails, which caused the cats to squeal at a certain pitch. Their dissonant cries of pain turned into a harmonious discord and supposedly cured a melancholy prince.

Before Google made it possible to find cat piano videos on YouTube and pictures in articles about "weird science," I learned of it the old-fashioned way: looking for images in Kircher's *Musurgia universalis*. In the flesh, the seventeenth-century version of this book is arresting, the kind of book that entices students to rare book libraries and special collections. There are drawings of human and animal ears to show the differences in sensory receptacles and transcriptions of nightingale songs juxtaposed with bird drawings. The sixth book, dedicated to musical instruments, has elaborate, fantastical drawings of hydraulic organs as minitheaters. The description of the cat piano has no pictures and is categorized not as an organ but as a percussion instrument. Kircher includes it at the end of an explanation of xylophones. The point of describing this machine (*machina*) is to demonstrate that "all earthly bodies can be arranged in such a way to make organized harmony."[1]

It's worth translating the whole title of Kircher's *Musurgia*:

> Musurgia universalis or the great art of consonance and dissonance, in ten books, in which are treated the whole doctrine and philosophy of sound, and both the theory and the practice of music, in all its various forms, are given: the admirable powers and effects of consonance and dissonance in the whole universe are explained, with many new and strange examples, and are applied to various practical uses for almost every situation, but especially in philology, mathematics, physics, mechanics, medicine, politics, metaphysics, and theology.

In this book the machine is a culminating rhetorical gesture merging ancient and modern. Though many of the objects in *Musurgia universalis* were in the museum Kircher made for the Collegio Romano and were part of live performances, the cat piano was always a ghostly entity captured on the printed page.

Gaspar Schott, Kircher's apprentice who worked with him on the museum at the Collegio Romano between 1652 and 1654, wrote his own enormous book called *Universal Magic*. He took Kircher's description of the cat piano almost verbatim and placed it toward the end of a long volume on acoustics. The illustration is on the same page as the donkey choir, a soundscape created by donkey urine to seduce male donkeys into making noises that could be construed as music. (Fig. Int.1).

> A certain instrument was built not so long ago, Kircher says at the end of book 6 of *Musurgiana*, paragraph 4, experiment 1, by an extraordinary and ingenious actor to drive away the melancholy of a certain great prince. He took living cats of all different sizes, and consequently of different deepness and sharpness of voices, which he locked up in a chest made specifically for this purpose and set them up so that the tail was sticking out through a hole, and the tail was attached to a sharp spike. The cats were aligned according to specific points of pain, as will be explained in the following Pragmatica. He arranged the cats according to the different magnitudes and tones, so that each key would correspond to the tail of a cat. And he brought the instrument prepared to restore the harmony of princes to a favorable location. With the key of the organ depressed by his fingers, as with sharp points they poked the tails of the cats, they became so raucous that, intoning with miserable voice, now low, now sharp, they rendered such harmony composed from the voices of cats, as could strive to hasten men to laugh or smile themselves into health.[2]

FIG. INT.1 Cat piano and donkey choir from Gaspar Schott, *Magia universalis* (Bamberg, 1677), pt. 2, p. 372. Courtesy of University of Sheffield Library.

The cat piano is a cruel amplification of musical instrument stories such as Orpheus's turtle lyre. A collective of animals are put into service rather than morphed into something beautiful; vibrating feline vocal folds act like vibrating strings. The most obvious link between the castrato and the cat piano might be pain: living beings mutilated for their sounds and altered to preserve their voices. After all, it makes modern readers almost as squeamish as does the castrato. But both also raise questions about what it means to manipulate matter. Such questions break apart interconnected binaries between voice and sound, animate and inanimate, mechanical and organic, enhancement and prosthetic. And they lead, at the end of this monograph, to unsettling the already tenuous spaces among machines, humans, and animals, which, in turn, dislodge the human from the center of the universe. As I alluded to in discussions of Orpheus and of the cicada, stories of humans metamorphosing into musical animals or musical instruments continued to reverberate through soundscapes and sound studies well into the Enlightenment, and the ontological distinction between them are forever unsettled. With the cat piano, autonomous beings morph into an assemblage that merges animal and technological, what Rosi Braidotti calls "becoming machine": the merging of beings with each other and the world around them and in a series of "enfleshed and extended bodies."[3] The cat piano, like the castrato, opens up a space to imagine organs as always on a continuum.

When I first read about the cat piano, I skimmed the large volume in search of details about the strange instrument. I missed the very important fact that Kircher and Schott were already part of the cat piano's imaginative afterlife. It is a machine that is always in the past tense, always something that someone else heard or felt "not so long ago" (*non ita pridem*).[4] I missed the fact that the cat piano was almost certainly a thought experiment, like Galileo's cicada vivisection. It is the afterlives of both the castrato and the cat piano that preoccupy this book. By locating the cat piano in the past, Schott and Kircher create an absence that as in the case of the castrato amplifies the challenges of writing about an embodied project where no bodies exist.

The cat piano is a repetition for which there is no original, a vignette that defies bibliographic control and whose wires cross in the vast Jesuit knowledge network. An elusive transhistorical idea tangled up in overlapping stories, it is filtered through multiple languages, genres, and centuries. When I put the cat piano in the title of this book, I thought of it as an object or isolated thought experiment. I didn't realize that it appeared in various histories at various times, often connected to writing about faraway places, as in, "that's the kind of thing they do somewhere else." The fifteenth-century pig organ made for Louis XI, which sounded by poking pig's butts, also has an extensive afterlife. This one appears in the CBBD television spin-off from the children's book series *Horrible Histories* in an episode called "Ruthless Rulers."[5]

The engraving (Fig. Int.2) description reads "There is no music sweeter to Midas's ear." When Pan challenged Apollo to a battle of the bands pitting panpipes against lyre, only Midas, who was Phrygian royalty, preferred the pipes. Apollo, angry at the betrayal, turned Midas's ears into those of a donkey. In Ovid the story of Midas comes just after Orpheus was silenced and killed by the noisy Bacchantes. In this emblem Midas's whistle overpowers strings and the chaotic noise of the animal kingdom. The engraving bespeaks an Ovidesque world where humans and nonhumans morph into one another. It is not anachronistic to think of it as a premodern episode of a posthuman present.

The engraving also has a history that maps on to the castrato's place in the zone of alterity. As I'll discuss later in the book, the castrato is of the Global

FIG. INT.2 Cat piano emblem. "There is no music sweeter to Midas's ear." Theodor de Bry, *Emblemata saecularia* (1596), 8th image. Courtesy of Robert J. Richards.

South, the sound of far away. Johan Theodor de Bry was the son of Theodor de Bry, who is responsible for some of the earliest European images of Indigenous Amerindians. The elder de Bry had fled the Spanish Inquisition, and his work oozes anti-Spanish and anti-Catholic sentiments. His New World engravings featured hermaphrodites and cannibals, and he and his sons reused the copper plate engravings so frequently that his images were implanted on the European imaginary for generations.[6]

Often discussions of the cat piano reference a cat organ played by a bear that the Spaniard Juan Cristóbal Calvete de Estrella described in his account of King Philip's entrance to Brussels.[7] Depending on which modern version you read, the organ either was something King Phillip II of Spain brought with him to Brussels, or it was an organ built to impress the king on his entrance. In June of 2020 *Psychology Today* referred to it as something that Phillip brought with him as the "most impressive part of the procession."[8] This statement reflects (and creates) its own bibliographic mess. It gestures toward a broader spectacular culture that is usually erased in discussions of this odd instrument. The most frequent source for Phillip's quirky keyboard seems to be Jean-Baptiste Weckerlin, the nineteenth-century French

FIG. INT.3 *La lecture du grimoire*. Engraving by Franz van der Wyngaert reproduced in Jean-Baptiste Weckerlin, *Nouveau musiciana: Extraits d'ouvrages . . . concernant la musique et les musiciens, avec illustrations et airs notés* (Paris: Garnier frères, 1890), 349.

composer, folk music collector, and writer who was head music librarian at the Paris Conservatory. Weckerlin concludes his book with a section on diabolic music that included the cat piano and the pig organ supposedly made for Louis XI (Fig. Int.3).[9] The image is a scene of dark magic produced by the Dutch printer and engraver Franz van der Wyngaert, who was a printer, engraver, and artist active in Antwerp in the seventeenth century. I leave it to the reader to make sense of the image. It looks to me like the after-party to Berlioz's black Sabbath. Women offer incantations from what might be chant books; an owl overlooks the whole thing. It is the stuff of Halloween but with no Orpheus to calm the savage beasts.

Weckerlin took his small coda from the French Jesuit Claude-François Ménestrier, who in 1681 published *Des représentations en musique anciennes et modernes*.[10] Ménestrier, who did a brief stint with the Barberini entourage, was a court chronicler, choreographer, and historian who also published emblems. He argued for spectacles as instruments for cultural messaging and for musical theater as transformative, and he insisted that all productions capitalize on the visual. He used Juan Cristóbal Calvete de Estrella's *El felicissimo viaje del principe Don Phelippe* festival book to prove his point about the power of image and spectacle.[11]

Calvete illustrated the entrances of Philip II as a performance of Hapsburg domination over current and future worlds. *El felicissimo viaje* (1552) was published after Calvete returned from accompanying the then relatively young King Philip on a festive trip through Italy, Germany, and the Netherlands (the Hapsburg northern territories) to visit his father the Holy Roman Emperor Charles V.[12] It was a wild ride of festival after festival with jousts, fireworks, horse parades, fire-breathing giants, Turks, and Indians. Calvete's book is as much creative propaganda as it is accurate depiction, and the quirky keyboard reveals the ontological impossibility of truth and language that preoccupy later sections of this book.

According to Calvete's account of the entry into Antwerp, the cat organ followed the mechanical trade guilds parade and a bull ridden by a devil spewing flaming rockets. A young man dressed as a bear sat on a chariot playing a strange organ powered not by pipes but by cats. Next came a dizzying array of live animals and humans dressed as animals, including a boy singing Flemish songs. The cat organ hardly stands out, but the description bears repeating anyway.

> Through clever artifice and placement, they made the tails accessible so that the bear playing the organ could pull the cat's tails appropriately and to the beat, some a lot and some moderately, and some a little. As each cat felt its tail pulled, it meowed according to its pitch, and together they made well-toned music.[13]

Calvete provided no image of this organ, and one wonders how many spectators actually witnessed it.

This bizarre history of the perhaps-imaginary cat piano is something I learned toward the end of writing this book, as I was nailing down images for my publisher. The rabbit hole of fact-checking early modern sources often reveals unexpected connections. The business of history, especially the history of a soundscape that can no longer be heard moves between looking for facts and suspending categories of time and place. By its very nature it becomes a project of unsettlement.

CHAPTER 3

# ORGANOSCOPE

## *Telescoping Sound*

The *Museo Galileo* in Florence houses and displays its namesake's middle finger, thumb, and tooth. The finger sticks out of a small glass cylinder and resides among telescopes and other instruments of science.[1] Deliberately or not, the exhibit powerfully links the telescope as a utopian bodily prosthetic to dismembered body parts as relics. And it presents both the body and the tool as assemblages: a living body manipulated and inserted into something else to do something. In the end, the Galileo Museum got it right with the finger, not because of the irony of preserving relic style the remains of a heretic but because for early modern thinkers, organs of the natural and human-made worlds occupied the same conceptual space.[2]

Historians surmise that Galileo made his first telescope using an organ pipe based on what looks like a grocery list. On the back of an envelope from the Veronese physician Ottavio Brenzoni, Galileo listed items he needed for the telescope project, including oranges, lentils, spelt, iron bowls, polished German glass, rock crystal, mirror pieces, Tripoli powder, Greek pitch and felt, artillery balls, and *canna d'organo di stagno* (a tin organ pipe). He recommended specific merchants from whom to buy the goods. The organ pipe is "verified" by the Roman satirist Gian Vittorio Rossi (who will appear extensively in chap. 9), whose snippy biography describes Galileo, upon hearing of a spyglass from a German student, immediately running home to "set some glass lenses of various kinds at specific distances in a leaden pipe which he took from an organ."[3]

In other words, the telescope was an assemblage; an extension of the human eye, it depended on the materials of a sound organ—a device that amplified sound. Artillery balls were used by optics makers to grind glass because they were perfectly round. Felt was used to polish lenses. The telescope was a collection of deliberately found objects put together to enhance the power of the human eye and to bring the planets closer to the viewer. It connects to the machinic castrato because it is an iconic early modern bodily extension, an artifice. To make a very blunt analogy, the telescope served as an ocular prosthetic set in motion by the hands that held it, revealing objects that had remained invisible to the naked eye. The trained voice itself was a prosthetic; an extension of the body and a reformulation of identity. The castrato was an extended human, a body whose constructed voice was, in turn, the tool for transmitting sound and music.

Natural philosophers explored echoes and resonances and worked with cunning devices to expand the sound world.[4] Composers and instrumentalists experimented heavily with instrumental sound. In each case instruments investigated the effects of sounds and harmony and made a statement about a larger philosophical, technological, or theological issue. The castrato embodied these philosophical and experimental efforts in the performance of vocal music that exceeded the capacities of the unaltered human voice, in his spectacular coloratura, in the immense length of his long tones, and in the startling breadth of his tessitura.

When Galileo died in 1642, his will stipulated that his body reside with that of his lute-playing, music theorist father in Santa Croce. But his followers worried that the church would ban the heretic from sacred ground and unceremoniously interred his remains in an unconsecrated box beneath the church's bell tower. The worry was not unfounded. Cardinal Francesco Barberini, of music patronage fame, explained that

> Having listened to the opinions of these Most Eminent people, His Beatitude decided that you, with your usual adroitness, should be sure to let it come to the Grand Duke's ears that it is not good to erect mausoleums to the corpse of someone who was sentenced by the Tribunal of the Holy Inquisition, and died while enduring the sentence: because decent people might be scandalized, to the detriment of His Highness's piety.[5]

Galileo languished there for almost a century until his reputation as a martyr of science became a catalyst for Enlightenment anticlericalism and a symbol of the need for the pope to release control of science. At 6:00 p.m. on March 12, 1737, ninety-five years after Galileo's death, his remains were moved from the unconsecrated box beneath the bell tower in Santa Croce to a memorial tomb in the church, just across from Michelangelo's fingers and other bones.[6] While no church figures attended this event, many important Florentines showed up, including Giovanni Targioni Tozzetti, a botany professor. Tozzetti, who would write in 1780 that bad fortune "caused much noise in the world," happened to have a knife with him—as one does.

With this instrument, Antonio Cocchi, Anton Francesco Gori, and Vincenzio Capponi removed three fingers, a tooth, and a vertebra. Most of Galileo's bones stayed memorialized in the church, but the fragmented bits then went their separate ways. Capponi took the thumb, index finger, and tooth and stuck them in a handblown glass base, which eventually earned itself a

FIG. 3.1 Galileo's finger displayed in the History of Science Museum in Florence. Museo Galileo, Firenze. Photograph by Franca Principe.

bust of Galileo. Supposedly he wanted the fingers that wrote the brilliant words. This disappeared in 1905. In 2009 two fingers, a vertebra, and a tooth turned up in a jar. (Before these parts were reunited with the middle finger, the museum conducted a DNA test to make sure they were authentic.) The vertebra had gone with Cocchi to the University in Padua, where Galileo started working in 1592 as the chair of mathematics. The middle finger has always been the property of the state, and it has always been an object of display. It first went to the Medici's Biblioteca Laurenziana, designed by Michelangelo. In 1841 it was moved to the Tribuna di Galileo (Tribute to Galileo) inside the Museum of Physics and Nature, now La Specola, a museum that constituted a Romantic celebration of genius. In 1927 the finger relic went to the Museum of the History of Science, which is now the Galileo museum. It has been giving tourists the middle finger ever since.

Today in Florence you can see this preserved body part sitting next to the instruments that those precious hands once touched (Figs. 3.1, 3.2). It is without a doubt a relic, and as such it was deemed complete no matter how fragmentary it is. In early Christian times removed body parts contained in them the whole body and remnants of the soul. Things saints touched could also be relics. Neither inert nor lifeless, relics were animate, with the potential to make miracles. And they inhabited an interstitial space between the saint's

FIG. 3.2 Galileo's first telescope in the History of Science Museum in Florence. Museo Galileo, Firenze. Photograph by Franca Principe.

earthly body and heavenly soul.[7] Although Galileo died a heretic, in 2014 the director of the Galileo Museum told the *New York Times* that "he's a secular saint, and relics are an important symbol of his fight for freedom of thought."[8] Note that the church didn't really rehabilitate Galileo until 1992, when they finally admitted that the judges who convicted him of heresy made a mistake.

Galileo's finger, the organ pipe, the dead cicada of the last chapter, and the castrato's voice all stand as early modern renditions of what Deleuze and Guattari thought of as organs without bodies. For them, each body has limits that can be extended to virtual worlds, and this collection of potentials becomes an organ without a body. They mean virtual elements of the body; not just the actual parts but their potential actions and affects. The human voice enacted a special version of this: an abstract machine that must always be disassembled. The moment the voice turns into music, it enacts a process of machining, of acting as an assemblage of parts.[9] The voice embodies cognitive and material means of reconstruction, and the castrato's voice messes with a normative assumption of the body as a unique organism. One could make a similar case for the telescope, an object that extends the human body and makes of it an assemblage of actions and sensations. The telescope connects the hand and the eye, sending the human gaze into unknown realms and possibilities. Galileo's telescope resounded with body parts: the hand that built the telescope, the eye that looked through it, the hand that drew the image on the page, the printer that set the type, the eyes of the reader, and maybe even the mouth of the reader who read it out loud.

In each of these early modern cases, the parts, pipes, keys, and bugs activate when they connect with other organs, and they do so in ways that create fantastical sounds. The body, or the instrument, is and was a space of potential. Galileo's preserved finger is far from useless. Rather, it "embodies" the immense potential of the human body to become an instrument of thought and creation. Tozzetti recalled Capponi explaining that he took two fingers "as relics because Galileo wrote so many beautiful things with them." Tozzetti in response "touched the corpse's skull" because he wanted what was inside the skull.[10] The organ pipe, likewise, repurposed, becomes an instrument that magnifies vision. In each case the organ or instrument defies categorization. The finger might stand as the ultimate body without organs: a body with unlimited potential and multiple unstable matters. The finger in the museum implies the muted organ pipe that made the telescope, the hand that played the lute, and the sensory organs that looked, listened, talked, and tasted.

Galileo articulated the telescope as an enhancement of the human eye. The eye for him was a natural organ, but it was also an operational instrument that worked by a series of intersections. When humans hold in their hands telescopes, microscopes, cell phones, cameras, binoculars, the natural world becomes magical, an ultra-high-resolution space that brings the faraway near and makes the small large. The viewer extends their sensorium. This happens with amplification devices as well; headphones put incredibly loud sounds directly in the ear, microphones make the breathing of a singer audible to the back of a concert venue.

Freud, who put sound at the center of the connection between instrument, body, and discovery, is a useful interlocutor here. In *Civilization and Its Discontents*, tools, which were developed as means to control and civilize the natural world, move easily between lines and categories, enhancing the sensorium and the body itself.

> With every tool man is perfecting his own organs, whether motor or sensory, or is removing the limits to their functioning. Motor power places gigantic forces at his disposal, which, like his muscles, he can employ in any direction; thanks to ships and aircraft neither water nor air can hinder his movements; by means of spectacles he corrects defects in the lens of his own eye; by means of the telescope he sees into the far distance, and by means of the microscope he overcomes the limits of visibility set by the structure of his retina. In the photographic camera he has created an instrument which retains the fleeting visual impressions just as a gramophone disc retains the equally fleeting auditory ones. . . . With the help of the telephone he can hear at distances, which would be respected as unattainable even in a fairy tale.[11]

Tools, instruments, and machines do not exist apart from the human body and the sensorium but, rather, connect it to the world with an extended capacity. New technologies—with their ability to enhance what humans can hear, see, feel, and reach—are modeled on the human body and its function and act as body-altering prosthetics. Freud reminded his readers that most technological innovations emerged as ways to enhance human functionality: Alexander Graham Bell invented the telephone in part to enhance deaf hearing, and early typewriter makers thought that they had a prosthetic for the blind. Edison imagined his phonograph as a device for reading to the blind, a phonographic artifact that could both bear voices into the future and make the technology of reading possible without sight. It could overcome the mortality and the limitations of the practice of reading and of eyes that didn't see enough to experience the printed text.

Galileo sold the telescope as a tool to see faraway ships and alert troops to impending invasions; the reimaging of the universe came later. And though he was far from the first person to make concave and convex lenses that enhanced human vision, he did offer a theatrical and sensational demonstration of their potential in 1609. On August 25, from the top of the bell tower in St. Mark's square, he presented an eight-times-powered instrument to the Venetian senate and called it a "nuovo artifizio di un occhiale" (new artifice of an eyeglass). Initially it was called an *occhiale* and related most to spyglasses and other lenses used to correct vision. In an early description, written in a letter to the Doge Leonardo Donato, Galileo highlights the prosthetic nature of the device,

> which brings visible objects so close to the eye, and represents them as so large and distinct, that something distant, for example, nine miles, appears to us as if it were only one mile away: a thing that for every maritime or land-based business and undertaking might be of inestimable advantage.[12]

When he introduced the instrument in print in his 1610 *Sidereus nuncius*, he referred to it as an organ seven times in the first two pages.[13] He described his discoveries as "of great interest, I think, both from their intrinsic excellence and from their absolute novelty unheard through the ages, and also on account of the *organo* by the aid of which they have been presented to my apprehension."[14] The discoveries are thus made possible by the organ, by which here he means the telescope, an extension of his own ocular organ. The origins he claimed for this were aural, and he begins with a sonic description of a visual event: "About ten months ago a rumor came to our ears that a spyglass had been made by a certain Dutchman by means of which visible objects, although far removed from the eye of the observer, were distinctly perceived as though nearby." In the 1610 *Sidereus nuncius* he described it poetically:

> unfolding great and very wonderful sights
> and displaying to the gaze of everyone,
> but especially philosophers and astronomers,
> the things that were observed by Galileo Galilei.[15]

He called it "artificial powers of sight."[16] Galileo described a vitally organic process that brought his eye to the lens. With the ability to make visible previously invisible objects, the telescope was a means to gain knowledge beyond the capabilities of human perception. Rossi's satire mentioned above gave a delightful rendition of the telescope prosthetic by explaining that with the tube Galileo pierced through the sphere of the Moon as if driven by a chariot.[17]

## EAR TRUMPETS

While working on the telescope project Galileo corresponded with his former student Paolo Aproino, who was working in Venice on an ear trumpet, an instrument designed to "multiply the sense of hearing." As Matteo Valleriani has argued, at the beginning of the seventeenth century the ear trumpet was an instrument designed to perfect and extend nature, but by the end of the century it was a mechanical tool in its own right.[18] There's a similar trajectory for the castrato: at the end of the sixteenth century they were men with high voices who could take the stage when women couldn't, but by the end of the seventeenth century they were cultivated for their mechanical extensions. Giambattista della Porta had already worked on a similar project in 1589, investigating the ears of various animals and suggesting shapes that mimicked nature. He began his explorations of the ear trumpet in order to design an instrument that could hear from a distance, an aural telescope with the potential to enhance the listener's ability to hear: "With the optical spectacles we have demonstrated how we can see quite far; now we shall try to make an instrument with which we can hear for many miles."[19] Della Porta's project was one of intensification of human processes, including a set of metal pipes that could be fitted together to transmit voices over great distances.[20] For della Porta, the ear trumpet did not alter the sound but, instead, changed the perception of the sound. Della Porta built on earlier writers such as Ctesibius, Phion, and Heron, whose works had been preserved in the words of Arabs and Byzantines and translated by Renaissance humanists. He explained that in ancient times there had been a brass colossus that "in violent tempests of wind from the nether parts received a great blast, that was carried from the mouth to a trumpet, that it blew strongly, or else sounded some other instrument, which I believe to have been easy, because I have seen the like."[21] Della Porta used the brass colossus to show the ways in which natural forces could be captured and turned into sounding instruments. From this he learned that engineered pipes could both trap sound and convey it through space.

The four letters written in 1613 between Galileo and Aproino, translated and interpreted by Matteo Valleriani, reveal both men's interests in sonic magnification as a process that, like the telescope, changed perception and expanded, literally and figuratively, the capacity of a human organ. Aproino's experiments aimed to show that the special sound magnification instrument could capture sound and, in effect, multiply it fourfold:

> It is sufficient that one makes what has been said: to show that a sound, caught with an artificial instrument, reaches the ear four times louder than when it is heard naturally. By showing this, which is what I wrote to you, not only will rational men have no chance to doubt, but they will also have more [material] to speculate about the nature of sound, more than what has been speculated upon until now.[22]

Aproino explained the importance of the experiment by discussing, among other things, the potential to hear church bells from far away as well as distant voices. And this had clear ramifications for music making:

> Concerning the music too, it seems to me that there is something subtle to speculate about as, when one is far away, the parts can be heard in a perfect blend of consonance, and the instrument allows the voices to be heard more vividly, as if they were closer.[23]

The speaking tube worked a bit like a mixer, something that could make the sounds crisper and, thus, perhaps influence tuning. Aproino also wrote of difficulties that are reminiscent of distortion and delay:

> As Your Lordship certainly observed, the distance decreases the voices but increases the sweetness of the consonance. As any other thing, also this instrument has its problems. One of them is represented by a little whirl. From this it follows that the words, at the beginning of their articulations, do not seem to be closer and that they then follow the proportion of the increase which they naturally make according to the essence of the sound. However, since such an accident does not follow from the instrument itself, as turbid sight does not follow from lenses, I hope to eliminate it trying with various experiments and with that patience and precision that this subject requires, which is made of *minimis naturae*.[24]

In the mid-seventeenth century, Athanasius Kircher did his own theatrical acoustics. He repeated Galileo's experiments and "invented" an ear trumpet, shaped like a cochlea, that gave the listener an experience of amplified sound. Throughout his *Phonurgia nova* he expressed fascination with the ways that sound was magnified in tubes and the ways that it could be strengthened by instruments and nature. He embraced uncritically synergies between optics and acoustics.

Kircher built on this work in Rome. The illustration of his 1672 demonstration of his speaking trumpet from the rock of S. Eustachio (Fig. 3.3), emphasizes the mechanical subjugation of the natural landscape to its circumscription. The flat landscape is divided with a linear grid system, and the barn and compass appear almost as big as the mountain, bringing out the man-made objects in the picture, the first used to house humans and the second to help them conquer the earth itself. Kircher explained later that he used this instrument for a four-part canon sung three miles from an illustrious personage's villa without his knowing where the concert came from.[25] Again,

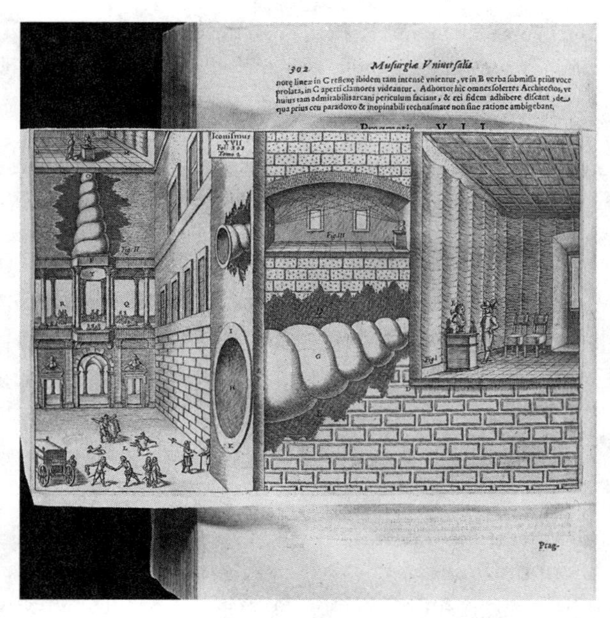


FIG. 3.3 Athanasius Kircher's speaking tube. *Musurgia universalis* (Rome, 1650), 2:303. Courtesy of the Albert and Shirley Small Special Collections Library.

the illustration oddly distorted the landscape with the trees planted in a linear pattern and the canon hovering in the heavens above. Kircher put speaking tubes to practical and performative use when he placed them in portals. Without raising his voice or changing timbre he could make disembodied voices sound in the courtyards.

> If our porters had to tell me about something, either of the arrival of guests or of anything else, so that the porters would not have the inconvenience of having to come to my museum through the confusing corridors of the college, they could talk to me while staying in the porters' lodge while I remained in the far away chamber of my bedroom, and, as if they were present, they could tell me whatever they wanted clearly and distinctly. Then I also could respond in the appropriate tone of voice according to the demands of the matter through the orifice of the tube. In fact, in that area of the garden no one could speak anything uttered in a slightly raised voice that I could not hear inside my bedroom, and this was a thing seen as completely new and unheard of by the visitors to my museum, when they were listening to the ones speaking but couldn't see who was talking. Thus also, so that I would not be suspected of any offensive art by

> the amazed people, I demonstrated to them the hidden structure of the device. It is hard to say how many people, even including many Roman nobles, were attracted to see and hear this machine.[26]

In this protointercom system a mechanical device enhanced the human voice so that to an unknowing listener the voice seemed to come from nowhere. One imagines a reaction similar in astonishment and wonder to those of early listeners to the phonograph who could not get over the novelty of hearing Beethoven symphonies coming out of boxes. Because the machine worked via a hidden structure that enhanced the natural sounds, it mystified and amazed its listeners.

## LYRIC EXTENSIONS: SCOPING IT OUT

By the time Galileo penned *The Assayer* in 1623, he saw the telescope as a prosthesis and pushed it and musical instruments together in an ontological argument about the limits of the body. The book stages a dialogue between Galileo and Sarsi, the pseudonym for Orazio Grassi. Galileo directly equated the length of the telescope with the length of the organ pipe.

> It is perfectly true that the lengthened telescope is a "different" instrument from what it was before, and this was essential to our point. . . . I ask Sarsi why it is that some organ pipes produce deep tones and some high. Will he say that this comes about because they are made of different materials? Surely not; they are all of lead. They sound different tones because they are of different lengths; and as to the material, this plays no part whatever in the formation of the sound. Some pipes are made of wood, some of pewter, some of lead, some of silver, and some of paper, but all will sound in unison when their lengths and sizes are equal. But on the other hand one may make now a larger and now a smaller tube with the same quantity of material, say the same five pounds of lead, and form different notes from it.[27]

Galileo wondered whether different pipe lengths made organs different instruments altogether. His concern with "deep" and "high" was about pitch and vibration:

> With regard to the production of sound those instruments are different which are of different sizes, not those which are of different materials. Now if by melting down one pipe and remolding the same lead we make a new tube that is longer, and therefore of lower pitch, will Sarsi refuse to grant that this is a dif-

ferent pipe from the first? I think he will not. And if we find a way to make this longer tube without melting down the shorter, would not this come to the same thing? Surely it would. The method will be to make the tube in two pieces, one inserted in the other. This may be lengthened and shortened at will, making diverse pipes which will produce different notes; and such is the construction of the trombone. The strings of a harp are all of the same material, but they produce different sounds because they are of various lengths.[28]

Galileo disagreed with Sarsi on two points. First, Sarsi wanted to call a long and short telescope by the same name; and second, Sarsi said the instrument was used differently to spy on the heavens and the earth. For Galileo the use was the same, as an extension of the eyes, but the measurement depended on the space between lenses—proportions like tuning an instrument. Galileo made an amazing analogy between strings, voices, and throats. In the passage below he ties musical practice to experimentation in natural philosophy and pitch to material or sound to organ:

On a lute, one string will do what many strings on a harp will do; for in fingering the lute the sound is drawn now from one part of the string and now from another, which is the same as lengthening and shortening it, and making of it different strings so far as it relates to the production of sound. The same may be said of the tube of the throat, which, varying in length and breadth, accommodates itself to the formation of various notes and may be said to become various tubes. Now since a greater or lesser enlargement depends not upon the material of a telescope but upon its shape, the tube constitutes different instruments when the same material is used but the separation of the lenses is altered.[29]

He then argued that use does not change according to the length of the scope:

The instrument is altered while its application remains the same. The same instrument is said to be differently applied when it is employed for different uses without any alteration; thus the anchor was the same when used by the pilot to secure the ship and when employed by Orlando to catch a whale, but it was differently applied. In our case the reverse is true, for the use of a telescope is always the same, being invariably applied to looking at things, whereas the instrument is varied in an essential respect by altering the interval between its lenses.[30]

The telescope as an extension of the human body and sensorium emerged even more explicitly in Giulio Strozzi's 1624 edition of *Venetia edificata*. The poem appeared in the same year that Monteverdi's *Combattimento* was performed in Venice. Based on Tasso's *Gerusalemme liberata*, it circulates among musicologists primarily because it contains an ode to Adriana Basile's voice.[31] It also has a scene in which the English magician Merlin invents the telescope and gives it to the one-eyed soldier Oddo, the commander of an army of former prisoners of war who were blinded by their enemy. The telescope is here a prosthesis:

> It has the form of a horn and a length of two
> cubits it extends; at the double holes
> two glasses sit, the one in its convex curves
> forms the beautiful work in the concave curves of the other.
> The eye is placed close to the concave; and then they show,
> where you point them, their effects
> and to make close a very distant object
> with immense magnification is the result.[32]

Strozzi is indirectly part of the telescope story through his *La finta pazza*. Francesco Sacrati took on the libretto and created an opera that was one of the greatest hits of the seventeenth century. The 1641 production celebrated the opening night of the Teatro Novissimo in Venice. The opera had an immense publicity machine, including a fifty-five-page volume called *Il cannocchiale per la finta pazza* (Telescope for the feigned madwoman), written by M. B. di G (Maiolino Bisaccioni, count of Genoa).[33] He wanted to make the production accessible to viewers, so that the "eyes of those even in the most distant and secluded foreign countries enjoy in these pages what eyes and ears have enjoyed in this city, which in its every aspect surpasses the bounds of the marvelous."[34] He aimed to make the technical details and feeling of wonder accessible but said that none had made the visual spectacle visible enough.

Bisaccioni's descriptions of the affective power of the production put the reader in the space, perhaps an early modern version of what the Metropolitan Opera HD productions do today. Using technologies of print and the idea of the telescope, he made a multimedia extension. The vivid verbal picture calls up the vibe of the crowd waiting for the symphony to begin and the open ears of the listeners who experienced, for instance, a castrato from Rome who sang "so delicately that the souls of the listeners, as if drawn through the portals of their ears, raised themselves to heaven to assist in the enjoyment of such sweetness."[35] The body of the listener is wide open for sensual reception.

Bisaccioni connected the opera production to the mechanics of the telescope and suggested that theater, print, telescope, and instruments all created sensory experiences:

> Throughout the centuries the discovery will be admired that, by means of two glasses, one convex and one concave, and by two contrary forces, the tiniest details never observed before are seen from the earth to the sky, not just of shadows, or recesses, or stars in the moon, but the little lights that serve as satellites around Jupiter and theirs in the Milky Way. I have been considering these days the composition the *Finta pazza* by Giulio Strozzi, with stage machines by Iacopo Torelli, and music devised by Francesco Sacrati, and they are a sky worthy of being contemplated by everyone, but being so far away from a majority of people, they would be deprived of a great value, having already come to see such a noble creation, if it were not made easy to everyone to see and admire it.[36]

## BEYOND THE THEATER

Francis Bacon's technological utopia *The New Atlantis,* written in 1624, describes a European ship lost somewhere in the Pacific. The fictional narrative presents an overwhelming array of ingenious devices designed to dazzle the reader.[37] The island's governor leads a tour of a strange palace called Solomon's house, which has among other things, a brew house, perfume house, engine houses, and a sound house. The book, which Bacon never finished, is a delightful mash-up of theatrical, literary, and philosophical techniques and tropes. Penelope Gouk has persuasively argued that the sound house demonstrated what musicians, architects, artisans, and engineers made happen in court masques.

> We have also sound-houses, where we practice and demonstrate all sounds and their generation. We have harmony, which you have not, of quarter-sounds and lesser slides of sounds. Divers instruments of music likewise to you unknown, some sweeter than any you have; with bells and rings that are dainty and sweet. We represent small sounds as great and deep, likewise great sounds extenuate and sharp; we make divers tremblings and warblings of sounds, which in their original are entire. We represent and imitate all articulate sounds and letters, and the voices and notes of beasts and birds. We have certain helps which, set to the ear, do further the hearing greatly; we have also divers strange and artificial echoes, reflecting the voice many times, and, as it were, tossing it; and some that give back the voice louder than it came, some shriller and some deeper; yea, some rendering the voice differing in the letters or articulate sound from that they receive. We have all means to convey sounds in trunks and pipes, in strange lines and distances.[38]

Scholars of sound studies, electronic music, and musical instruments like this passage, which is one very small paragraph within a creative work that exudes mechanical wonders.[39] Indeed, Bacon goes on to describe carriages that make themselves go, machines that fly like birds, and ways to preserve food. For Bacon these ingenious devices, all of which exist in the modern world, "deceive the senses." In the sound house, voice was one sound among many others. And "articulate sounds and letters" sat alongside the "voices and notes of beasts and birds." For Bacon, in other words, what mattered was the raw materials of sound and how to produce and reproduce it.

This passage gets attention as one of the first documents in the European tradition to privilege instruments and as one of the first to understand instruments as constitutive of sound. And it is portrayed by modern scholars most often as prophetic, with its extension of sounds that nature provides into an acoustic virtual reality. But it was not as fanciful as it often seems to modern commentators. Most of Bacon's tricks came from medieval and early modern natural philosophy. His description embodies the work of composers, engineers, musicians, and artists. Already air and water had powered hydraulic organs; fantastic keyboards had animated an assemblage of other instruments in a Deleuze-on-Boulez sort of process.

Bacon had previously made clear that instruments were extensions of the human body and that the concept of organ was vexed in ways I will elaborate in the next chapter. In 1620 Francis Bacon said in *Novum organum*,

> Neither the naked hand nor the understanding left to itself can affect much. It is by instruments and helps that the work is done, which are as much wanted for the understanding as for the hand. And as the instruments of the hand either give motion or guide it, so the instruments of the mind supply either suggestions for the understanding or cautions.[40]

The complete title of Bacon's work is generally translated as "New Instruments of Science." But for Bacon, the word *organum* refers both to material technologies and to spiritual essences: stuff you can't feel.[41] He meant mechanical instruments—instruments created by nature and bodily organs—*and* he meant reason and logic. For him, *instruments* already occupied an interstitial space between subject and object.[42] That Bacon saw no ontological difference between organs endowed with sense and those without shows distinctly that the modern notion of machine as dehumanized and soulless did not apply in the early modern world.

CHAPTER 4

# ORGANS AND ORGANS

In 1707 the Huguenot lawyer Charles d'Ancillon wrote a book that offered multiple arguments against marrying a castrato. The English translation, published anonymously in 1718, inserts a sound bite supposedly from d'Ancillon's 1705–1707 sojourn in Rome.

> But if Pasqualini was allowed to be the greater Master, Paoluccio was allowed to have the finest Voice. This Eunuch who was then about 19 Years of Age (and now about 30) was indeed the Wonder of the World. For besides, that his Voice was an Octave, at least (and I speak within Compass) higher than any ones else, it had all the Warblings and Turns of a Nightingal[e], but with only this difference, that it was much finer, and did not a Man know the contrary, he would believe it impossible such a Tone could proceed from the throat of any Thing that was human. "Jeronimo, (or Momo,) had a Voice so soft, and ravishingly mellow, that nothing can better represent it than the Flute-stops of some Organs."[1]

The description references three castrati to telescope the uncanny timbre, virtuosity, and wonder of the castrato voice all told. Pasqualini, who had been dead for fourteen years by the time d'Ancillon looked for castrati in Rome, was the master of all things. Paoluccio—Paolo Pompeo Besci, who served Queen Christina and sang for Scarlatti at San Giacomo degli Spagnoli—had the voice of a nightingale, a figure long associated with troubled expressions of interiority and abstracted voice. The third singer, "Jeronimo" is an interpretive mess. He sounds like the flute stop that in French organ music of the late seventeenth century was a unique otherworldly timbre in the midst of a highly virtuosic and eclectic organ sound palette. The castrato occupied an imaginary space between nature and artifice and divorced sound from the body and mortality of the individuals who made it.

The word *organ* is fabulously ambiguous, etymologically and philologically tangled. In the period of the castrato it could mean the organ as a specific musical instrument, the organ as an instrument, the organ as a body part, and the organ as an Aristotelian property.[2] Because instruments, tools, and organs did not occupy separate and autonomous spheres, the strikingly similar vocabularies used by sixteenth- and seventeenth-century commentators to describe organs and singers is more than metaphor and linguistic play. Word choice reflects on the one hand, a viscerally material understanding of the human voice as a substance to be manipulated and, on the other, a curiously animate notion of the organ as a machine that sang. To take seriously the idea of musical instruments and the human voice as organs is to hear both substances as products of nature ripe for manipulation by human hand and will. On the one hand, as scholars working in the seventeenth century, and especially Rebecca Cypess, have beautifully shown, musical instruments take on lives of their own, change bodies, change minds, and more.[3] And on the other hand, some musical practices deliberately make the voice sound so much like an instrument that it's hard to know how the sound is made; think beatboxing.

These questions around the concept of *organ* build on Feldman's archaeology of castrato vocal production and engage an interdisciplinary conversation around issues of music technology. They also refuse received philosophical distinctions between those who work with their hands and those who work with their minds. The fluidity of *organ* in its early modern messiness is key to the resonance between castrati and machines. And to understand what *machine* and *organ* meant requires putting aside contemporary assumptions. Cecilia Bartoli, for instance, says she sings against the idea of the castrato as machine: "The mistaken belief that castratos were nothing more than heartless, brainless singing machines needs to be quashed once and for all. Immediate evidence to the contrary can be found in these highly diverse arias."[4] The problem with this statement is that early modern machines were not in fact heartless and brainless; the contrast between expressivity and the idea of a machine is a post-Enlightenment concept.

This chapter collects organs: vocal organs, keyboard organs, body parts, and the word itself. Dwelling in the long seventeenth century during which artists, writers, musicians, and artisans played extensively with the potential liveness of matter, I begin by using the castrato voice to critique the notion of organology and move on to the making of voices and Michele Todini's museum of musical instruments as a space that made instruments living beings. Leaning into the messiness of the term *organ* emphasizes the voice as a technical practice inextricably intertwined with practices of instrumental music.

## ORGANOLOGY OR NOT

Translations of most early modern musical and philosophical treatises translate the word *organo* as "instrument." *Instrument* has, since the nineteenth century, connoted a mechanical thing used for some purpose. And since Guido Adler's 1885 discipline-founding taxonomic scheme of music study, music scholars have located organology under the rubric of history.[5] In the Western ethnomusicological tradition, organology refers to the cultural study of instrumental classification, with schemes that go back to the Sachs-Hornbostel system initially published in 1914.[6] Curt Sachs and Erich von Hornbostel worked at the Berlin Phonogramm-Archiv on a system that in part mimicked late nineteenth-century biological species classification and understood instruments as key to understanding material culture. They were interested in how instruments made sounds, not humans as instruments.

It worked differently in the premodern world. Instruments, tools, and organs never existed in a separate and autonomous sphere. Latin and Romance languages intertwine the words *instrument* and *organ*. Organ derives from the Latin *organum*, which means instrument, and which comes from a Greek word that means tool, instrument, or organ of the body. Soft, pliable, and alive, organs do things. Aristotle used the word to refer to plants and animals as well as abstract parts of philosophical systems. In modern terms one might think of the notion of *organ* as encompassing bodies, technologies, and artifacts—like the telescope, record player, and camera—or as symbolic forms or modes of perception—such as perspective, theater, and words.

Before the eighteenth century, writers in music and natural philosophy who discussed sound, singing, and playing instruments drew heavily from the traditions of Aristotle, Greek Neoplatonism, Galenic medicine, early Christian writers (like St. Augustine), and the medieval learned Latin tradition. This entanglement of organ, instrument, and body emerged in the foundational texts of Aristotle and Galen, which were continuously refueled by anatomy, physiology, and theology.

In the ancient world, the body and bodily organs contained and animated the soul; internal structures were not based on a coherent plan as in the nineteenth century but instead made up an animating essence. Aristotle understood the body as an organic substance composed of malleable instruments with the body serving as an instrument of the soul. Similarly, Galen understood the body as the proper instrument of the soul and the parts as organs: "I define *organ* as a part of an animal that is [the] cause of a complete action, such as the eye for vision, the tongue for discourse, or the legs for walking." Here, Galen connected human-made bodily extensions to the body:

> But man is neither naked, nor vulnerable, nor without protection, nor barefoot, for he can have an iron *thorax* whenever he wants, an organ more protective than any skin, and he has every kind of sandal, of weapon, and of armor. And not only does man have the *thorax*, but he also has houses, walls, and towers as protection.[7]

Music making involved doing things with hands, with voices, with brains. The hand and the voice both connected to the soul and were special instruments. Galen explained that "the soul is analogous to the hand; for as the hand is a tool of tools, so the mind is the form of forms and sense the form of sensible things."[8] Aristotle positioned the voice as an instrument of the soul, a vital function, and an operation. Only human beings had voices, for "what produces the impact must have soul in it and must be accompanied by an act of imagination, for voice is a sound with a meaning, and is not merely the result of any impact of the breath as in coughing."[9] But instruments could give voice, as the pipe, lyre, and other sounding bodies have range, melody, and articulation. Galen, in *On the Usefulness of the Parts of the Body*, expanded on Aristotle:

> Hands are an instrument, as the lyre is the instrument of the musician, and tongs of the smith. Hence just as the lyre does not teach the musicians or tongs the smith but each of them is a craftsman by virtue of the reason there is in him although he is unable to work at his trade without the aid of his instruments, so every soul has through its very essence certain faculties, but without the aid of instruments is helpless to accomplish what it is by Nature disposed to accomplish. . . . For though the hand is no one particular instrument, it is the instrument for all instruments because it is formed by Nature to receive them all, and similarly, although reason is not of the arts in particular, it would be an art for the arts because it is naturally disposed to take them all unto itself. Hence man, the only one of all the animals having an art for arts in his soul, should logically have an instrument for instruments in his body.[10]

For early church thinkers like St. Augustine, who wrote at the turn of the fifth century and remained influential through the Middle Ages, *organum* referred to all musical instruments and to the musicians themselves: "The whole body, inside and out, can be looked upon as a kind of organ with a music all its own. The beauty of this music no one has discovered because no one has dared look for it."[11] Music was made of bones, flesh, skin, and organs, by voices and instruments: "That not the voice alone may praise, but the works too. When timbrel and psaltery are taken, the hands harmonize with the voice." In praise, "Thou hast taken to thee an instrument, and thy fingers agree with thy tongue." Augustine went on to conflate the crucifixion with the stretching of strings: "Let him not set down his psaltery, let him not set down his timbrel, let him stretch himself out on the wood, and be dried from the lust of the flesh. The more the strings are stretched, the more sharply do they sound."[12]

Augustine sounds materialistic and embodied at the same time. His text, which placed very little space between musical instruments and singing voices, was foundational to theological speculations around music and the body through the Middle Ages and early modern period. This sensuality permeated writings that conflated instrumental strings and viscera, drums, and skin. This conflation was not, however, a matter of organology where, like some gruesome horror story, you can make an instrument out of the parts of a corpse. Instead, the human body worked like an instrument, and music harmonized the soul.[13]

The most influential medieval commentary on Aristotle was Albertus Magnus. Around 1260 he published a book called *Questions Concerning Aristotle's* On Animals, which explores a series of contested passages. His discussion of voice as instrument built on explicating the difference between the voices of men and boys: "a vocative potency is natural, seeing that its instruments are natural."[14] The voice, in his reading of Aristotle, is made by nature and controlled by the soul.

When medieval thinkers and practitioners read Aristotle alongside anatomy texts and dissection accounts, this led to a move toward specific physiological explanations for bodily functions and sensory activities. And, by the sixteenth century, literatures that we would describe as philosophical, psychological, and anatomical saw a convergence of responses to Aristotle and an increased emphasis on individual organs. Sensation, for instance, depended on organs inside and outside the body, and attempts to understand how it worked veered toward the physical. It was not just how do you hear but how does the ear work. In music making the body was understood as a site of musical production. So, a text about hearing might refer to the anatomy of the ear. Fabricius ab Aquapendente, of the vivisection voice studies, explored individual body parts across different animals. His interest in body parts was about the relationship to the whole, and the animal was a composite form. Investigating vision in *De visione*, he looked at the function of the eye, drawing on his dissections and vivisections, as well as Aristotle's *On the Soul* and *On Generation* and Galen's *On the Natural Faculties*.

Writing around the same time as Fabricius, Zarlino explained *organ* as an instrument that moves matter. In querying whether the organ should be understood as an ancient or modern invention, he pursued his answer in a genealogy of the word itself. He explained that the name *organ* emerged not in relation to a particular instrument but because "it also behooves all of the mechanical instruments that are useful in any of the arts and sciences through the help of which one can bring any work to the desired end."[15] Organs, in his terms, constituted anything that can change matter or immaterial forces. An organ or a tool is, in this model, both material and immaterial:

> The hammer that the ironsmith uses to make nails and the saw that the carpenter uses to saw and cut a plank are called tools or instruments. Also money, with which we buy the things that are necessary to human life, is called a tool. It's not just the material things that have a permanent form; but those without form such as logic that we call a tool.[16]

By the early seventeenth century, the Aristotelian notion of logic as an organ was connected directly to new modes of inquiry. Galileo's understanding of natural philosophy as a pipe organ brings this to the fore.[17] He makes the connection in a correspondence concerning sunspots with the Jesuit Christoph Scheiner.

> I do not want on this account to despair and to abandon the enterprise; on the contrary, I would hope that these novelties might serve me wonderfully to adjust a few pipes of this grand [but] discordant organ of our philosophy, which, in my view, many organists labor in vain to tune to perfection. And this is because they go about leaving and preserving three or four of the principal pipes out of tune, such that it is impossible for the others to respond in complete harmony.[18]

In 1609 Adriano Banchieri, who would start the Accademia dei Floridi in Bologna in 1614, published the first book of instructions for church organists. The *Conclusioni nel suono dell'organo* uses the word *organ* in very slippery ways. He had dedicated the work to the patron saint of music, who was imagined by some to have invented the organ, and he reminded readers that *organ* could mean "organo della voce humana."

> Alexander of Alexandria says that the Virgin Cecilia repeatedly sang only to God with music of the organ, as we have it in the appropriate liturgy for her feast day. By this is meant the unarticulated organ of the human voice, which was her manner of singing, and no other sound. For further confirmation of this we have it in the Hymn of St. John the Baptist, *Organa vocis*, and in Job, *Organum meum in vocem flentium*.[19]

For Banchieri an organ could be the musical instrument he wrote music for, or it could be the human voice. Note that the organ was still the primary instrument for most keyboard players who played a range of musics, from sacred surround sound to dreamy improvisations and basso continuo accompaniments for singers.

## MAKING THINGS

In 1592 Lodovico Zacconi explained that singers had a harder time making large intervals than keyboard players because they could not use fingers. Half a century later Athanasius Kircher focused on the tremulant, a mechanical device installed on the organ's windpipe that rhythmically modulated the airflow to the wind chest and produced a vibrato with an almost constant frequency. Giovanni Battista Marino's 1614 *Dicerie sacre* made these technical points in a more fantastical manner.[20] Marino will come up later for his writings on the magic of stagecraft and his vivid description of looking at a Titian painting.

His discourse on music used words to make the human body into a music machine and described using knife and tongue to dissect human members. He anatomized the mouth as a pastiche of things: printing press, bell, paintbrush—an assemblage extraordinaire.

> I certainly do not see any sense for whose operation nature provided more machines and marvelous tools than for the mouth, which it made specially for music and in which it employed all its ingenuity and effort. The instruments are worked with so much care and subtlety, and they have been brought from such a distance, that all the limbs of the universal body seem to have been created only to serve music, such that the brightest intellects, in their philosophizing, and the most learned hands, in their writing, exhaust themselves in such considerations.[21]

An animated lyre, the mouth is a voice-making machine tied to the birth of the pipe organ. He did not focus on the spirit or the breath but, rather, links the organ to a body part you could touch, feel, taste, and hear.

> This organ is likewise found in the mouth of man; the voice takes the place of sound; the lungs substitute for the bellows, which are compressed by the chest in order to process the air it receives; the throat is like the cannon through which the spirit flows; the varied disposition of the teeth resembles the various reeds that have the task of fragmenting and shaping the voice and dividing the articulations of the song. Do you want the artisan or the musician? Here is the intellect that uses the tongue instead of the hand, corrects the disordered breath, and gives norm and form to the voice that comes, unruly and lawless.[22]

Meanwhile musical treatises embraced the parallel mechanics of the voice and organ; organs were described in embodied terms and voices in material terms. The human voice was animated by the breath, which pushed air from the chest through the throat and into the mouth, where tongue, lips, and teeth gave it articulation as speech or song. Pitch and volume depended on how swiftly the air moved through. In 1588 Zarlino explained that the organ acquired its name because it is constructed in the manner of the human body:

> Thus I say this, that the proposed "organ" acquired this universal and common name as its own particular name from a certain excellence of its natural parts that form the voice, which are called the natural instruments. Because [the organ] was constructed in the way of the human body, with the pipes corresponding to the throat, the bellows to the lungs, and the keys to the teeth and the part that sounds to the tongue, and so on with other parts that correspond to the same ones in man. But to tell the truth, our Organ as a material structure is not very old; on the contrary it is rather modern, inasmuch as the bellows were added to the modern [organ]; the bellows from the box that used to contain the water, which now we call the "Sommiero," create the air that goes into the pipes, as Vitruvio paints it, from which it acquired the name of "Hidraulica" (Hydraulic); and thus we can see that our organ is not a modern instrument, except in its alteration of its first form; the air that now is produced by the bellows is put in the place of the [air] that was made using water.[23]

Giovanni Battista Aleotti, Fabio Colonna, della Porta, and others described more specifically the process that Zarlino theorized, articulating a practice used in Italy's iron foundries, using only water, air, and a recording barrel. As flowing water entered a long vertical tube, it mixed with air. When the water flowed out the end, the pressurized air separated to feed the organ pipes, and the water drove a paddlewheel. The wheel in turn drove a large barrel with pins that opened the valves of the pipes that created pitch. Because they seemed to power their own sound, such mechanical organs blurred the line between man and machine.

Describing his experimental attempts to build a hydraulic organ, della Porta wrote about a process that mirrors that of the human voice. He mixed air and water either in the end of a pipe or in his mouth. After many failed attempts, he found a way to create a "warbling sound and keep the tune" with the instrument by forcing the air from the bellows to bubble in the chest.[24]

> Let there be made a Brass bottomed chest for the organ, wherein the wind must be carried; let it be half full of water, let the wind be made by bellows, or some such way that must run through a neck under the waters; but the spirit that breaks forth of the middle of the water, is excluded into the empty place:

> when therefore by touching of the keys, the stops of the mouths of the pipes are opened, the trembling wind coming into the Pipes, makes very pleasant trembling sounds, which I have tried and found to be true.[25]

Some of the effects were the same. The vibrato effect della Porta wanted to create for the organ was next of kin to the singers' tremolo, and the shared emphasis on tremolos from writers on singing and on organs highlights this kinship.

Through the seventeenth century, the tremolo was perhaps the most basic singing technique. The tremolo sounds like a tremble. It is a throat articulation, not a vibrato. In his treatise for singers, the Venetian teacher and falsetto singer Lodovico Zacconi, who wrote extensively on voice quality, included it in a chapter on ornamentation and grace and imagined it as a gateway to ornamentation all told.

> The tremolo in music is not necessary, but performing it, in addition to showing sincerity and audacity, makes tones more beautiful. . . . I say again, that the tremolo [that is, the tremulous voice] is the true opening to the *passaggi* and way to master the *gorgia*; just as the ship moves more easily when already in motion and the dancer leaps better if he moves into the leap. This tremolo should be slight and pleasing, for if it is exaggerated and forced, it tires and annoys. Its nature is such that if used at all, it should always be used, so that it will become a habit, for this continuous motion of the voice helps and readily encourages the movement of the *gorgia* and facilitates admirably the beginnings of *passaggi*.[26]

When Monteverdi first heard Pasqualini sing in 1628 at the Farnese-Medici wedding, the young singer had apparently not yet reaped the benefits of such training programs; the composer remarked despairingly to Alessandro Striggio of the boy who was about eleven that "he seems not to have a pleasing voice either—he can do little ornaments and something of a trillo but everything is pronounced with a somewhat muffled voice."[27] But this is a skill that he would acquire.

## MUSEUM OF LIVING THINGS

In chapter 3 I mentioned Kircher's speaking tube that spat words from his bedroom to his garden. He eventually moved the speaking tube to the museum in the Collegio Romano where it lived among other automatic machines. The museum, a focal point of the global Jesuit project, hosted visitors from all over the world. In the museum Kircher reincarnated the tube as the Delphic oracle. The tube was a secret actor.

> Now it wanders about with a low and hidden voice, in ludicrous oracles and invented consultations with such an artifice that no bystander can perceive anything of the secret technique of the reciprocal murmuring of the discussion. And it is shown to strangers even to this day not without the suspicion of some hidden demon among the ones who do not understand the machine, because the statue opens and closes its mouth as if it were speaking, and moves its eyes.[28]

In this case Kircher's handiwork breathed life into a statue by making it speak. He did so by placing the tube in the mouth of a statue created in the image of the famous Delphic oracle. The Delphic oracle, like the castrati, made a sound that listeners could not quite figure out and whose inner mechanisms remained mysterious to those who enjoyed it.

Zooming in on museums shows yet another space in which the lines modernity draws between animate and inanimate were much more fluid in the era of the castrato. Early modern museums felt very different from the collections of static exhibits that we know today.[29] Often called "theaters of nature," they were resounding spaces that served as important sites for performance and display. And they often engaged a broad public through books with *theatrum* in the title that worked like virtual tours.

Kircher's museum was used as a theater for ceremonial visits. In 1622, for example, the Collegio Romano sponsored a series of festivities to celebrate the canonization of Saints Ignatius and Francis Xavier, for which the college was turned into a simulacrum of ancient Rome. The college museum hosted stagings of plays representing events in the lives of the two new saints, which included an opera—with a libretto by the mathematics professor Orazio Grassi and music by Kapsberger—that featured many castrati and elaborate stage machines.[30]

The museum endeavored to inspire wonder in spectators in part by playing in the space where art and nature met. Naturalists, like musicians, took pride in their ability to amaze. The machines, instruments, and experiments housed within the museum walls occupied a space bordering on the theatrical. Kircher's assistant Gaspar Schott clarified this in the preface to his descriptions of Kircher's machines:

> There is, in the much-visited Museum (that we will soon publish in print) of the Most learned and truly famous Author mentioned above [Kircher] a great

> abundance of Hydraulic and Pneumatic Machines, that are beheld and admired with enormous delight of their souls by those Princes and *literati* who rush from all cities and parts of the world to see them, and who hungrily desire to know how they are made.[31]

Schott described Kircher's mechanics as creating an abundance of delight and desire

> so that I can satisfy their desire to know the construction of the machines, I have undertaken to show the fabric, and almost the anatomy of all of the Machines in the said Museum, or already shown elsewhere by the same author.[32]

Note his use of the words *fabricam* (fabric) and *anatomiam* (anatomy), both of which recall contemporary anatomy treatises and serve to bring the machines closer to their human observers. The word *anatomy*, used here to describe both machines and humans, serves conceptually to link the two. For Schott the machines aim to please princes.

> Neither will we be content with delighting only the eyes, we also prepare pleasures for the ears, with various self-moving and self-sounding organs and instruments, that we will excite to motion and sound only by the flow of water and the stealthy approach of air, with no less ease than skill.[33]

Just a few blocks away from Kircher's global enterprise, Michele Todini made a museum of musical instruments in his house. His machines, especially his keyboards, mediated on the entanglement of humans and machines.[34] Todini was a trombone player, instrument maker, and keeper of instruments at the Congregazione di Santa Cecilia. He articulated the museum as a space of living, breathing musical organs, a space where musical instruments as tools, living organs, and thought processes did their work.

His instruments did not exist apart from nature or from players. Instead, they, like singers, stood as continuous mechanical systems that extended natural materials. Todini explained that making machines meant "talking with the hands." He also published a *Dichiaratione della galleria armonica* (1676), designed more to demonstrate his own virtuosity than to take visitors on a virtual gallery tour.[35]

Like Kircher, Todini wanted to make inanimate objects talk and insisted that to do so would involve the motions of natural organs. The key to such vocal action lay in making use of organ pipes made of soft, malleable marble.

> The sound of a pipe of an organ or another concave thing, passing through a twisted duct, made out of soft materials, like animal skins or a similar thing, would be enough to make that with which such great art, and even with miraculous art, is scarcely distinguishable from the natural organ, guided by the will, which makes every part move, for whatever effect is necessary.[36]

The sweetest organ came, according to Todini, from Indian wood. In this description, Todini articulated in some form what Diderot would say a century later.

> Imagine a harpsichord with sensation and memory, and then tell me whether it will not repeat by itself the tunes you play on its keyboard. We are instruments endowed with sensation and memory. Our senses are merely keys that are struck by the natural world surrounding us, keys that often strike themselves.[37]

For Diderot the harpsichord was, as Roger Moseley put it, already a media system capable of transition. Read backward, though, Diderot, like Todini, embraces the potential of an organ to be alive and unalive, organic and material all at once.[38]

This is especially the case with the instrument called the "grand machine": a compound keyboard that put on display the whole matter of instrumentation and instrumentality. It stood as an elaborate and ornately decorated conglomeration of seven instruments—harpsichord, spinets, organ, violin, and lyra viol—that could be activated in any combination from a single keyboard. (Fig. 4.1) Todini built the machine in phases. In 1650 he assembled the keyboards, and they were in place by 1656, but then he had to seal off the room because of a plague outbreak.

The fourth chapter of his *Galleria* describes the keyboards in the third room. Evidently, he toiled for years on this instrument, struggling to make the instruments work together remotely. After describing the many mechanical challenges, he explained the layout of the instrument. The keyboards sat away from the wall, touching nothing. Most visitors, he said, assumed that more than one person played the instruments.

> There are three quilled instruments put inside the room; a small spinet [*Spinettina all'ottava alta*], a *spinettone* [large harpsichord],[39] and a small theorbo/small keyboard,[40] facing opposite the person who plays them and situated so effortlessly that it would seem impossible that they might have any communication with the intrinsic machinery. Without knowledge that [this machinery] exists

FIG. 4.1 Michele Todini's *Galleria armonico*. Plate from Athanasius Kircher's *Phonurgia novo* (1673), vol. 1, sec. 7, p. 168. Courtesy of the Albert and Shirley Small Special Collections Library.

and how the instruments operate, the effect makes it seem almost invisible. The instruments are separated from one another by more than a palm on their tables and each on its own stand. They don't touch walls, nor anything else, and the whole thing appears to have nothing stable. There is an Organ, as already said, that also plays with the said *tastatura* and that has many registers, and one in particular is rare and sweet because it is made of Indian wood.[41]

Todini meticulously detailed the operation of the machine in part through a gripping description of a demonstration for princes during which he removed all twenty-three parts of the instrument that "control the operation," threw them on the floor, and then put them back together again. He then numbered each part and drew a diagram that would in theory allow anyone to take apart the instrument and put it back together again. Todini insisted that the making of his instruments required technical skill and used this to deliver warnings for youths who come to Rome hoping for an education. He advised that they embrace the technical, not just books:

[They] must reflect on the fact that practical, mechanical mathematics [i.e., engineering] is much more difficult than theoretical mechanics [i.e., physics], be-

> cause for the latter, one just has to read books about these subjects and know how to talk or write about them, which probably can make them seem simpler, but in the former case, you must reconcile the weight, the number, and the measurements, which cannot be falsified in any way.[42]

The second room in the museum contained the gilded harpsichord known as "La Macchina di Polifemo e Galatea," the only surviving instrument from this collection. In this room Todini turned a musical instrument into a minidrama and put into the hands of mythological figures instrumental sounds ventriloquized by instruments imitating the sounds of other instruments. The room embodied the idea of metamorphosis and is yet another space where vibrant materiality lives. A keyboard instrument became a fanciful marine scene. A combined organ and harpsichord imitated the sound and timbre of the *sordellina*, a keyed bagpipe instrument that Polyphemus played in a futile attempt to attract his nymph love object.

> In the second room you see the story of Polyphemus with a number of statues covered in gold, and among the others is Galatea, who is pictured moving through the ocean carried by two dolphins harnessed by a cupid, and sitting on a seashell, courted by sea nymphs and served by life-size tritons who carry a harpsichord, the case of which is rich with carvings representing, in bas-relief covered with gold, the triumph of the above-mentioned Galatea, with sea monsters that offer her various sea creatures. Polyphemus is seated on the slope of a mountain, where he lived, as the fable relates, in the act of playing a sordellina or musetta in order to please Galatea; and within this said mountain is found the device to make the tones of the sordellina, which sounds with a keyboard placed under that of the already mentioned cembalo. The statues are made by worthy men, as are also the other materials, which were used to represent the sea, the mountain, or the air. This machine takes up the space from the floor to the ceiling; the difficulties [of its creation] were many, and will be described at the end, so as not to impede the brevity of this account.[43]

As Galatea, holding a lute, moves through waves on a dolphin-drawn carriage, Polyphemus plays a sordellina. Creatures blow horns, and a strange figure plays the lute. A harpsichord stood in front of a mountain that contained a small pipe organ. The organ was the voice of the pipes and the harpsichord of the lute. Todini describes a set of *linguelle* (tongues) activated by rubbing fingers on the harpsichord keys and six knobs that activated certain pipes. The same kind of organ pipe that made Galileo able to see planets here activated the voice of a sordellina played by the mythological Polyphemus.

## AFTERLIVES

In 1677 more than thirty creditors sued Todini, and it did not end well. Rather than sell the machines, as the court mandated, he dismembered his own creations and fled with key parts to Santa Maria dell'Anima. Almost as soon as they were made, Todini's magical keyboards became ghosts in a failing machine, body parts disassembled and reassembled in the Verospi family palace. Todini was in perpetual debt to the Verospi.[44] Like the cat piano and even the castrato, Todini's instruments may do even more work in their afterlife than they did in their historical moment. His instruments were ghostly almost as soon as they appeared, quickly morphing into specters of themselves.

The only picture of Todini's grand machine is in Kircher's *Phonurgia*.[45] Kircher attached it as an appendix to the first volume, the one about the influence of music on human beings. Todini later said the illustration was based on fantasy. Kircher describes the demonstration that "with each key strike not only seduces the listener's ears but also astonishes their eyes."[46] The machine provides a wide range of emotion: soft tones, excitement, fury. Kircher was not willing to reveal "how the harmonious instrument was built or the hidden art" of its construction.[47]

A generation later, Filippo Bonanni, a Jesuit archivist and curator, published an illustration in his 1722 *Gabinetto armonico pieno d'istromenti sonori*. The author was over eighty at the time. The book, a gorgeous pairing of texts and prints, was made by Arnold van Westerhout. Todini's machine here sat alongside ancient instruments, Turkish instruments, automata, and more. It was already, in other words, a fantasy. The book as a whole stands less as what would constitute an accurate description in modern terms and more as an intertextual and visual synthesis of a series of ideas. The pictures, even according to Bonanni, were representations. The illustration looks different from Kircher's and seems to be an inscription of memory; the series of illustrations removes the instruments from their natural habitats. Charles Burney also wrote about the museum's decay.

> All the accounts of Rome are full of the praises of this music gallery; or, as it is called, gallery of instruments; but nothing shows the necessity of seeing for one's self, more than these accounts. The instruments in question cannot have been fit for use these many years; but, when a thing has once got into a book as curious, it is copied into others without examination, and without end. There is a very fine harpsichord, to look at, but not a key that will speak: it formerly had a connection with an organ in the same room, and with two spinets and a virginal; under the frame is a violin, tenor, and basse, which, by a movement of

> the foot, used to be played upon by the harpsichord keys. The organ appears in the front of the room, but not on the side, where there seems to be pipes and machines enclosed; but there was no one to open or explain it, the old *Cicerone* being just dead.[48]

For Burney the gallery was lifeless, and he gestured toward a repetition with no original, a thing that is always in the past tense.

To end on a castrato note; recalling the same trip to Italy, Burney, in his 1789 *A General History of Music*, penned a description of Farinelli's voice that was a stylized instrumental hearing. As I will argue later in the monograph that Burney's literary recollections of castrato sounds were as much about Burney's ideas about voices and Italians as they were about whatever he heard.

> There was none of all Farinelli's excellencies by which he so far surpassed all other singers, and astonished the public, as his *messa di voce*, or swell; which, by the natural formation of his lungs, and artificial economy of breath, he was able to protract to such a length as to excite incredulity even in those who heard him; who, though unable to detect the artifice, imagined him to have had the latent help of some instrument by which the tone was continued, while he renewed his powers by respiration.[49]

This description, performative as it is, highlights the tension between material and organic that was embodied in the singer. His lungs were natural, but his breath was artificial; he somehow seemed to possess the secret help of some instrument. The castrato appears to be an enhanced human being, a natural body modified by artificial means. Burney's description offers an eighteenth-century rendition of a worldview that did not draw hard lines between organs and organs.

Burney responded to the pyrotechnical style of castrato singing that modern listeners tend to know best. For example, Senesino (Francesco Bernardi), for whom Handel wrote many operatic parts, possessed a vocal instrument that could shift inaudibly from head voice to chest voice and hold fourteen-measure phrases. This technical mastery over the body was by then associated as much with instruments as with singers. Handel's aria "Perfido," from the *Radamisto* performed in London first in April of 1720 and then in December of the same year, is a well-known example of late eighteenth-century castrato vocal control. The opening motive of the aria, written for Senesino to replace a less virtuosic aria sung by Margherita Durastanti just eight months earlier, demanded that the singer switch from head voice to chest voice, showcasing the castrato's peculiar talents. There are long trills and melismata that high-light agility and make language into sound.[50]

D'Ancillon's description of the castrato voice that this chapter opens with goes on to say,

> You think you are almost satisfied with those Luxuriancies of Sound, you are most agreeably charmed anew with the soft Strains of Jeronimo (which I have sometimes almost imagined have been not unlike the gentle Fallings of Water I have somewhere in Italy often heard) lulling the Mind into a perfect Calm and Peace.[51]

The voice signified something in Italy; maybe fountains, maybe the Mediter-ranean Sea. The garden is our next stop.

CHAPTER 5

# INTO THE GARDEN

When Charles d'Ancillon described a castrato as reminiscent of falling water in Italy, he dropped a sound bite that signaled the imagined sensory experience of castrati and of gardens and a temporal threshold of a mythological past and hybrid future. To walk through a villa garden today is to step on and through the materials of history. Existing in an afterlife of virtual reality, these gardens were, and are, palimpsests of representations and sensory experience. Like the spectacles that preoccupy the next chapters and like singers of all body types, they possess a textual life that transcends the natural world. From the late sixteenth century on, Italian villa gardens featured prominently in guidebooks, machine books, and horticulture books, elaborately illustrated and described. The images engraved in books were not photographic; they were meant to incite a sensory experience inflected with the presence of ruins and, like descriptions of the cat piano or castrato voices, must be read with many grains of historical and philosophical salt. Like *Il cannocchiale per la finta pazza*, the descriptions invited readers from far away into the immersive experience and reminded those who might see it what and how to feel.[1]

In the sixteenth and seventeenth centuries, villa gardens were imaged as immersive performance experiences. Anton Francesco Doni, in his literary description of villa life written in 1566, constructed the villa as a space for feasts and other rituals animated by music and nature.[2] The Jesuit historian Daniello Bartoli described water bringing statues to life and creating sounds of moaning, sorrow, singing, and anger.[3] These ephemeral, multimedia, ghostly works of cultural expression functioned as what Foucault called heterotopic spaces, separated from their surroundings with controlled movements and systems. Or, in different terms, they made theaters out of nature and were theaters of nature. And in still different terms, they embodied the performative culture that was a hallmark of late sixteenth- and early seventeenth-century Italy, especially Rome. And they embodied the generative property of matter in that they were neither art nor nature. Bartolomeo Taegio used the term *third nature*.[4] Human artifice shaped natural materials into imitating human artifice. With marble statues, automata making noise, figures made out of plants, and advanced hydraulic engineering, the gardens played with art and nature. Artificial grottos seemed natural; hydraulic devices imitated bird sounds, which in turn prompted real birds to sing; botanical structures looked like humans; and creatures seemed to lactate, vomit, cry, and sweat.

The Villa d'Este at Tivoli, nestled in the hills nineteen miles outside of Rome, invites musings on gardens. Once the summer residence of the second-century emperor Hadrian, it is now a UNESCO world heritage site and tourist attraction. Just a short train ride from Rome and a fifteen-minute walk down a hill, across a stream, and up an old staircase, the gardens, fountains, and sculptures even today seem to emerge out of rocky hills. The path of a hundred fountains consists of three levels of gushing water, sculptures of birds, and spitting animals. Supposedly the hydraulic organ plays every two hours, its eerie sound embodying a dialectic of generation and decay (Fig. 5.1).

In 1572 or 1573, depending on whom you ask, Pope Gregory XIII visited the villa, in part because the air made it a health retreat and in part as a diplomatic mission. Giovanni Maria Zappi, a chronicler and editor, described it:

> That organ is found inside a large niche. When someone gives the order for it to play, first two trombones are heard, which play quite a bit, and afterward follows the harmony, generally so well ordered and with moderate pace that [the organ] of San Salvatore del Lauro in the city of Rome cannot play more

FIG. 5.1 Hydraulic organ at Villa d'Este, Tivoli. Photograph by author.

> graciously than this with good measure. And so, people believe that when His Holiness, Our Lord Pope Gregory XIII, found himself in the city of Tivoli in this year 1573, the order was given to make it play, and its playing gave His Holiness such satisfaction and amazement that he returned not only one time to hear it but two and three times, and he wanted, in any case, to speak to Claudio, its inventor, and having seen and heard such rare magnificence in the presence of numerous cardinals and important men, a countless number of gentlemen could not believe that this organ played the registers tunefully with water by itself, but they believed instead that there was someone inside there, and consequently they stayed to confirm [that there was in fact nobody inside]. After the organ played, the order was given at the same time to give water to fifteen little jets of water, which was scattered with such great force and violence that the water rose so high that it was called the Deluge.[5]

This performance of eerie, automated music, water, sound, and special effects occurred either just before or just after the pope was elected. Workers hastily completed a dragon fountain for the pontiff's visit, a reminder that this garden, like so many, was always in production and always decomposing. Cardinal Ippolito d'Este, who commissioned it, died before the massive reconstruction concluded. In 1582 someone broke in and destroyed the organ, and by 1586 the whole garden languished under the custody of the College of Cardinals. Nevertheless, these extravagant gardens emerged eventually as metonyms for the threshold between the mythical past and the imagined future and were part of what made Italy on the Grand Tour an interpretive playground. This garden and organ are both well documented and hard to pin down. "Eyewitness accounts" may or may not have been eyewitnessed.[6]

Hydraulic organs have inspired marvelous descriptions since the ancient world. The Roman poet Claudian wrote of the musician "whose light touch can elicit loud music from those pipes of bronze that sound a thousand diverse notes beneath his wandering fingers and who by means of a lever stirs to song the laboring waters."[7] Claudian's enchantment with the organ player's ability to create sound from waters was repeated frequently in early modern descriptions of hydraulic organs. There are some obvious parallels between garden hydraulic organs and musical practice. Both depend on sound, and late Renaissance commentators referred to the music of falling waters and discussed at length its musical properties. Claudio Tolomei, who was deeply invested in Vitruvian architecture, focused on the sounds of fountains.[8] Marin Mersenne, a French mathematician and music theorist who was heavily influenced by the Italians, connected the voice and waterfalls. He wrote in his *Harmonie universelle*, published first in 1636, that

> there is still another vocal quality that renders the voice full and solid and that augments its harmoniousness. One can describe it by comparing it to a canal that is always full of water when it flows or by the comparison of a body and face plump and in good health; as opposed to those voices lacking in this quality, which are analogous to a trickle of water flowing in a large canal and a face that is thin and emaciated.[9]

Like the humanist musical practices of Zarlino, Monteverdi, and Galilei, these organs and the gardens they inhabited stood as physical attempts to rewrite antiquity through iconography and engineering. And, like the music dramas, lyric poetry, and operas that castrati envoiced, early modern gardens frequently embodied Ovid's stories, putting into early modern practice the near obsession with material transformation and movement between kinds of matter. Ovid's lyric depiction of ubiquitous metamorphosis from animal to vegetable and back again found animate reality in such gardens. The theatrical hydraulic organs that wowed Pope Gregory were sites of sensory overload and magic making meant not to sound like acoustic instruments or even a church organ. Instead, their affective potential was in the idea, image, and feeling of sound coming out of marble and horticulture without the aid of the human hand. They made acousmatic sounds: sounds that have no visible source, like Pythagoras, a frequent figure in these gardens who was said to have taught his students from behind a screen so that the visual did not distract them from absorbing his sounds.

Hydraulic organs exist predominantly in textual abstraction, appearing in centuries of books called theater of nature.[10] Kircher described hydraulic organs at Villa Aldobrandini, the Palazzo del Quirinale, and the Villa d'Este in book 9 of his *Musurgia universalis*.[11] The Palazzo del Quirinale, built by Lucca Biagi in 1598, included a mechanical representation of the legend of Pythagoras. Kircher's illustration anatomized the whole extraordinary machine, showing exactly how the cylinder worked by means of perforations that allowed poles to slip through and open the pipe. The blacksmith's hammer is a metronome and animator. His prose renders the machine a readable text, a kind of tablature that makes the magic graphic. He graphed the cylinder and put in musical notation an eight-voice song (Fig. 5.2).

## TIVOLI

The hydraulic organs at the Villa d'Este in Tivoli and the gardens are exceptional in their history, documentation, and afterlife. Here they provide an interpretive window into the vibrant matter that connected art and nature,

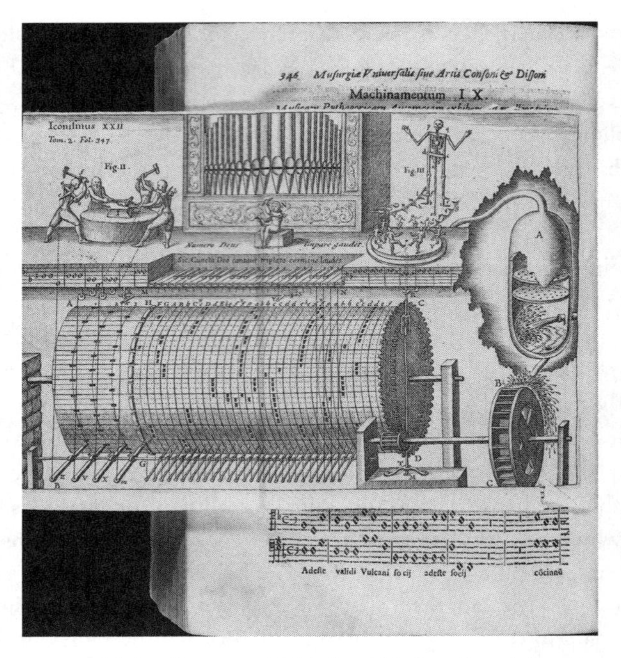


FIG. 5.2 Hydraulic organ with blacksmith's forge from Athanasius Kircher, *Musurgia universalis* (Rome, 1650), 2:347. Courtesy of the Albert and Shirley Small Special Collections Library.

theater and life, voice and instrument. The Villa d'Este water organ, the first such organ in a Roman garden, and its sensational emergence on the spectacular scene incited the construction of other such marvelous machines, leading to what would be called today a "competitive villa scene." A tourist site from the start, the garden cost Ippolito over five thousand *scudi* and required him to sell large quantities of silver, flatten hills, and move aqueducts.[12] The garden was the brainchild of Pirro Ligorio, the official antiquarian to the Este Court, and it was realized by a massive team of architects, artists, fountaineers, and craftsmen. Ligorio and his team imagined themselves creating a paradise composed of ancient relics and modern marvels. Not surprisingly, they were ruthless and had no problem flattening one church, pillaging others, and taking over land. In 1564 fifty "schiavi turchi" (Turkish slaves) staged a revolt and threw their handlers/masters into the fountains. Ligorio quickly replaced them with fifty more purchased in Venice.[13]

The mythological program embodied in the garden emerged over decades, and the result is a who's who of Ovid's *Metamorphoses*. Orpheus and Apollo, for instance, were added in 1609, just after their appearance in multiple music dramas. Stone animals spitting water and making music standing alongside mythological figures marked an ontological continuity among animal, human, and Earth. Preparations for Pope Gregory's visit in September of 1572 (mentioned above) included the hasty construction of the fountain of dragons. Nicolas Audebert, who traveled to Italy from 1574 to 1578 and chronicled the visit in journals, described a fountain of four dragons with scales and a series of sounds and explosions designed to evoke fire breathing and fireworks.[14]

The organ, built by the Frenchmen Luca Clerico and Claude Venard, was the organ that Zarlino claimed indirectly inspired his *Sopplimenti musicali*. Responding to a provocation by Leonora d'Este to explain what she, following Vitruvius, called a hydraulic machine, he explained the machine.[15]

> But the similarities that emerge between this description of the hydraulics and our organ led the Illustrious Sister Leonora d'Este in November of the year 1571 to ask me through Francesco Viola, my remarkable friend, whether that organ was ancient or modern, and where it got its name. After I had first answered her, I was intrigued by the request, and I decided to write the present *Sopplementi*.[16]

A small theater in itself, the organ displayed marble hybrid sculptures. Orpheus with his lyre and Apollo with his viola da braccio are in residence with their respective stories of art ending with dismemberment: the severed head of Orpheus and the flaying of Marsyas. In 1581 Montaigne described it like this:

> The music of the organ, which is real music and a natural organ, though always playing the same thing, is effected by means of the water, which falls with great violence into a round arched cave and agitates the air that is in there and forces it, in order to get out, to go through the pipes of the organ and supply it with wind.[17]

Montaigne highlighted the organ as performance. It started with trumpets, moved into an organ madrigal accompanied by automaton birds, and ended with a loud dramatic deluge. Montaigne explained the workings of the organ as giving agency to the water itself, and he made clear that the gentlemen who, according to Zappi's 1576 account, thought the organ contained a man inside it were not privy to the workings of the magic.[18]

> Another stream of water, driving a wheel with certain teeth on it, causes the organ keyboard to be struck in a certain order; so you hear an imitation of the sound of trumpets. In another place you hear the song of birds, which are little bronze flutes that you see at regals; they give a sound like those little earthenware pots full of water that little children blow into by the spout, this by an artifice like that of the organ; and then by other springs they set in motion an owl, which, appearing at the top of the rock, makes this harmony cease instantly, for the birds are frightened by his presence; and then he leaves the place to them again. This goes on alternately as long as you want.[19]

In Montaigne's description, nature and artifice imitate and control each other until it becomes almost impossible to distinguish between the two. The water-powered machines imitate the sounds of the man-made trumpet while birds sing in notes and harmony molded into sounds that only a human could generate. The artificial owl, created to look like a real owl, looked like it was conducting the little birds, stopping them with its presence. And most bizarrely, gunfire and artillery cascaded through the otherwise pastoral scene.

The inner mysteries of the organ were anatomized in an incredibly detailed description written by Audebert, who arranged for a secret tour by the organ maker himself, Claude Venard. He found the fountains most fascinating and was especially drawn to their elaborate and mostly invisible mechanisms. Audebert's behind-the-scenes tour sounds incredibly fun; it involved climbing ladders leading into small chambers and crawling behind organ pipes. His descriptions of water and air moving by way of springs, gears, and wheels moves almost seamlessly into descriptions of specific musical mechanics. Different levers make breves and semibreves, sharps and flats.

Audebert depicted sound as the mechanism through which art and nature merged and through which listeners became unable to tell real from imaginary. The organ played

> with such a melody that it is difficult to tell artifice from reality. However, an owl appears slowly between the feet of one of the statues and the muzzle of the boar, and just as it arrives, all of the birds fall silent. As long as the owl is there, they are quiet, but once it leaves, they begin to sing again. This occurs not all at once and all together, but in a sort of round. First begins the ringing of the lark. Then, another murmurs something between a staccato chirp and a deeper sound as if it fears still that which it has just seen. A third begins to whistle. When all of them sing together, they start to chatter like before, continuing until the owl returns.[20]

Paralleling Montaigne, Audebert used musical language describing sound to fuse artifice and reality. He described climbing into the fountain, revealing the innards and inner workings and performing for his readers a revelation of the ways that art and nature play with one another as if showing the inside of a singer's throat or the strings of a harpsichord.

> The manner of doing this is as follows: there are a number of whistles that ornithologists are accustomed to using in order to imitate the sounds of various birds. These correspond with a canal that fills with wind along with the water, that makes the whistles create each individual bird's song. But when the owl arrives, it finds itself in a place where it suddenly corks the conduit from which the wind enters. Then, in returning to its former place little by little, it first lets loose the pipe that corresponds only to the lark, then that allows another to chirp very low, not having enough wind, then it does the same to another that whistles, then, the owl having completely departed and the conduits remaining entirely open, all of the birds restart their original chirping with such a volume that they drown each other out and create the sound of a large number of birds in a flock.[21]

His description of the organ likewise begins very much like Montaigne's until the secret entrance. He reveals the human intervention.

> These organs, without any human intervention play a musical song with everything that comes with it, with beat and trills, by no means any less pleasing and melodic as the most excellent player would perform. However, to make the ingenious invention seem even more admirable, most visitors are not permitted to view the underlying mechanical construction. Unlike them I had the honor of a detailed explanation and was permitted to access the most secret places, where I curiously observed all movements and means of this admirable invention. I will describe them here as I saw it.[22]

Audebert went on to describe the relationship between the birds and the organs:

> While this music is playing, there is also a Nightingale singing endlessly by means of a tube under the water that receives air from one end of this cavern, its other end dipping into a little pot full of water. These organ pipes play just one single long and musical piece of music, which is for five voices.[23]

And for the final cadence, when the music finished, a machine operator pulled a small cord that made a large cascade of water and a deafening sound.

## GARDEN VARIETY

Thus far I have focused on gardens in the country, gardens where princes went for healthy air, and I have dwelled on the garden as performance. But Rome had gardens, too, and those gardens were part of the city's spectacular viscera that preoccupies the next chapter. In the seventeenth century, gardens in the city of Rome became more and more performative and increasingly served as living cabinets of curiosities. Flavio Chigi, the nephew of Pope Alexander VIII, presided over gardens and had a massive and notable collection that included rhinoceros horns, portraits of famous women, long telescopes, human skins, and ancient statues. In 1668 he staged a garden entertainment that was a musical comedy in honor of the Feast of the Assumption in the Garden of the Quattro Fontane near Santa Maria Maggiore. Carlo Fontana designed the whole event, Bernardo Terrenti composed the cantata, and Gianlorenzo Bernini created an "azione scenica," an authentic work of theater. The ten musicians included, among others, Arcangelo Corelli on violin.[24] Chigi hosted elaborate garden parties frequently. A play within a play, with a garden within a garden, it displayed the business of theatrical illusion and assimilated the practice of technological scene change into the sculpted garden of nature.[25]

In the evening, a long meal with interspersed performances began with a singer dressed as a gardener meeting guests, inviting them into the theatrical production. Unlike almost every other production in this book, this one delightfully involved a female virtuoso. Giulia Masotti sang a concluding recitative. Make no mistake, this was an exceptional event that took months to prepare. In the final days two harpsicords were tuned and musicians got new shoes. The gardener initially appeared surprised by the incoming guests but promised to put something together pronto. Pomona and Flora sang, and Bacchus, who got mad when the guests seemed underwhelmed, kicked the table in frustration. The noise served as distraction for a quick scene change. A fountain and a theater emerged seemingly out of nowhere. Lighting came in the form of branches and leaves illuminated by one hundred torches. As per the following description, each course seems to have had a scene change, and for the final cadence, lightning and thunder struck, followed by a rainstorm of scented water.

> Rome, August 18, 1668. The aforementioned banquet given by the Chigi for the Rospigliosi lords and ladies followed on Tuesday morning in the garden of Santa Maria Maggiore. The princesses of Rossano and Farnese attended as well. The sumptuous things prepared included that as the guests entered through the garden gate they found inside, with lovely perspective, another door, where, after

> knocking, a musician dressed in rustic clothing appeared and, in song, asked what they wanted, and, after they responded that they wanted to see the garden, not only was this granted to them, but also a table laid out with *ricotte* and other rustic foods. Then, four forest nymphs, who were patrons of the place, quickly rushed up and, after many reproaches against the rustic guardian, turned the table over onto the ground. After this, another table appeared, regally laid out, where, as soon as the guests assembled, came a rain of perfumed waters and a hailstorm of sweets. After this had finished, the banquet began with playing and music. After this, an echo of song from different birds. The whole thing succeeded perfectly because it was prepared by the Cavalier Bernini.[26]

The production was an early modern virtual reality. More about this later. It didn't end well, by the way. Somehow in all of the elaborate stage mechanics the stoves didn't work, and the guests were evidently quite hungry.

There are still some magic gardens. The garden of Bomarzo, commissioned by Pier Francesco Orsini and designed by Ligorio, about an hour's drive from Rome, is now billed as a monster park. Visitors can still walk into a giant screaming mouth and take a meal at a table in the shape of an erect tongue. The tongue comes out of a large nasty face with flared nostrils, and the space creates massive reverberations. Even today if you stand just outside, it sounds like the mouth might really be a roaring monster. Called the Sacro Bosco in the sixteenth century, it features horticultural giants viscerally tearing each other apart.

This short chapter has highlighted resonances between castrati and spectacular hydraulic organs located in gardens, both of whom reverberate with the vexed but inextricable intertwining of human, machine, and nature in the early modern period. Each can be constructed as experimental mixes of nature and techne that tended to obscure human labor. The organs, even at their heights, were essentially creatures of an imaginary that now, centuries past their prime, cannot possibly exist such that the modern experience of them almost always depends on reconstruction. As Vernon Lee wrote, "Roman villas are really a kind of architecture cut out of living instead of dead timber."[27] For her, gardens are haunted not by individuals but by ghosts of experiences that now "exhale in the breath of the honeysuckle and the murmur in the voice of the birds in the rustle of the leaves."[28]

# ROMAN REVERB AND SEA CHANGES

CHAPTER 6

# WHEN IN ROME

## *The Castrato as Special Effect*

I thought I was starting a project about the place of the castrato in Roman spectacles, so I read through dozens of printed festival accounts looking for musical references or names of castrati. Textual shards of these ephemeral papal coronations, royal births, piazza dedications, and feast days exist now only in bound volumes of small pamphlets, usually just a few pages each, whose titles reference "real accounts," "triumphal possessions," "solemn processions," "celebrations." They describe "delightful music," "marvelous fireworks," "beautiful horses," and more. These are not eyewitness accounts, and that the author may not have actually "seen" the event hardly got in the way of transmitting the ideological position of the sponsor and the interest of the writer. To feel more of the story, I walked cobblestone streets looking for traces and realized just how little ground this city that thought it centered the world occupied, how sound bleed must have been the norm.

Four hundred year old pamphlets blurred Ferdinand II and Ferdinand III and rendered piazzas whose names changed over the centuries difficult to identify, but traversing Rome on foot brought these things to life and turned the city back into the organism about which I had read. I wandered from Santa Maria Maggiore looking for the Garden of the Quatro Fontane. Rome still hosts spectacles and rituals on an almost daily basis. Watching stages and props go up and come down, I could imagine early modern Rome as a space where opera, spectacle, and city were connected in spiritual and pragmatic realms. Luigi Allegri showed that often the same craftspeople made props and machines for indoor theatrical productions and outdoor spectacle. I'd like to say I was embodying the eighteenth-century notion of embodied practice as key to the production of knowledge and embracing a sensory connection between past and present. And I would like to say that these walks were feminist resistance, escaping the monastical scholarly ideal. That would all be an ex post facto justification for gelato, wine, and movement.

One cold February I wandered from Santa Maria dell'Anima to what is now called Palazzo Taverna, following the steps of elaborate festivities staged for the election of the Holy Roman Emperor Ferdinand III in 1637. Luigi Manzini's one-hundred-sixty-page festival book with elaborate engravings by Luca Ciamberlano helped conjure arresting fireworks that celebrated

FIG. 6.1 Luca Ciamberlano, *Victory over Ottoman Turks*. Engraving in Luigi Manzini, *Applavsi festivi fatti in Roma per l'elezzione di Ferdinando III* (1637). Internet Archive.

victories (Fig. 6.1). The image of fireworks representing victory over the Ottoman Turks and descriptions of fire-breathing dragons to the tune of trumpets and drums is more vivid when you walk the streets.[1] On a brutally hot July day a distant parade conjured the coronation of Leopold I (Fig. 6.2) in which, "to awaken the pens of the most erudite to celebrate the glories of this house, just as the drummer, the most vile among soldiers, wakes up not only the most generous warriors, but all the leaders of the militia."[2]

Sounds called everyone to action:

One heard in every quarter the drums used to gather the citizens, who assembled in squadrons in different piazzas and especially around the gates of the

RELAZIONE
DE' FVOCHI ARTIFICIATI,
E FESTE FATTE IN ROMA
Per la Coronazione del nouello Cesare
LEOPOLDO PRIMO
Dall'Eminentissimo, e Reuerendissimo Sig.
CARDINAL COLONNA
Protettore del S. Romano Imperio, &c.
Colla descrizione delle Cirimonie, e Solennità fatte in Francfort
nella Coronazione di sua Maestà CESAREA
il primo di Agosto M. DC. LVIII.
All'Illustrissimo Sig. e Padron mio Colendissimo il Sig.
DON ANTONIO COLONNA ROMANO
de' Duchi di Montalbano in Sicilia.

IN ROMA, Appresso Francesco Caualli. 1658. Con licenza de' Sup.
Ad istanza di Giuseppe Elmi da Spoleti.

FIG. 6.2 Title page of Giuseppe Elmi, *Relazione de' fuochi artificiati e feste fatte in Roma* (Rome, 1658). Biblioteca Casanatense.

Piazza del Senato, between which and the church was erected a wooden bridge covered with 4,000 *braccia* of black, yellow, and white cloth (among which is the royal gold [emblem]).[3]

Festival descriptions often mention sound and singers. They don't always identify singers as castrati, but in papal courts women didn't sing in streets and squares, which means that festive soundscapes with high voices included castrato voices.[4] They don't have musical notation, but if you can hear a mash-up of Corelli, Monteverdi, and a touch of Gesualdo with extra rich orchestrations in your head, that might give you an idea of the composed music, and for the streets think loud voices, trumpets, drums, big wheels, fireworks, and sometimes animals. Castrati singers were part of the material production of an ephemeral festive culture, embedded in a sensual space that moved between natural and artificial. To put this in theoretical—and especially Deleuzian—terms, the castrati were the viscera of Roman spectacle culture: the organs, blood, and guts that animated experiences.[5] Seventeenth-century theatrical practices aimed to move the affections through material practices. The castrato was one of them, existing alongside fireworks, fountains, piazzas dressed up as theaters and battle grounds, crowd control, human-made thunder, and lighting effects. In each case, bodies moved natural force—air, fire, water—to entice affect.

Castrati animated the performative structures that instantiated the absolute power of the papal courts and the connection of ruling parties to a divine order. For the Barberinis especially, Jesuit notions of theater as persuasion mixed with ritual performance of civic power about which I'll have much more to say in chapter 8. This claim builds on Martha Feldman's arguments about opera and the performance of sovereignty in eighteenth-century absolutist opera.[6] "The social order was—by ideology, not in reality—meant to exist naturally, inevitably, and endlessly. The opera that represents it merely turns the pages of eternal time, its messages and denouements hardly susceptible to validation through inspections by earthly mortals."[7] Likewise, characters in spectacles embodied figures embedded in transcendent truths. And as Olivia Bloechl has argued for French opera of the eighteenth century, these productions enact a political theology.[8]

This chapter explores the power of stagecraft and mechanics to create sensory apparatus and positions the castrato voice as a special effect. Descriptions of stagecraft and stage sets as marvelous and miraculous are ever present, but they are not empty clichés. The practice of wonder did important affective work, and stagecraft was as much a part of ensuring Aristotelian unities of space and time as singing. If, as Simon Werett has argued, pyrotechnic feats were laboratories for technical knowledge, so, too, were productions that featured singers such laboratories.[9] In other words, as singers increased the capabilities of their voices to delight and incite wonder, so, too, did engineers and stage designers increase their ability to create wondrous artificial figures and devices that extended and replicated human and natural processes. Angelo Poliziano's 1488 *Orfeo* was also a lab for Leonardo's flying machines. And if anatomical theaters, public autopsies, and proto-Cartesian ideas about the body played on the conceit of the body as a machine, then operas and spectacles used the castrato as both a machine and a malleable object.[10]

The material production of music, of theater, of piazzas, of art was at once permeable and operational. By the middle of the seventeenth century, the idea of *machine* in theater referred to stage sets, instruments that moved sets, chariots, and special effects. Theatrical and spectacle productions created an immersive, affective experience for viewers in part through mechanical arts, including stagecraft that deliberately tricked audiences. The designer of stage machines was, in the sixteenth and seventeenth centuries, called a *macchinatore*, meaning a trickster or smooth operator. Torelli, the engineer who turned to stagecraft in Venice, earned himself the name *il grande stregone* (the grand sorcerer) because his creations animated unexpected changes in space and ignited a collective metamorphosis. We live in an age that loves cyberspace, one that William Gibson famously described as consensual hallucination.[11] Early modern Romans were frequently transported by theatrical spectacles that transformed entire city spaces, turning them into raging floods, fireworks, and sites of flying gods and goddesses. Their virtual reality, like ours, was not unreal. Martha Feldman has argued in the context especially of eighteenth-century opera that urban ephemeral practices frequently dressed and redressed entire cityscapes in multimedia systems of sound and light.[12] By the eighteenth century, events would more and more have definitive constructed settings, and civic festivities would become things to view rather than things to experience.[13]

Rome as theater of the world is a ubiquitous assertion, and as Valeria De Lucca among others has shown, it meant something particular with respect to seventeenth-century musicking.[14] Architectural historians of this era articulate a concept of scenographic public space in which piazzas were records of spectacle performances and in which ephemeral cultural practices reached permanent marble form.[15] The built world of early modern Rome drew explicitly on memories of festive decoration and served as technical laboratory for baroque festival experiments and stage scenes often featured scenes of urban Rome.[16] Painted wood, stucco, papier-mâché, clay, canvases, and more created fictions of illusions with raw materials. Human-made structures turned into miniature cityscapes, and fantastic creatures from myths and fables roamed the stage as enormous automata. In stage sets, statues coexisted with live singers and actors, while in parades, floats and people were made to look like statues. The seventeenth century also saw an increase in designated performance spaces. In Rome the Barberini family, Queen Christina, and others built designated theatrical spaces. By the eighteenth century, as Feldman argues, theatrical culture increasingly captured and controlled the chaos of outdoor space. Outdoor balconies turned into theater boxes as the ritual space was moved inside. But, in the seventeenth century the distinctions between outside and inside were still very fluid.[17]

Rome's soundscape and music moved inside and outside and through sacred and secular spaces. In recent decades, scholars from across the disciplines have brought that sound world to modern ears. Valeria De Lucca and Christine Jeanneret's essay collection *The Grand Theater of the World: Music, Space, and Performance of Identity in Early Modern Rome* offers a scholarly sound walk.[18] Much of the story comes from seventeenth-century guidebooks and musical descriptions, such as Grazioso Uberti's *Contrasto musico*, a musical dialogue that comes off as a sound walk through baroque Rome.[19] And to judge from the descriptions in primary sources as well as the invaluable work of Margaret Murata and Frederick Hammond on Barberini Rome, baroque Rome was a noisy space of barely controlled chaos.[20] Church feasts, diplomatic events, and dynastic celebrations provided massive possibilities for scenographic, musical, and creative marvel. Building on that work, this chapter focuses primarily on outdoor festivals and their intersections with the operatic stage and for the most part does not attend to the interconnected sacred rituals that occurred inside and outside of churches.

This chapter offers a virtual tour of Rome, lingering in Piazza Navona and Piazza Barberini as key scenographic spaces in the city's political ecology. This thick description includes snapshots of Rome and specific festivals. I discuss Marc'Antonio Pasqualini and his appearance at massive spectacles mounted in 1634 that included a joust and the sacred opera *Il Sant'Alessio*. These productions did their work through stage magic that is illuminated by the much-discussed *Il Corago*, an anonymous theater treatise published in 1630 and other contemporaneous documents. The tour concludes with Bernini, festivities for Queen Christina, and the Triton fountain in Piazza Barberini.

## ROME: A SPECIAL CASE

The history of the castrato in the streets and in Roman civic theater intersects in the late sixteenth century with the short but transformative reign of Pope Sixtus V. His actions and choices greatly affected the civic performative fabric of the city and the emergence of castrati as key instruments of civic and sacred performance. Castrato enthusiasts know his name from the 1587 papal bull *Cum frequenter*, which forbade eunuchs, spadones, and castrati from marrying because they could not procreate.[21] He also attracts attention from music scholars because in the late nineteenth century Alessandro Ademollo falsely accused him of banning women from the Roman stage.[22] Outside of musicology he gets attention for his deliberate urban planning: irrigation systems that would allow resettlement of the hills, street networks connecting churches, and aesthetic unity.[23] He left his mark on Rome through a deliberate revival of obelisks; he installed four in important locations and added crosses on their peaks, which symbolically linked imperial Rome and the Catholic Church.

The years in which castrati ascended to musical prominence in Rome were the years in which the papacy made itself into an absolutist state. In Rome, the pope served a dual role as patron of the Roman people and bishop of the Roman state. Paolo Prodi argued memorably in 1982 that the papacy in the sixteenth and seventeenth centuries was intimately tied to the formation of modern Europe.[24] The pope had two souls: one that was the father of Catholic Europe and global converts, and one that was the prince of the Papal States, the bishop of the city church of San Giovanni in Laterano, and a patron to the people.

These were also years of rapid papal turnover. Between 1555—when Pope Paul IV was elected—and 1655—when the Chigi pope Alexander VII came to power—fifteen popes came and went.[25] Paul IV, the earliest pope associated with the castrati as a musical institution, imported two Spanish castrati from Naples to audition for the papal chapel. Urban VIII lasted a comparatively long time. He reigned for twenty-one years of the Thirty Years' War, invested heavily in his own military might and consolidated a creative program of which castrati were key instruments.

Even within the absolutist state of Rome, the city constantly seethed with conflict or the threat of conflict and with tension between moralistic prescriptions and carnivalesque transgressions.[26] This is because it was run by a succession of elected rulers from rival powerful families. The death of a pope, or even the rumored death of a pope, was unpredictable and disrupted the orchestrated ritual time of sacred and secular calendars. Between popes, Romans entered an interregnum or *sede vacante*, a period of massive misrule in which the city devolved into power struggles. Violence erupted between noble families and on the streets.[27] Carlo Ginzburg writes of ritualized pillages after the death of an old pope and during the election of a new one. He focuses primarily on the belongings of dead church officials. But John Hunt argues for a more overarching period of misrule in which illicit activities of all kinds rose to the surface.[28]

Each new pope felt compelled to prove their own might. Papal dynasties created new power and spatial structures overnight, and the law of the previous pope ceased to matter. Meanwhile, Catholic rulers from all over Europe sent entourages who built their own churches. Theatrical diplomacy, complete with jousts, processions, and music dramas that conflated ruling families with gods and goddesses existed alongside a rampant criminal street culture. Rivalries and contests for power occurred in the connected theaters of the piazzas, villas, and palazzos.

Rome hosted multiple international communities and tourists. Thanks to a robust print culture, visitors could choose from a wide variety of accounts and guides ranging from small pocket books to illustrated volumes that helped plan and remember travel and that suggested ways to plan voyages around particular spectacular performances.[29] During the 1650 jubilee a Good Friday celebration drew pilgrims from all over the world to San Trinità dei Pellegrini. According to the Roman diarist Giacinto Gigli (whose records of Roman life from 1608 to 1670 included dozens of these spectacles), pilgrims held torches and chanted the liturgy, each in their own language, which sounded like "musica dissona et stravagante" (dissonant and bizarre music).[30] The Spanish colony in Rome celebrated with ravishing spectacles in their many piazzas.[31] On Easter Sunday of that jubilee year, Carlo Rainaldi designed a festival with music and fireworks. Gigli wrote that the Spanish Easter celebration cost twelve thousand scudi and included eight choirs of musicians. He also explained that some people stayed away for fear of being injured by the fireworks.[32]

Rome's rich musical culture, cosmopolitan vibe, and history of cultural exchange made it a natural castrato hub. Young castrati went to Rome to study, and dignitaries from other Italian courts looked to Rome to fulfill their singing needs. Like natural wonders, property, and books, castrati were subject to exchanges between dignitaries based on complex negotiations. The activities of Paolo Faccone, a Mantuan emissary and recruiter introduced to musicologists by Susan Parisi, confirm the importance of Rome and shed light on the nature of contractual negotiations over castrati.[33] A prominent musician employed by the papal choir first as a bass singer and then as the chapel master between 1586 and 1615, Faccone earned stipends and other rewards in exchange for bringing musicians to the Gonzaga court. In June 1615 he wrote of his failed attempt to recruit a singer at the court of Cardinal Borghese, so he turned to the court of Cardinal Arrigioni.

> If the castrato who serves Cardinal Arrigioni were free or if it could have been arranged with his parents, by this time My Lord I would have taken that solution, if it had been possible. But because as I wrote to Your Highness the young man is obligated to the said cardinal who had him taught and clothed at his own expense as well as giving him a benefice and provision regularly Your Highness will need to write a word to the cardinal. Or if you order me on your behalf, I will ask him. For in this case we have to go to the top.[34]

In 1618 Girolamo Fioretti wrote from Rome to Enzo Bentivoglio, who was in Ferrara about some castrato scouting:

> For sopranos: Signor Francesco Severo, musician of the Chapel, and currently in the service of his Lordship the Cardinal Borghese, Patron, Signor Lorenzo

> Marrobino, musician, likewise, of the Chapel, and who has a most powerful voice. Both are castrati.
>
> For contralto: Signor Ferdinando, musician of the Chapel who although he has only one arm yet has an iron one with which he performs so that his defect is not noticed in many actions—and he would be perfect for he has a most powerful voice, and skill. Don Pietro Bolognese, musician of His Lordship the Cardinal Montalto and he is the one who served Your most Illustrious Lordship in the tourney as Mercury.[35]

## PASQUALINI IN THE PIAZZA

The beginning of the 1630s did not go well for the Barberinis. They endured a plague, a war, and a lot of bad press generated by their successful prosecution of Galileo as a heretic. The visit of Prince Alexander Charles Vasa of Poland in early 1634 provided an opportunity for early modern damage control. The grand procession, opera, and a theatrical joust left no room to question their greatness (although the prince was in fact no longer in Rome when this event occurred, having already left for Florence). The culminating event occurred in Piazza Navona.[36] Throughout the festivals Pasqualini animated a series of transformations. He exemplified the ways in which the castrato voice and body were implicated in the spectacular. Cardinal Guido Bentivoglio, the brother of Enzo Bentivoglio, also produced a fabulous festival book that sold for over six hundred scudi.[37] It was an expensive, dramatic, printed performance of Barberini greatness.

Pasqualini, whom Bentivoglio called "celebre Musico del Sig. Cardinale" (the celebrated castrato of the cardinal), made a dramatic entrance into Orazio Magalotti's house in a chariot drawn by a giant metallic eagle and "drawn above four wheels decorated in gold."[38] In the role of Fame he introduced the joust that would occur later. The illustrations positioned Pasqualini's performance as the goddess Fame in a prized place, and the text described the castrato's impact as a special effect.

> Fame, on a lovely chariot, pulled by a large eagle and drawn above four wheels decorated in gold, presented herself at the center of the room, where the ladies and various other noblemen had assembled themselves. The body of the cart was divided by many carvings adorned with leaves and golden decorations that stood out even more against the green background. But from the body of the same chariot over two harpies of silver rose the seat of Fame, which was also supported from behind by a giant silver harpy. One went up to the said seat

> through two silver steps all worked with numerous arabesques and carvings, and on the edge of the footboard, where the Eagle had the ropes to pull it, two lovely silver vases adorned the floor of the carriage. Fame, who sat majestically on top of it, appeared superbly dressed, and her gown, which was multicolored and woven with gold, was also studded with a multitude of eyes, mouths, and ears. She held a golden trumpet in her hand, and on her shoulders, she unfolded two wings that were also covered with eyes, ears, and mouths. The chariot stopped when needed, and while the people were waiting to hear what Fame would bring, she was accompanied by a harmonious consort of instruments and in these notes with a very sweet song explained the reason for her arrival.[39]

Pasqualini's appearance as Fame was one of many instances in this festival, and in seventeenth-century Rome, in which humans functioned as special effects (Fig. 6.3). Fame's wings, and those of the eagles drawing her chariot, suggest an animal, while the wheels, which powered machines, stand as artifacts of human industry. The wheel, a marker of human technology, moved everything from chariots that carted popes through processions to water that powered hydraulic machines. The silver harpies that carried Fame's seat, half woman, half bird, were themselves liminal creatures. Fame wields a golden military trumpet and bears responsibility for upholding the power of her patrons. The metallic sheen of Pasqualini's costume enhances the image of a constructed instrument, built, like the trumpet, through a human process while the multiple eyes and ears suggest a human body reduced to its component parts and reconstructed.

Pasqualini's Fame echoes ancient mythology and the costume nearly exactly recreates Virgil's description of Fame, in which she spreads her terrible wings and each of her many eyes comes with a tongue, a voice, and an ear, repeating everything, first in a whisper to a few and then louder and stronger with each breath:

> Pinioned, with
> An eye beneath for every body-feather,
> And, strange to say, as many tongues and buzzing
> Mouths as eyes, as many pricked-up ears,
> By night she flies between the earth and heaven
> Shrieking through darkness, and she never turns
> Her eye-lids down to sleep.[40]

Already in this epic written between 30 and 19 BCE, Fame is a fragmented and an extended being. The costume also recalled Ovid's description of Fame's

FIG. 6.3 Ramona Martinez, *Pasqualini Channels Fame.*

home "built of echoing brass."[41] Like these mythological forbears, Pasqualini-as-Fame stood as an embodied reverberation of sound, always resonant. Resonance, when the vibration of one object causes another to vibrate with it, is something that occurs widely in nature but is also recreated with human-made devices. And it matches the image and description of Fame in Cesare Ripa's *Iconologia*, a dictionary of iconological forms published first in 1593 and again in 1603. Ripa, quoting Virgil extensively, described Fame as a woman with many eyes and feathers holding a trumpet.

As Fame, Pasqualini was an acoustic instrument that sang with a "harmonious concert of instruments" as she announced her own arrival. She arrived from on high to call the audience to order. While the score has been lost, we have the words written by Fulvio Testi. Frederick Hammond points out that the alternation of seven- and eleven-syllable lines suggests a Monteverdian prologue.[42]

> Io che sol frà le bocche
> Invisibile altrui,
> Sù le lingue mortali
> Vò dispiegando l'ali
> Qui vengo, e co'l sembiante,
> Quella sarò io, son'io che le grand'alme, e l'opre
> Ignote al cieco Mondo
> Fò note, e col mio volo
> E' termine al lor grido il Mare, e'l Polo.
>
> I, alone among other invisible mouths
> spreading my wings
> on mortal tongues
> [I] come here.
> I will be the one
> who makes known the great souls and works
> unknown to the blind World
> and with my flight
> Their praise will be heard all the way to the sea, to the [farthest] pole.[43]

Fame, in the body of the Sistine Chapel singer, stood under the protection of another castrato wearing the arms of an important Roman family. Her costume resembled the trumpet she wielded. The metallic gold color further enhances the image of Pasqualini as a constructed figure composed of materials that are not entirely human. The costume of multiple eyes and ears reduced the human body to the component parts and reassembled it, like the castrato who is so often reduced to his parts and reassembled as a voice. Pasqualini's own existence as an invention shines even farther here because he appears on a piece of strange machinery, itself working through the most up-to-date innovations.

The events celebrating the 1634 visit of the Polish prince centered on the Saracen joust. This joust reenacted the 1492 victory of the Spaniards and the expulsion of the "infidel Saracens" from Granada. The battle had been recreated on a yearly basis in the piazza, but 1634 was exceptional. Festival architects turned the piazza into a stage, with bleacher-type seating on three sides and private viewing stations in palaces on the fourth. The audience was almost as impressive as the performers, according to accounts. Two detailed paintings now on display in the Museo di Roma record at least the fantasy of the event.[44]

The spectacles concluded with a tremendous boat of musicians gliding through the Piazza Navona (Fig. 6.4). Bacchus piloted the boat, which had elaborate silver and taffeta sails. One thousand torches turned the night sky into dawn. Artificial waves camouflaged the boat's low wheels, underscoring the ability of humans to imitate and extend nature:

> As the sweet sounding instruments began, every murmur in the theater ceased immediately and it was soon filled with angelic voices. The first to sing was Bacchus, followed by a chorus of nymphs and shepherds, and for the finale, Laughter, with superhuman grace, finished the music, which was, however, interrupted by a delightful *balletto* danced by the Shepherds and accompanied by a good instrumental consort. This, while delighting the eyes and pleasing the ears, insensibly stole the hearts of the spectators.[45]

Accompanied by music, the boat with cannons sticking out of the gunports sailed around the square, pausing at the boxes of various dignitaries. The public was desperate to see how it all worked, so the parade of ships kept going, inspiring wonder and awe.

FIG. 6.4 *The Musicians' Boat*, in Vitale Mascardi, *Festa fatta in Roma* [ . . . ] (Rome, 1635). New York City, The Metropolitan Museum of Art.

As in most court spectacles, the music, which Mascardi described as possessing superhuman grace, worked as a special effect: an illusion and a theatrical trick. There is of course no score, but music asserted itself most powerfully at moments when it assaulted the senses not just aurally but also visually, in this case through the accompanying spectacle of the brightly lit sky. This concluding parade featured liminal creatures, including a strange and monstrous fish plated in gold that sat on the bow of the ship, a siren with a double tail, bacchants, satyrs, and sixteen fishermen clad in blue robes covered with silver scales who ran alongside the boat carrying torches. Each of these creatures—like the castrati themselves—had long-held associations with sound, excess, and the limits of humanity. Half human, half animal, and in possession of unrestrained passion, satyrs with their ever-present panpipes had long been associated with a kind of liminal music. Sirens represented dangerously seductive bird women. Juxtaposing these mythological figures with—and embodying some of them in the bodies of—castrati on elaborate stage machines completely confused the already fluid boundaries between human, animal, and machine.

## THE PIAZZA

The idea and practice of Rome as a civic theater goes back to the founding of the city and is built into the ancient ruins. The body of the city was inextricably intertwined with the body of the Roman empire, a point Vitruvius made in the first century in his ten-volume architecture text, which receives musicological attention because of his work with acoustics and theater.[46] But he also wrote importantly on city streets, lead pipes, the body, and more. As Rome was revitalized in the thirteenth century, jubilees inflected those spaces with Christian ritual. The seventeenth century marks an important transition for Rome. Spectacular culture moved from what Luigi Allegri has theorized as object apparatus to backdrop apparatus.[47] Object apparatuses are large spaces that are full of things as opposed to backdrop apparatuses where objects are mounted; an object apparatus puts things in a place, while a backdrop apparatus make the place a setting, such as turning a whole city into a ritual stage.

Even within the context of Rome as a spectacular cosmopolitan theatrical city, and even with the understanding of Italian piazzas as vibrant sites of civic and personal performance, Piazza Navona stands out. In the ancient world it housed the stadium of Domitian. From the years 86 to 96 CE, it hosted games and gladiator contests. In 1477 it became the site of Rome's largest food market.[48] During the seventeenth century the piazza transformed from a large civic space that needed temporary structures for any festival to a festive scenographic space ready for action.

When Bernini designed the Fountain of the Four Rivers in Piazza Navona and the fountain in St. Peter's Square in 1656, he set in motion a process of carving in stone the ephemeral festival culture of Rome. Maurizio Fagiolo dell'Arco pointed out the direct relationship between the festivals for Innocent X and the Four Rivers fountain itself, highlighting visual connections between the four river gods and the temporary structures featuring four female gods personifying the four continents that had been put up in Piazza Borghese for the *possesso*, the pope's symbolic possession of the city.[49] Kircher, in the 1650 jubilee year, published an elaborate celebratory book titled *Obeliscus Pamphilius* that demonstrated the obelisk as an amazing artifact and highlighted it as a prop in civic theater.[50]

Gigli's account of the 1650 jubilee in Piazza Navona makes the theatrical nature of the piazza and of the ceremony clear, especially the details of Easter Sunday (April 17).

> Ornaments were made in Piazza Navona, as had been done before, but even grander this time. The two fountains that are in the piazza were enclosed within an arch with four facades of very high columns, and above the arches were towers and domes that all looked as if they were made of stone and colored marble. Inside these were stands from which choirs of *musici* sang during the procession. In the middle of the piazza where there is now an obelisk (the decoration of which is not yet finished) there stood a large wooden enclosure covered with marvelously painted canvases, and in the four corners there were four towers with balconies inside for the *musici*, and in a straight line flush with the obelisk, towards the middle of the piazza, stood other painted obelisks and other machines, all filled with fireworks. The entire theater of the piazza was surrounded by arches of painted wood, all filled with lit lamps, as were all the towers and other ornaments. In front of the obelisk, where the Church of Sant'Agnese is now, there was made a very beautiful altar, with columns and a cornice above, painted and gilded, on which the Holy Sacraments were to be placed.[51]

Fireworks, painted wood, and other ornaments pushed together natural and artificial materials and displayed the convergence of creative practice. The fountain was designed to display the greatness of the Pamphili dynasty and was understood as a "gift" to papal subjects that had, in effect, converted a useful space to a scenic spectacle and a wonder of the world.

There is a darker side to this story of turning a communal square into a stage set for elite productions and the continuous reproduction of the elite. Markets and drinking tubs became scenic palace facades for papal families and other dignitaries. Vendors were evicted from the piazza, and Jewish sellers had to return to the relatively nearby ghetto.[52] The transition between civic useful space and scenic authoritarian theater came with violent contests between the absolutist papal authority and the consequently unruly street culture. In describing the 1650 jubilee, Gigli railed against the Spanish, who spent over twelve thousand scudi on the festival; he also noted that very few people attended because of fear of violence, worry about fireworks damaging nearby buildings, and general fear.[53]

Between 1646 and 1648 Pope Innocent X, Giovanni Battista Pamphili, bankrupted himself building the Fountain of the Four Rivers. To compensate he raised taxes and made bread smaller, which led to rampant protests in the streets. Gigli reported that Pasquino, the talking statue who lived just around the corner, declared: "Noi volemo altro che Guglie e Fontane. Pane volemo: pane, pane, pane!" (We don't want fountains and spires. We want bread: bread, bread, bread!).[54] As evidenced above, the piazza was a frequent site of giant spectacles. Gigli took some delight in the public's fury: "While the pieces of the obelisk arrived in Piazza Navona, people began to say that it wasn't time for incurring such expense, when there was a shortage of bread and grain and various words were attached to these stones."[55] Pasquino provided the counterpoint—or perhaps ground bass—for the orchestrated celebrations planned by the princely family. In effect then, the square staged a theater of theaters, official choreographed celebrations, and bawdy satirical responses.

The fountain, a creation of marble and water, dramatized four figures circling an obelisk, with each figure representing a river.[56] An example of public scenography that deliberately mirrored and made permanent the ephemeral spectacles, processions, and feasts of Roman elite culture, the fountain featured lifelike exotic creatures imbued with the motion of life, such as serpents writhing and lions blowing through their nostrils. A marble theater, it inspired wonder and was described in lyric poetry and guidebooks as a cross between a mechanistic processional float and a wonder of nature.

Athanasius Kircher translated the text of the Obelisk with, as he explained, great effort. The emperor Domitian, who ruled from 81 to 96 CE, brought the obelisk to Rome. It stayed at Circus Maximus for centuries, where it was broken into pieces. When Pope Innocent X had it moved, oxen pulled wagons with the smaller pieces, and horses pulled carriages carrying the largest bits. Even its movement was a technical marvel. The installation of the obelisk in the fountain came with a festival as elaborate as any put on for human dignitaries and prints of moving obelisks circulated all over Europe.[57] Some of these prints painted a picture of the fountain itself as a kind of procession. A pamphlet by Maria Portia Vignoli (Fig. 6.5) painted the fountain

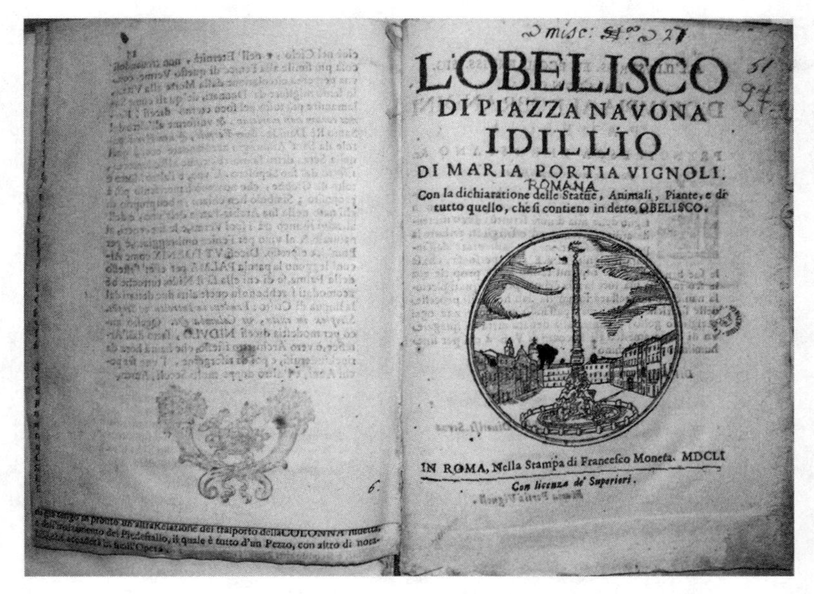
L'OBELISCO
DI PIAZZA NAVONA
IDILLIO
DI MARIA PORTIA VIGNOLI,
ROMANA
Con la dichiaratione delle Statue, Animali, Piante, e di tutto quello, che si contiene in detto OBELISCO.

IN ROMA, Nella Stampa di Francesco Moneta. MDCLI
Con licenza de' Superiori.

FIG. 6.5 Title page of *L'obelisco di Piazza Navona idillio di Maria Portia Vignoli* (Rome: Francesco Moneta, 1651). Biblioteca Casanatense

as a procession machine moving in and out of the sun and celebrated Bernini as a spectacle engineer.[58]

> But who then was the artisan of such a beautiful work?
> It was the expert, Bernini,
> Who like a new Archimedes
> With his valor exceeds all value.[59]

Fame started the celebration, holding a trumpet in one hand and an olive branch in the other. A second figure dressed as Curiosity followed, exhorting the people to move toward the piazza. Fame sounded her "sonora tromba" (sonorous trumpet) as she passed through the streets of Rome. She orchestrated the hordes of people moving through Rome toward the piazza:

> In this way, a great whisper having arisen among the citizens, each was more astonished than the next, while none of them could investigate the cause of the prodigious event, and in this way, realizing that the souls of the admiring crowd were all in suspense, suddenly they saw another woman appear whose robe was ablaze with two vibrant colors, red and blue.[60]

Eventually people realized that it was indeed Fame and Curiosity calling their attention and their bodies to the piazza:

> "What are you worrying about? Come on, everyone, move your feet toward that place that you once called the Circus of Alexander, because there you will see, revealed, the most sublime marvel that the entire world can boast of in our time. It is created, with great magnificence, by the Innocent Dove of the House of Pamphili, which guards and defends the great machine, despite its magnificence and loftiness, under its lovely foot." At those words, the souls of the astonished people were awakened, and recognizing one woman as Fame and the other as Curiosity that spurred them, all the people, eager to see such a wondrous thing, with precipitous speed, began to walk in that direction, where each person that arrived became a spectator of a construction so majestic that they stopped there in admiration, so that one could scarcely perceive the difference between them and those revivified pieces of marble that they stared at with such fixed and intent eyes.[61]

Sound animated the people toward a spectacle. The spectacle survived in the fountain and in the many festival books that described it.

## PRACTICAL MAGIC ON THE OPERATIC STAGE

During the carnival for the Prince of Poland, the one with the joust in Piazza Navona, Cardinal Francesco Barberini staged *Il Sant'Alessio*, with music by Stefano Landi and text by Giulio Rospigliosi, seven times. Pasqualini moved from his role of Fame to the wife of Alessio. The production occurred in the Palazzo Quattro Fontane in a medium-size room that could seat about two hundred:

> On Sunday evening and again on Wednesday, the production of *Sant'Alessio*, with music, was performed again in the Palace of the Most Excellent Prefect of Rome [Taddeo Barberini], for the satisfaction of the lords and other nobility that could not see it the first time, and it was a very beautiful thing, both because of the performers, each of whom played his part excellently, and also because of the loveliness of the costumes and the variety of the scenery and the *intermedi*.[62]

Even the seating chart was a power struggle. Count Fulvio Testi, an ambassador for the d'Este family, was mad because his assigned seat was "not of a reputation which is fitting to the duties I hold." To get a more prestigious seat he faked a stomachache.[63] The opera, which had first been performed in 1632 during the Thirty Years' War, told the story of the ancient Roman martyr who tossed aside his family's wealth for a life of penance. Women were identified by their relationship to Alessio: *madre* and *sposa* (mother and wife). The first performance was dedicated to a Holy Roman ambassador and focused on a saint already deeply associated with Rome.

In 1634 the Barberini commissioned Francesco Guitti, a student of Giovanni Battista Aleotti, to make five new stage sets. But Rospigliosi was not impressed.

> There is no doubt that when performed for the Lord Prince of Poland, the machines could have worked better, but there was very little time. And I feared worse, for they had hardly ever been tried out. In addition to the multitude and size of the machines, the space where everything operates is so narrow that it would be a wonder if a thousand disorders did not always result.[64]

The papal family published an edition of the opera as well as a series of eight engravings by François Collignon that memorialized the sets featuring, among other things, a cityscape, the mouth of hell, a palace gate, and a forest. The 1634 version is a textbook presentation of tropes and a symbolic mess. Comic figures (*zanni*) caused complicated trouble in the household, and messengers disguised themselves and their messages. The devil at one point turns into a bear.

The opera has striking vocal ensemble writing that showcases nimble voices. Some of it is noticeably conventional: a chorus of soprano angels hanging out on clouds accompanied by the full instrumental ensemble. Sometimes jarring musical contrasts are enhanced by thick orchestration; keyboard basso continuo gets back up from theorboes and harps throughout, and repeated ritornellos are jaunty and memorable. The violins have the melody much of the time, which makes it sound at times closer to the eighteenth century, and some of the orchestration is chordal and idiomatic to the instruments (Fig. 6.6). Note the Barberini coat of arms; the iconic bees between sections of the score.

The vocal range is, to modern ears, overwhelmingly high: the score calls for four tenor or bass roles and ten soprano and alto roles (when William Christie put on a production in 2007, he used nine countertenors). The roles went to singers from the papal chapel and the Barberini courts. Angelo Ferotti, who had trained and boarded with the composer, likely sang the role of Alessio. The comic roles of Martio and the mother went to a soprano from St. Peter's and an old castrato, respectively. The production began with a symphonia for three violins, theorbos, lutes, harps, keyboard, and bass violins with a basso continuo for harpsichord.

The published version includes a dramatic illustration of the opening

FIG. 6.6 Sinfonia per introduttione del prologo, Stefani Landi, *Il Sant'Alessio* (Rome: Paolo Masotti, 1634). Beinecke Library.

scene that focuses on the chorus of slaves sitting in shackles at the feet of Rome, who is personified as a woman. The slaves are among other "spoils of war," and the score describes the scene as

> Rome, sitting on top of a pile of trophys of war and surrounded by slaves, having heard the praise of Prince Alexander Charles of Poland and the communal jubilation over his arrival, resolves to perform for him the fortunes of St. Alexis, who, among her citizens was no less famous for the glory of sainthood than many were for their valor in arms. And to demonstrate that she values her role as queen of hearts above all other domains, she orders that the slaves be freed from their chains.[65]

The chorus of slaves is followed by Rome descending on a cloud and singing of her own glory (Fig. 6.7). Rome in the body of a castrato released the slaves as they explained while she used to rule their bodies, she now rules their hearts; absolute power in the form of control of body and soul was on aural and visual display.

In the second scene of act 1, Alessio sang a prayer that contrasted with the upbeat aria of the pages. Their tunes sound like secular street music. Meanwhile the demon choir reifies papal conquest. The Devil makes his first appearance in act 1, scene 4, with infernal shades carrying fiery torches dancing a morisca. Act 2, scene 6, shows the confrontation between Alessio and the Devil, who sounds a bit like Monteverdi's Caronte, the bass moving by large leaps, but in this case the Devil sounds more extravagant than Alessio, who sings higher as he calls for God's help (Ex. 6.1). Harmonically the part moves all over the place. Finally, seemingly defeated, Alessio sings alone on stage of his own willingness to die in a strophic aria in which rhythmic flexibility disrupts the words. The phrase "O more gradit" (I welcome death) repeats throughout the aria in ascending harmonic sequences; the vocal part seems to sigh.

The commemorative score included an account by someone Landi called

FIG. 6.7 *Il Sant'Alessio*, first engraving, chorus of slaves, called "spoils of war" (Rome: Paolo Masotti, 1634). Beinecke Library.

EX. 6.1 Landi, *Il Sant'Alessio*, act 2, scene 6, mm. 1–12

a man of letters, whose "pen made without colors a portrait of the opera."[66] The writer described the ways the drama conformed to Aristotle's rules and praised the singing and acting, noting that "also the gestures and movements seemed as harmonious and consonant as the voices."[67] The costumes evidently fit the singers perfectly and were stunning in and of themselves. And the bodies of the singers and dancers executed grand control.

The learned man praised the Aristotelian depiction of heaven and hell, the scene changes, the artistry of the depiction of Rome, and the concluding garden.

> The first introduction of new Rome, the flight of the angel in the clouds, the appearance of Religion in midair, were works made with ingenuity and machinery, but they rivaled nature. The sets were most artful, the appearance of Heaven and Hell marvelous, the transformations of the wings and the perspective scenes more and more beautiful; but the last scene, with its receding perspective, and the darkness illuminated by the portico and the appearance of the garden in the distance, was incomparable.[68]

According to the score, a sinfonia accompanied each scene change. The Barberini reign in Rome coincided with the advent of quick and easy scene changes. This technology drew heavily on the theatrical culture of Florence around 1600.[69] Musicologists have mined the cohort formerly known as the Florentine Camerata for promoting sung dramas. But the composers of those dramas also served as impresarios, *macchinatori*, technicians, and builders. Gagliano's preface to *La Dafne*, published a year after *L'Orfeo*, depicts theater as an illusion that depended on human ingenuity. Gagliano worried about the effectiveness of the scene in which the singer who played Apollo had to fight a python before singing an aria:

> But very often a singer is not capable of this attack on the serpent, needing for this effect, ability leaps, and wielding of the bow with the appropriate poses, traits that appear in a good fencer and dancer as well as a fine singer. So finding himself having to combine both these aspects it would be difficult to sing after the combat because of the fatigue from the previous action. So let two Apollos be dressed alike, and the one who sings shall enter in the place of the other one after the death of the python, with the same bow in his hand or another like it, and he sings as we have described above. This exchange works so well that no one ever perceived the deceit during the many times it was played. The fighter must of course fight in time with the music and the scenographer ought to make the python breathe fire and the wings move.[70]

This was standard practice through the seventeenth century. A documentary record of stage practices, the anonymously written *Il Corago*, published in 1630, is essentially a how-to book for producers of musical theater. The author presented a picture of events that included poetry, music, acting, dancing, instrumental playing, costumes, props, stage sets, and more.

> The corago's art will here be taken to mean that competence which enables a man to lay down all the needful ways and means by which a drama once it has been written by a poet may be staged with the perfection it requires. . . . It seems that in modern times he should be more trained and proficient in the arts of stage machinery and acting.[71]

The author devoted many pages to explaining how to mechanically create the scenic artifice that made the theater art.

The document details the means of manipulating the sensory effect of theatrical performance, and it encourages the use of entire theatrical spaces for special effects, including water, fire, and smoke. It describes two methods for representing the sea: the first involves cutting planks the shape of a wave and covering them with silver fabric that moves, and the second involves spits that rotate and represent waves. It was also evidentially challenging to create illusions for viewers close and far from the stage and to keep actors in view of the audience,

> because they must raise their eyes more than the natural and comfortable way we look at things. This doubles the lack of satisfaction of the view because the

> appearance of the object is higher than it should be to look at it in a perfect way. But this inconvenience is unavoidable on the modern stage unless we build it in a bigger proportion than we usually do. Because if the actor wants to remain far from the viewer, next to the perspective [set], immediately he looks much bigger, not only than the doors of the nearby palazzi but also than the roofs and chimneys of the houses that are painted in the background. This appears very wrong to the eyes of the viewers and reveals too clearly the fiction of this artifice.[72]

The author went on to compare the modern stage to an optical illusion:

> It cannot be denied that, as far as the [perspective] view is concerned, our scenes recreate much more than the ancients, although they do appear somewhat incorrect to the naked eye, just as views, even false ones, seen through the artifice of crystals give more pleasure than simply looking at real objects directly.[73]

In 1638 Nicola Sabbatini published his *Pratica di fabricar scene e machine ne' teatri*, a handbook of theater engineering. It is a series of directions to an architect whose job it is to turn a space into a theater complete with stage, lights, and effects. It gives a snapshot of late sixteenth- and early seventeenth-century stage practices including the creation of stage effects for, among other things, fire and fireworks and the sound effects of thunder. For thunder he made a device that threw cannon balls down wooden steps.

Arguing for human intervention in the process of wonder, the manual teaches the architect how to astound his audience through a vocabulary of pulleys, levers, weights, and tricks of perspective. The second book, devoted to scene changes and the vanishing of stage sets, explains that the changing of scenes arouses great wonder and delight, especially when it is done fast enough that—with the aid of attention-distracting tricks—viewers cannot tell how it happened.

> Of these artifices the best, in my opinion, is the sounding of a trumpet or some other instrument because the representation of a brawl or the collapse of a staircase brings many dangers, just like the instigation of great tumult, which is not easy then to quiet down. But on the contrary, having heard only the sound of the instrument, as we said above, once it finishes, the audience immediately pays attention to the scene, quieting down as they were before, ready to relish with wonder and gusto the new scene represented before their eyes.[74]

The Barberini used these theatrical techniques. Antonio Sacchi added to the Palazzo Barberini a permanent theater that could seat 3,500 to 4,000 people.

In 1642 the theater staged a performance of *Il palazzo incantato*, with theater sets designed by Sacchi and a story based on Ariosto's *Orlando furioso*. The seven-hour production displayed his theater, his machinists, and his castrati. Star positions went to Loreto Vittori, Pasqualini, and Lorenzo Sances. An *avviso* from Paris describes the opera as a spectacle off the stage as well as on. On the stage, the comedies worked well except that they were too long. Off the stage, Paolo Sforza "was almost forced to engage a Frenchman who wanted to harm Marc'Antonio il castrato in a duel." The cardinal himself also apparently engaged in the fights.[75]

> The same cardinal (Antonio) by his own hand beat a few of the servants who were outside the gate who had made too much noise while waiting for their masters. He threatened others who told the performers that they should speak more loudly, and continuously inserted himself in matters that were not appropriate to his rank.[76]

This incident suggests the potential for manipulation of power and misrule in a performance. And it shows the ways that Barberini power was always in danger of collapse.

The opera was a display piece for Pasqualini's vocal talents and the voice. Though rarely performed or recorded today, the music is beautiful and endless with crisp solo vocal parts and luscious choirs. To modern ears recitative, aria, arioso, and interlude blend into one another. And Rossi, a virtuosic harp player himself, used musical instruments as part of the dramatic structure, with a big string section and lots of brass. Based on an episode from *Orlando furioso*, it explicitly combined magic and artifice. It begins with a symphonia with notation only for basso continuo, and then Painting, sung by Pietro Paolo Visconti, sings a duet with a chorus of streams with an eight-part accompaniment. The two highest parts remain tacit when the soloist sings. It almost sounds like a madrigalism. Music, Painting, and Poetry argue over who is the most powerful. In the end, Magic flies in, finishes the scenic construction, and declares herself the unequivocal winner, disciplining the others along the way. She argues that her effects are new and different, and her spells ultimately triumph over all the other arts. And then in a fabulous effect, the magic castle explodes out of nowhere, ready to host the shenanigans to follow. Magic allows Poetry, Painting, and Music to do their work because all of them rely on deception to seduce and entertain their audience.

The castrato Angelo Ferrotti, who had sung the role of Alessio, sang the role of Magic, giving voice to the altered reality he occupied. Submerged in artifice, his body reminded the audience that in the end the castrato's voice retained the supreme power to extend and alter nature. And by setting up the castrato's voice as the embodiment of magic, he created a place for Pasqualini and the others to do their musical thing.

## BERNINI

Bernini still inhabits Rome. He relished the relationship between audience and spectator and designed immersive experiences to inspire wonder and deceive the audience. He made a fire that looked and felt real enough that the audience screamed for their lives. He made architecture that looked like movement and built theater sets that looked like the built world spectators lived in.[77] Building on the stage machines of Sabbatini and the hydraulics of Aleotti, he used hydraulic mechanics, trap doors, flying actors, movable sets, and just about every technology of illusion available at the time.[78]

Floods seem to have been a favorite immersive experience. Cardinal Pamphili and his closest confidants called a design contest for the construction of the fountain of the four rivers. After the cardinal examined Bernini's miniature model for quite some time,

> he asked Bernini if he could turn the water on so he could see it. Bernini said it wasn't possible. After benediction as everyone was leaving he had not yet gone out of the enclosure when he heard a loud sound of water. Turning back, he saw it gush forth on all sides its great abundance. The Cavalier had, at the crucial moment, given a certain signal to the person whose job it was to open the water ducts, and he quickly had it coursing through the pipes to the mouth of the fountain. Bernini knew that the more unexpected it was, the more pleasing it would be to the pope.

The pope was "overcome by such originality and gladdened by so beautiful a sight."[79]

This was not Bernini's first flood. In 1638, a year after a massive Tiber flood, he staged one that terrified spectators sitting in the theater in Piazza Barberini. Just as people ran screaming from what felt like a flood, the waters receded. Massimiliano Montecucoli, in a letter to the Duke of Modena, recorded the event:

> As the curtain fell there appeared a miraculous scene with relief that showed distant buildings, particularly the Church of St. Peter, Castel Sant'Angelo, and many other buildings well known to those who live in Rome. Closest to view was the Tiber, which seemed to be rising by means of artifice and rare invention,

> as the Cavaliere wanted to display the unfortunate events that occurred the year before, when the Tiber almost flooded the whole city. Closer to the stage, where the performance was taking place, was real water contained by levees that had been set up for that purpose all around the entire stage; and one could see real men who ferried people from one side to the other, as if the river, occupying the lowest parts of the city, had made transit impossible, just as had truly happened the year before. While everyone was astonished by the spectacle, several workers were reinforcing the riverbank, adjusting beams and structures so that the river would not submerge the city. But all of a sudden the bank collapsed, and water flooded the stage, rushing furiously toward the audience, and the people closest to the stage, afraid of the water, rose to their feet to flee but, when the water was about to drop on them, a levee rose suddenly at the edge of the stage, and dispersed the water without harming anyone.[80]

Since the effect was so real that spectators feared for their lives, they were swept into the production. Later in the production, a scene appeared that again seduced spectators into the stage set itself:

> In the middle of the comedy there was seen another marvelous scene that started with hearing music; attached to the stage where the singing was occurring a street appeared full of carriages, horses, and real people who stopped to listen to the music. At a distance one could see noble buildings and large piazzas, which were frequented by fake carriages that moved and fake people. There was also the sky with the moon and the stars that at times were moving, and a donkey who brayed appropriately was introduced into the scene.[81]

Bernini's stagecraft for *La fiera di Farfa*, the act 2 *intermedio* to the 1637 carnival production of *Chi soffre speri*, created a fire that terrified the audience, a thundershower with red-colored rain, and working fountains. With improvised music, a duel, and braying donkeys, the fair felt real to the audience. It took a month to build, and the effects, which required twenty-four men, were rehearsed almost as much as the singing.[82] Multiple accounts highlighted Bernini's phenomenal depiction of the rising sun and especially his ability to imitate nature and create illusions that remained inexplicable to the uninitiated observer.[83]

Bernini's 1644 comedy *La Fontana di Trevi*, also known as *The Impresario*, took as its central topic these kinds of effects.[84] It contains a play within a play and is a comic satire on the theater, Papal Rome, and art in general. The character of Aldoro, a satiric representation of Salvator Rosa, begged to learn the secrets of the stage machine that had made Graziano—the satirical mouth of Bernini—famous. In addition to the usual story of cross-class love conflicts, the play put the business of stage artifice on display and came complete with countless moments at which the magic of the stage failed to work. After much fuss over clouds that first failed to rise properly and then plopped to the ground, the comic manservant Zanni announced, "Oh! I see beautiful clouds floating in the air. In truth, where there is naturalism there is artifice."[85] Naturalism always depended on artifice. The materials of machinery and levees and pulleys created images from nature. The character Graziano explained that

> the inventiveness, the design, is the magic art through which one can trick the eye in order to astonish, and to make a cloud appear on the horizon and have it advance, always clear and with a natural motion, so that little by little as it gets closer to the eye it enlarges, appearing bigger. To show the wind rising and transporting it away here and there and then to have it go up, and not to have it come down because of the counterweights.[86]

## FESTIVAL FOR A QUEEN

In conclusion, let's return to the fountain in Piazza Barberini. My first sighting of the Triton fountain in Piazza Barberini was historical buzzkill. In seventeenth-century sources, I had read about the majestic, superpowerful merman in the piazza. But after a long, hot, uphill walk, I arrived at a busy metro station, looked across a terrifying Roman traffic scene to see an almost impotent merman perched on a giant median. The muscular merman, who once glorified a sea god and a pope, seemed more like a fish out of water than an artificial structure designed to manipulate and move the natural force of water. In his original natural habitat, the Triton spewed a hydraulic vision of Neptune's powers to make a sound colossal enough to reach the four corners of the world. These days he spits little streams of water that, thanks to regulations imposed in the 1960s, fail to rise to anything resembling the remarkable heights they reportedly reached in the seventeenth century. My twenty-first-century pilgrimage was the visual equivalent of hearing Alessandro Moreschi's records after reading eighteenth-century accounts of castrati, a voice that sounds not resonant and smooth but small and scratchy.

The fountain, commissioned by Urban VIII and orchestrated by Bernini in 1642 and 1643, deliberately brought a garden-style spectacle to an urban setting by staging a merman who has a humanlike body and a fish tail. It was a marble celebration of Pope Urban VIII as a scholar and as an investor in the Thirty Years' War. By 1642 Urban's popularity had faltered thanks to his increasingly oppressive governance and because of heavy taxes he set to finance his family's personal war with the Farnese family: the Wars of Castro. He levied over fifty-two taxes on food, including an aggressive wine tax, and even managed salt, meat, and bread taxes from his sick bed.[87]

The sea god is on an open shell held up by four dolphin tails; the dolphins have wide-open mouths as if they might drink up all of the water in the earth.[88] The merman raises a conch to his lips. We know that Urban sent Bernini a verse of Ovid in which Triton, summoned by an angry Neptune who had previously incited the waters of the entire world to catastrophic flood, rose from the depths of the stormy seas to blow his echoing conch and give the rivers and streams the signal to return:

> And the sea's anger dwindled, and King Neptune put down his trident, calmed the waves, and Triton, summoned from far down under, with his shoulders, barnacle-strewn, loomed up above the waters. The blue-green sea-god, whose resounding horn I heard from shore to shore, wet-bearded Triton set lip to that great shell, as Neptune ordered, sounding retreat, and all the lands and waters heard and obeyed.[89]

The fountain served as a cross between an actor and a special effect in the drama of urban space and Barberini glory.[90] Neptune and Triton presided over a variety of Roman spectacles and events long after the Barberini reign.

In 1655, Queen Christina of Sweden made her triumphant entrance into the holy city. Her iconoclastic persona, deep voice, conversion to Catholicism, and other "peculiarities" rendered her an object of historical fantasy even in her own time. Her much-touted arrival in Rome inspired festivities, including parades, a joust in Piazza Barberini, and an opera in the palazzo itself; the palace was both a temporary and a permanent structure in this instance.

The festivities for Queen Christina lasted a couple of months, beginning with her entry into the Papal States at Ferrara on November 22, 1655, and concluding with a joust on February 28, 1656, at the Barberini palace. Along the way, she visited numerous papal cities where she saw musical entertainments; it reads as if she were handed off from one cardinal to another. In Venice there were ceremonial barges with military bands. In Ferrara she entered a lit-up city in a parade and enjoyed musical entertainments.[91] These events were described in several printed pamphlets by Gualdo Priorato, who joined the queen's team in 1657. His chronicle reads like a compilation of other smaller sources.[92]

The queen arrived in Rome in an ornate carriage designed by Bernini himself and drawn by six horses. With silver decorations and pale blue upholstery, it pulled up to the Porta del Popolo, where the cannons of Sant'Angelo saluted the queen. The procession included young boys wearing white stockings, and the streets were adorned with flowers.

> With the incessant sounds of trumpets, drums, and bells in the background, and with festive salutes, the queen passed through the infantry in two ranks.... At the gate of the city, among the shimmering of candles and torches, Signor Fabio Compagnoni, the commissioner in charge of the military, presented himself and dedicated to the queen for royal service a group of beautiful young boys, who for reasons of birth or of character were most remarkable. The pages wore a uniform of black velvet trousers embroidered with silver, silver coats of plate, and white stockings, all dressed in a most elegant and harmonious style. The royal carriage emerged from amid the pages and passed through the city streets, decorated with greenery, and by the windows, decorated and ornamented, and overflowing with waves of people.[93]

The Barberini family, just back from exile in Paris, spent tons of money on their contributions; they funded three operas in the palace theater and a massive joust in the piazza. Giovanni Francesco Grimaldi, who had designed the 1634 festivities, designed the outdoor theater for the joust, the costumes for the horse riders including their massive hats, and the triumphal carriages. The celebratory joust featured a fire-breathing dragon that had just been conquered by Hercules and triumphal carriages filled with singers dressed as gods, goddesses, and nymphs.

The queen saw Marco Marazzoli's and Giulio Rospigliosi's *La vita humana* (Human life, or, The triumph of piety) at least three times, including January 31, 1656, just after visiting Kircher's collections at the Jesuit college. The opera, an allegorical celebration of her Catholic conversion, displayed Queen Christina's favorite castrato, the soprano Bonaventura Argenti, impersonating the title character Vita. Argenti was one of five castrati released from Sistine Chapel duty for this production.[94] He sang in an operatic world that looks, from a modern vantage point, like a microcosm of baroque spectacle culture, with stage sets that featured fountains, fireworks, and gardens. Festival records open with accounts of the palace plumber and his supplies for making the fountain "real." They also paid the tinsmith for lighting. The opera and the spectacles surrounding it also surely served political and moral ends; the presence of a feisty libertine arts patron queen in Rome made the papal court nervous, to say the least.

This Opera, with its parade of singers from the papal chapel, emphasized the material ways in which the castrato voice was displayed. Giovanni Battista Galestruzzi made engravings of Grimaldi's sets for the printed edition of the score. The drama began with Aurora descending, not surprisingly, on a machine, singing and throwing small pieces of silver paper meant to look like dew. Act 1 opened with a trumpet and drum fanfare that set up the opposing forces of virtue and vice. The swordsmith provided swords and daggers for musicians and dancers. And the tailor constructed costumes and furnished gloves.

The singers, like the stage effects of water and fire, functioned as special effects created by ingenious devices. Act 3 was magical. A garden erupted overnight, and the act opened with Pleasure working Armida style to bring Guilt and Understanding to his evil enchanted garden (Fig. 6.8). The garden fountain was decorated with tritons whose water gushed from a double scallop shell and overflowed onto the floor. Georgina Mason argued in 1966 that the fountain was real because Giovanni Battista Bolla, the plumber at the Pantheon, used "a large basin (Vaso) of lead and a socket with a large veil of lead to make the fountain of the cascade."[95] He provided twenty-four lead pipes to move water from the palace garden to the theater and made a spray at the "foot of the stage, eight small round fountains and eight small sprays."

FIG. 6.8 *Il trionfo della Pietà ovvero La vita humana*, engraving of a scene from the third act, Giovanni Francesco Galestruzzi (Rome: Mascardi, 1638). Biblioteca Casanatense.

In this space the fountain manipulated water and the singer manipulated the air he breathed.[96] The visual assemblage of the spectacle was inextricably intertwined with the sonic force of the singers. It was a staged immersive experience.

The opera concluded with a *girandola* or Catherine wheel firework rising above a representation of the Castel Sant'Angelo. Festival architects drew on one of the most famous fireworks displays in history, the annual Girandola shot off the Castel Sant'Angelo that can still be seen today. The direct architectural reference to the city further blurred lines between the street and stage. For this effect, Bolla provided a reflector in the form of a sheet of pierced tin that was six meters long. The drama ended with a multimedia show of dancers, fireworks, cannon noises, and singers, with castrati surrounded by the noises of guns and fireworks. Outside the opera the Girandola lit the castle with torches, firing cannons and mortars, and firing rockets from palaces across the Tiber.

The festivities concluded at about 9 p.m. on February 28 with a joust, *La giostra dei Caroselli,* in the Piazza Barberini. Grimaldi turned the piazza into a stage with a face-lift that knocked down a few buildings and added a triumphal arch. This was the most dramatic piazza redesign since the 1634 joust in Piazza Navona. While the Barberini used public space in 1634, the productions for Christina occurred on ground attached to their family land and was perhaps a play to remind Rome of Barberini power. The audience of over three thousand saw some of the most fabulous triumphal carriages and costumes ever seen:

> In between them rose a magnificent gate, and at the place where the queen was to arrive, it was embellished with various figures, which, around Her Majesty's coat-of-arms, adorned a display of other noble arms. On the top of this gate, in four large windows, with false shutters, were choirs for the musicians, who played exquisite melodies with various instruments.[97]

The joust began after dark. The two squadrons, preceded by eight trumpeters on horseback, met and began first with musical banter: "The chorus of musicians located as I said above the arch that was erected for the queen from time-to-time played harmonious symphonies. Then they gave up to the sound of the trumpets, which urged them all to battle."[98] They shot pistols at each other until Hercules, riding a fire-breathing monstrous dragon, interrupted.

For the joust itself, rival teams of mounted cavaliers and Amazons battled it out, with the Amazons winning. Matteo Barberini and two other cardinals dressed as Amazons. The fire-breathing dragon who had been conquered by

Hercules shot rockets and flames from his mouth. Both sides had to battle it out with him. Finally, a cart carrying Apollo, the four seasons, and twenty-four representations of hours created a visual cadence. Seventeenth-century Roman theater and the city itself embodied what Deleuze said of baroque theater, all told: "The essence of the Baroque entails neither falling into nor emerging from illusion but rather realizing something in illusion itself, or of tying it to spiritual presence that endows its spaces and fragments with a collective unity."[99] For Deleuze, a baroque aesthetic was defined by a unity of form and coalescence of means; it doesn't matter if the creator uses marble or sound. It was a theater of matter in which the castrato, the fireworks, the foundations, and the machines all worked toward a sensory delight and persuasion. And the spectacle is meant to live on in print and in the imagination of the viewer. The work of that aesthetic lay in the hands, mouths, and feet of the technical players: the stage designers, the instrumental players, and the singers.

CHAPTER 7

# ON THE CUSP

In December 1645 Cardinal Jules Mazarin, the Sicilian cultural broker formerly known as Giulio Raimondo Mazzarino, staged a performance of *La finta pazza* at the court of the then-seven-year-old King Louis XIV of France. It featured many of the exiled Barberini's favorite stars. Two years later he mounted Luigi Rossi and Francesco Buti's *L'Orfeo*. Both productions were sensational in every way. But, a year after his operatic triumph, there was a massive pamphlet attack on the Italian cardinal. Over five thousand pamphlets appeared, which included calls to have his balls cut off and pamphlets blaming him for the moral degeneration of France. The Italophobia was not new; the French had long called sodomy "le vice italien" and dubbed practitioners "la secte de Rome." The English associated France and Italy with the Catholic Church and with sexual deviancy.[1]

The cardinal, as diplomat and impresario, is a pivot-chord modulation in the story I have been telling, a pivot that lands on a historiography of the ways Anglo-American music scholars have heard the castrato phenomenon. To get from Mazarin to the twenty-first century involves thinking about how the castrato, imagined as a premodern technology, has been heard predominantly through twentieth-century readings of eighteenth-century texts. This chapter is less about the Pasqualinis than it is about the idea of the Pasqualinis, and it is less about the castrato on the eighteenth-century stage than it is about bringing twenty-first-century theoretical questions to bear on the castrato phenomenon and acknowledging resonances of Enlightenment thought and practice in current interpretations of the castrato. This means reading between the lines of fact and fiction and reading received Enlightenment understandings of the castrato not as epistemological truths but as iterations of an exoticized Italian sound.

In the pages that follow I position the castrato as living sonic borderlands, singing on the cusp of the West and the rest, the Enlightenment and the unenlightened, what Mary Louise Pratt calls a "contact zone."[2] A contact zone is one where dissonant cultures meet and often clash, and sound is always a primary point of contact. In the Western European and Anglo-American imaginary, the castrato, who migrated north and west from southern Italy, was always the sound of somewhere else. In this imagined myth of origin, the somewhere else was southern, a place that, according to Aristotle and Pliny, was the uninhabitable torrid zone between the Tropic of Capricorn and the Tropic of Cancer. So, for instance, the associations of castrati with animals and beasts that Martha Feldman explores were tied also to climate taxonomies.[3] The castrati hailed from the edge of the European and English geographical imaginary and a space associated with an idealized pastoral prehistory.[4] And they occupied the place of a nonnormative Other who shored up the hegemonic normativity of the explorer. In modern terms, this place has been the zone of the third world, the developing world, the Global South. As Serena Guarracino writes, "Europe, English-speaking Europe in particular, situates this castrated voice in the South—a geographical South located in Italy and in its southern regions, but also a bodily South where the *castrati*'s mutilated sexual organs become the Other of Western reason and modernity."[5]

Attending to this narrative decenters the erotic politics of the castrato that have been so much at the center of Anglo-American recoveries of castrato song in favor of taking seriously eighteenth-century European fascination, fears, and desires with song as key in ideologies of speaking and writing, reason, and civilization. It also recasts eighteenth-century accounts of the castrato that have been already overread by Anglo-American scholars as part of a literary and representational tradition, not "evidence" or "fact." Such a recasting reorients a musicological cartography. Music scholars, especially those who inherit a German and Anglo-American intellectual tradition, tend to think of the Italian Renaissance predominantly in the context of Florentine humanism, Roman papal courts, Venetian nascent capitalism, and the codification of new musical practices. But there is another side to the story. Already the South was a subaltern space, and as Walter Mignolo described in *The Darker Side of the Renaissance*, these same humanist texts justified colonial expansion, and the early stages of modernity and emergent eighteenth-century aesthetics were instrumental in classifying humans.[6]

Castrati rose to musical prominence in a geopolitical sea change; a moment when the axis of global power was shifting radically. The dire consequence of the Italian wars, the Spanish colonial project, and the Reformation led to the disempowerment of the Italian states. By the time writers affiliated with the papal chapel began to both search for and justify the practice of castration, Italy, and especially the south of Italy, was already on the way to being subaltern. As Martha Feldman has argued, the economic depravity of the 1570s and the severe economic decline of the seventeenth century led to an entrenchment of patrimony amid the increased production of castrati singers.

By the mid-eighteenth century, the Italian peninsula was decidedly marked as Other, an exotic stop on the Grand Tour, and the castrato, located predominantly in the Papal States and the Kingdom of Naples, was a prime attraction.[7] In 1714 Austrian Hapsburgs replaced Spanish Hapsburgs, and in 1734 they combined as the Kingdom of Sicily. There were, of course, gradations of geocultural otherness that inflected the ways the English and the French saw Naples in relation to Paris but also to cities that were decidedly outside the West. Nevertheless, when Auguste Creuzé de Lesser traveled from France to Italy in 1807, he put in writing a common idea: "Europe ends at Naples and ends quite badly. Calabria, Sicily, all the rest is part of Africa."[8] And in 1924 Walter Benjamin, writing about Naples, said,

> what distinguishes Naples from other large cities is something it has in common with the African kraal: each private attitude or act is permeated by streams of communal life. To exist—for the Northern European the most private of affairs—is here, as in the kraal, a collective matter.[9]

In telling the story of the castrato as a technological heir of the pre- and early modern periods, I have eschewed the eighteenth century. But the avoidance creates its own lack. It is the century of Farinelli, and music written for castrati in the eighteenth century reverberates through twentieth- and twenty-first-century popular and scholarly accounts. Moreover, the Enlightenment and especially its sonic ideologies lie at the heart of Anglo-American music scholarship. It was in some fundamental way the eighteenth century that created the castrato and the Italians as Other. Indeed, the castrato might even stand in for the baroque all told. *Baroque*, for the Portuguese, meant misshapen pearl. For Rousseau it was confused harmony: "Baroque music is that whose harmony is confused, overburdened with modulations and dissonances, with a stiff and unnatural melody, a difficult intonation, and a forced movement."[10] This was in effect music that sounded harsh to Rousseau's ears: dissonances, phrases that lasted too long. For Charles de Brosses, whose writings on castrati permeate music scholarship, it was "ridiculous," excessive, and ornamented to the extreme.[11]

This chapter is framed by two familiar eighteenth-century accounts of castrati that are embedded with inscriptions of Italy and the South as zones of alterity and endless deferrals. They exemplify the ways that Enlightenment travel narratives put the castrato at the center of an aural search for a nostalgic past and as a sonic site of northern assertions of an unruly south; an errant voice that by nature disrupted. Castrati, with their nonnormative bodies and voices, sat on the cusp of anxiety and desire, or a nexus of what Felicity Nussbaum calls the intertwining of imaginative power and physical defect.[12] As long as the practice remained in the faraway southern and Catholic zones of tarantellas, endless Catholic feasts with extravagant massive musical productions, and native Italian speakers, it did not cause trouble. When castrato singers escaped the pages of the Grand Tour and put their live bodies on the English stage, their physical presence threatened the natural order of English society and caused trouble that still echoes in Anglophone scholarship.

The two stories articulate the castrato's migrations within the context of the Italian and Mediterranean fall from power and grace. First, Charles Burney's futile quest to locate someone who made castrati is a tale of endless deferral that fits squarely into the narrative of the English traveler who journeys to the far reaches of Italy in search of an exoticized European Other and the primordial roots of his own "civilized" culture. The earliest European encounters are peppered with ravishment by incomprehensible savage sounds that oppose classical order of harmony imposed by the Enlightenment; civility that opposed noise.

The second account, Johann Wilhelm von Archenholz's description of a castrato called Balami who sang so spectacularly that his testicles popped out, is set in Naples. Balami was a voice from the torrid zone, a kind of essentially erogenous zone that oozed desire and sex. This account located Naples in an exotic past far from the center of Europe in time and space. And as an example of the ways that travel narrative and history merge into one another, the castrato in that reading vibrates with otherness and stands as a human contact zone who embodies Fanon's notion that otherness can't just be a fixed dichotomy between self and other or the West and the rest.[13] I will for the most part defer the erotic and sexual element to the castrato's alterity. But it's there. Foucault, in his field-defining *History of Sexuality*, made a distinction between the *ars erotica* of the East, grounded in pleasure, and *scientia sexualis*, grounded in truth and self. In this narrative, it is sex and the body that distinguishes between us and them, here and there.[14]

As in much of this book I want readers to hear the music, but score examples likely won't do the trick even for music scholars, especially because the accounts are far from phonographic. The music is in my head from the very deep work of especially my colleagues in the world of seventeenth-century Rome like Roger Freitas, Margaret Murata, Valeria De Lucca, and Amy Brosius.[15] For what listeners heard in Paris I could offer a duet between Achilles and Deidamia in *La finta pazza*, an erotic song for two high voices. For what they heard in eighteenth-century Naples I could show a score of "In braccio a mille furie," a pyrotechnic spectacle written for Caffarelli that sounds like a lot of fast notes. For street music I could show a seventeenth-century guitar song book, an anthology that looks like a fake book.

But even recordings won't really give a soundscape or, more importantly, the feel. This is the imagination part of history. When castrati sang in European cities, they sang in counterpoint with chaotic loud zones. Locals and travelers in Rome, Naples, Paris, London, and elsewhere wrote with wonder, fear, disdain, and anger about bells, screaming vendors, singers, rats, reverberating coaches, and barking dogs. In seventeenth-century Paris, for instance, the Pont-Neuf was a chaotic soundscape, a paved stone bridge that hosted singers, marionettes, street vendors, and oral news.[16] Eighteenth-century London was a mess of loud street criers, crashing chains, and screeching animals.

Even in the opera house, music itself was hard to hear. Burney complained about the noise of Italian theaters, and audiences didn't grow silent until well into the nineteenth century. But here are some sound bites. For the operas Mazarin brought to Paris, imagine close dissonances in duets and jaunty rhythms in dances. For Naples in the eighteenth century, imagine a string section with eighteen violinists on each part. And for something perhaps more recognizable, Handel wrote a version of "But Who May Abide the Day of His Coming" for the alto Gaetano Guadagni; Guadagni could with his voice make the sounds of violin études up and down his range and seems to have sung very in tune.

## BURNEY'S GOOSE CHASE

Charles Burney' was looking for the castration procedure, but he never found it. In 1765 he wrote,Y"

> I was told at Milan that it was at Venice; at Venice, that it was at Bologna; but at Bologna the fact was denied, and I was referred to Florence; from Florence to Rome, and from Rome I was sent to Naples. The operation most certainly is against the law in all these places, as well as against nature; and all the Italians are so ashamed of it, that in every province they transfer it to some other.[17]

This sound bite exemplifies the ways that taking texts out of context erases crucial narratives. Burney looms large in the musicological imaginary and is still touted as an important historian of music: "impressive if inconsistent" says the *New Grove Dictionary of Music*.[18] The anecdote has often been read as indicative of Enlightenment ambivalence about genital mutilation; no one wanted to admit that they knew anything about such an unseemly procedure, goes the logic. But it's really not a case of pre-Freudian castration anxiety. Everyone wanted to blame the imagined introduction of purposeful castration into Christian Europe on someone else. Burney and other eighteenth-century authors, whose facts and fictions sit at the origins of music scholarship, do not just speak from the emergent Enlightenment; instead, they continue and culminate early modern thought, particularly humanism. Like thinkers of the sixteenth and seventeenth centuries, Burney answered questions about music through ancient philosophical traditions and through travel to faraway places. And musical sound always signified something larger about society. Cultivated music reflected cultivated societies, disordered music reflected disordered societies.

Sonorous accounts of the Grand Tour reveal the centrality of music in the constructions of alterity.[19] Descriptions of southern Italy include outsized sounds of nature: water, cicadas, rain, and Italians humming and singing everywhere. Piazzas across Italy were theaters of sung verse, nasty satire, and medical advice, often in the form of recipes: it was like live TikTok, Pinterest, and infomercials. Tourists went to Naples to hear the "people" sing. The Irish tenor Michael Kelly, who toured Italy from 1779 to 1780, was fascinated with Venetian gondoliers' singing and reported in Naples "listening to some ragged fellow near the Mole, who recites lively stories from Boccaccio, in the Neapolitan jargon, and perhaps sings the verses of Tasso, or Ariosto."[20]

Travelers on the Grand Tour landed in Naples partially to experience the music of theaters and the streets, and they almost always mention a castrato performance—Naples being a primary site for the production of singing voices, castrato ones in particular. Three of the four major singing schools that had emerged from orphanages in the mid-seventeenth century were, by the eighteenth century, sponsored by the King of Naples. Castrati often received preferential treatment. Matteo Sassano was trained by Giovanni Salvatore at the Conservatorio dei Poveri di Gesù Cristo. Santa Maria di Loreto was the oldest of the four. Nicola Porpora taught Farinelli, Caffarelli, and Salimbeni at the Conservatorio di Sant'Onofrio.[21] And the streets were their own cacophonous soundscape. In 1816 Goethe published accounts of two years in Italy (beginning in 1776), which included vivid descriptions of street sellers loudly peddling lemons, liquors, and more.

In literary renderings of the Grand Tour, southern Italy was a site of natural beauty, sensory pleasure, an immersive experience in ancient civilization. It was an authentic sensory experience of Virgil, Dante, and Petrarch, among others. Goethe's quest for natural wonders of the south and the traces of ancient civilization included in rural areas buzzing cicadas and women dancing with tambourines. The Marquis de Sade, who fled France because of a sexual scandal, was fascinated with southern Italy. But he condemned the castrato in Rome and called Naples degenerate.[22] When Stendhal traveled in 1816, he combined his fascination with the sounds of southern Italy with a study of the morals of the Italians. For him, civilization ended at the Tiber River in

Rome. And as many scholars have argued, sometime before the second decade of the nineteenth century, the axis of musical power shifted, with German instrumental music as the teleological dominant end.

## TORRID ZONE

Historians of Western music typically pick up the history of castrati in the middle of the 1560s, when they appear in Sistine Chapel diaries. Giuseppe Gerbino traced their emergence in the papal chapel, in part, in the context of Spanish-Italian diplomacy. On March 4, 1558, Pope Pius IV requested that two Spanish singers—Francisco Bustamante and his nephew Hernando Bustamante—come from Naples to audition. They were at the time employed in the chapel of the Spanish Viceroy of Naples. Hernando did not stay in the papal choir for long and, by 1561, was installed at the court of Ferrara.[23] As I will discuss more fully in the next chapter, Pope Sixtus V's 1587 papal bull *Cum frequenter* that regulated castrato marriage begins with Cesare Speciano, the Spanish nuncio (papal ambassador) worrying about what he saw as an alarming uptick in eunuch/castrato/*spadone* marriage: "With frequency in these regions there are eunuchs and spadone" (Cum frequenter in istis regionibus Eunuchi quidam, et Spadones).[24]

Early modern learned Europeans inherited from ancient and medieval thinkers an understanding of human diversity as depending on place and climate. The climate taxonomy built on literary and philosophical traditions that tied moral superiority to geographic superiority. The north was the frigid zone of people with character strength and weak intelligence; southern Europe had people with moderation, capable of thought and self-governance; and the torrid zone was full of lusty savages. Aristotle had explained that cold nations had courage; nations in Asia did not have courage or thought and could thus be enslaved. Albertus Magnus tied politics directly to place in *De natura loci* (*The Nature of Places*) and believed that hot climates produced inferior minds. He suggested in the thirteenth century that the "torrid zone" could accommodate human lives but not a civilized human life. For him, perfect bodies had perfect souls, and bodies from the torrid zone could never be perfect.

Sixteenth-century European and English geographical consciousness was shaped by these earlier texts. Travelers and explorers alike found what they were looking for and articulated it in the language of ancient and medieval writers. So, when Amerigo Vespucci wrote about the "new world," he used the language of Dante and Petrarch to project a magical space of pierced and magical natives, "innumerable serpents, and other horrible creatures and deformed beasts."[25] Columbus articulated himself as a cross between a contemporary Aeneas and an agent of the Catholic king's manifest drive for a universal empire. Writing about Columbus, Nicolás Wey Gómez suggests that already naturalized imaginings of the South

> enabled [Columbus] to believe not only that he was about to venture into the most sizable and wealthiest lands of the globe, but also that the peoples he would encounter in those latitudes were bound to possess a nature—ranging from "childish" to "monstrous"—that seemed to justify rendering them Europe's subjects or slaves.[26]

To Columbus, the Global South was not the New World but rather an "other world" or *otro mundo*.[27] *Otro mundo* means something like different places outside of the known, inhabited lands.

It was a zone of marvels populated by people, animals, and vegetation that he described as "like us" or "not like us." It was inhabited by monsters who existed outside received ontological categories. Using myth and ancient philosophy, Columbus described sirens in what is now Haiti, "not as beautiful as they are painted, though to some extent they have the form of a human face."[28] And in a 1503 letter to Lorenzo Pietro Francesco de' Medici, he depicted a race of women whose excessive desire caused them to magically enlarge their men's genitals, which sometimes resulted in castration.[29] On February 15, 1493, he penned a letter that described Hispaniola as marvelous ("La Spañola es maravilla"). He went on to say, "You can't believe the seaports without seeing them" (Los pouertos de la mar, aquí no havría crehencia sin vista).[30] A *marvel* meant something arresting, something on the cusp of the real and the imaginary, taking the looker and listener out of the ordinary.

Printed performances of Columbus and other navigators indelibly imprinted the monstrosity of New World inhabitants on the European imagination. The de Bry family, who made one of the much-circulated cat piano images discussed earlier, never crossed any ocean. But between 1590 and 1634 they put out a twenty-five-volume collection based on dozens of travel accounts. The engravings were designed to illustrate "the known world." The image of a massive crocodile printed in India Occidentalis II (Fig. 7.1) demonstrates the indelible imprint of monstrosity of what is now Florida. It takes many men to kill one massive crocodile, a beast that is dragon-like and amphibious. The scene condemns animals and humans to alterity. The image of an Indian chief in Virginia (Fig. 7.2) oozes animality: a tail hangs between his legs, and his skin has animalesque tattoos. Thomas Jefferson described the de Bry images to John Adams as a mingling of fact and fable.[31]

Texts in a variety of genres about the torrid zone questioned the humanity

FIG. 7.1 Theodor de Bry, *Killing Crocodiles* (1591). State Archives of Florida, Florida Memory.

FIG. 7.2 Theodor de Bry, *A Weroan or Great Lord of Virginia*. Harvard Art Museums/Fogg Museum, Gift of William Gray from the collection of Francis Calley Gray, Photo ©President and Fellows of Harvard College, G567.

of the inhabitants. The Scottish surgeon Peter Lowe, whose description of the castration procedure has been cited frequently, explained that the people of the south are mad, furious, and lustful. Lowe studied in Paris and served under Philip II in the Spanish siege of Paris and wrote on surgery and venereal diseases. According to Lowe,

> The people towards the North are cold and humide, neyther so wicked, nor deceytfull, they are faithfull and true, yet because they are of a grosser wit and more strength, they are more cruell & barbarous, they haue greater force and are stronger by reason of the thicknes & coldnes of the blood, they be verie courageous for the great abundance of blood and smaler judgment, they haue great heate in their interior partes and therefore eate well and drinke better which is an unhappie vice. . . . But principally the people of the South, from whom is come the use to geld men, whome they call Eunuches to keepe their wiues. Moreouer, they who are towards the North are more laborious and giuen to artes mechanicks, & more proper for wars, than sciences.[32]

Lowe located the business of castration squarely in the south.[33] If monsters are defined by what humans are not, then the castrato was a kind of monster, a living thing that obliterated the normalcy of the category of human. Not surprisingly, the Barberinis, whose artistic program depended on castrati, were also fascinated with monsters and curiosities.

Climatic and temporal differences made their way into debates on the nature of men and women, debates that, as I will argue in the next chapter, greatly inflected the ways the castrato body was understood. Lucrezia Marinella's 1600 *La nobiltà et l'eccellenza delle donne, co' difetti et mancamenti de gli uomini* (*The Nobility and Excellence of Women and the Defects and Vices of Men*) made one of the first female contributions to the vernacular debates on the status of women's bodies.

> Why should some people be unstable, others great gluttons and guzzlers, others lively and audacious, and others unbridled and given wholly to concupiscence and pleasure? I believe, as all writers who describe people's customs confirm and as can be seen by experience, that in general the origin and cause of these matters are one's country of birth and one's bodily temperature.[34]

For Marinella temperament ties to temperature and location on the globe.

Seventeenth-century Roman thinkers mapped this climate taxonomy directly onto musical practice. In the process of creating universals centered around European music, Kircher described "the spirit of the place and natural tendency, or from custom maintained by long-standing habit, finally becoming nature," and commented:

> The peoples of the East [Orientis populi]—Greeks, Syrians, Egyptians, Africans sojourning in Rome—could hardly endure the refined music of the Romans. They preferred their confused and discordant voices (you would more truly call it the howling and shrieking of animals) to said music.[35]

The climate taxonomy played out in debates around temperament as well. In 1640 Giovanni Battista Doni, a Florentine, wrote that in Rome, organ tunings were lowered around 1600. An expert in ancient music, he had been part of the Barberini court and was fascinated with creating new instruments to demonstrate his harmonic theories. Referring to organ tuning he used the contemporaneous word *tuono*, which can imply pitch, mode, and tone in the mode. He linked and blamed this, in part, on castrati and suggested a pitch-latitude correspondence that tracks with castrato prominence. His discussion of organ pitch came in a long section devoted to correlating the *tuono* of artificial instruments with innate attributes of singers. He explained that if you lined up organs in Naples, Rome, Florence, Lombardy, and Venice and played the same note on each, it would play a scale that ascended in semitones that could be visualized as a graph of castrato prominence; they were most associated with Naples and Rome, where pitches had to be lowered the most: "and so, beginning with Naples, we know that the *tuono* of the organ there is lower than that of Rome."[36]

Doni argued that Roman pitch-mode correlation was the most aligned with the Dorian of the ancients:

> I have heard experts discuss these matters regarding the *tuono* of Rome in various ways. Some attribute its lowness to the weakness and slothful nature of the singers; others attribute it to the large numbers of castrati who, once they are of advanced age, are no longer able to sing with the same accurate voice as those formed by uncastrated boys; and finally still others to the large number of *bassi profondi* found here more than elsewhere."[37]

In technical terms, a lower pitch might have helped castrati sing entire pieces without the break in register that occurred when they shifted from chest to head voice. Doni's words exude meaning: the weakness and sloth of singers, a geographical insistence that equal temperament would never work, and a dig at aging castrati.

## AS THE WORD TURNS

Over the course of the sixteenth century, as the axes of power shifted from the Mediterranean to the Atlantic, Italy lost its center of gravity. The sea change came in part from the Atlantic slave trade and the Protestant Reformation. As Aníbal Quijano writes,

> A historically new region was constituted as a new geocultural identity: Europe—more specifically, Western Europe. A new geocultural identity emerged as the central site for the control of the world market. The hegemony of the coasts of the Mediterranean and the Iberian Peninsula was displaced toward the northwest Atlantic coast in the same historical moment.[38]

Naples occupied a particularly vexed place. As a center of Spanish power, it linked Italy to the New World, and early modern naturalists read the plants and animals they saw in the New World in terms of what they saw in southern Italy. The Jesuits even referred to Naples as "the other Indies."[39]

Italians like Amerigo Vespucci, Columbus, and Giovanni Caboto (John Cabot) dominated navigation and narratives of explorations, but they did not work for the Italians; Cabot worked for the English, Columbus for the Spanish, and Vespucci for the Spanish and Portuguese. For centuries Italy had been the primary legacy of the Classical tradition in which the Greeks and Romans stood against barbarians. By the time of European exploration and expansion, however, Italy began to be imagined as barbarous even as its explorers were instrumental in a cartography of difference. The presence of Christopher Columbus in the Gallery of Maps in the Vatican Museum is indicative of particularly the papal relationship to exploration.

In 1578 Pope Gregory XIII, the pope who supposedly inspired the dragon fountain at Tivoli, commissioned the Gallery of Maps with thirty-two panels to demonstrate his global dominance (Fig. 7.3).[40] The allegorical Columbus looks like one of the castrati of the last chapter: think Pasqualini riding into Piazza Navona on a stage. Columbus sporting a Roman emperor-style toga rides a chariot seemingly powered by Neptune standing on a seashell. A giant merman blows through a conch trumpeting the voyage. The banner says, "Christopher Columbus of Liguria, Discoverer of the New World." This imagined cartography asserted a Catholic domination that was already fantasy and nostalgia. In the early modern period, the Gallery of Maps was updated frequently to emphasize ecclesiastic possession and conquest and to proclaim Italy as the center of world history. While Urban VIII built the

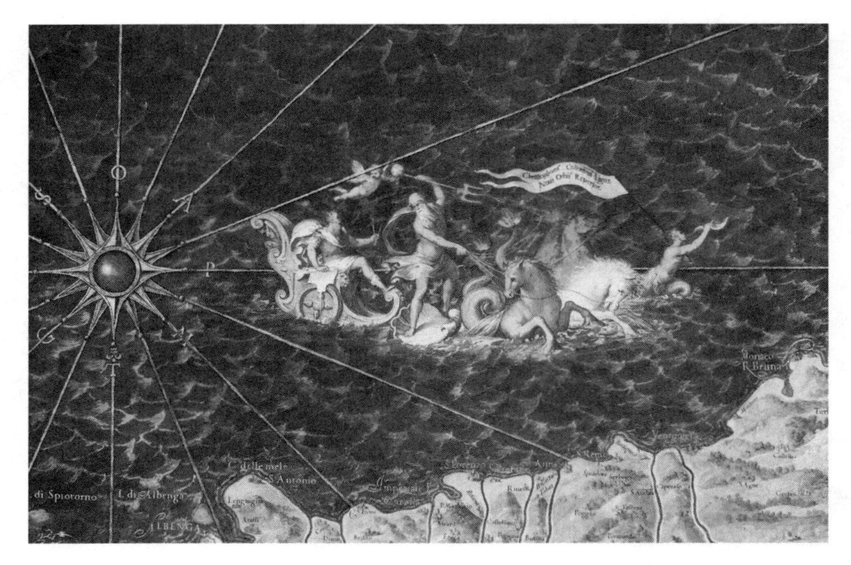


FIG. 7.3 Christopher Columbus and Neptune. Detail, Vatican, Gallery of Maps. Adam Eastland / Alamy Stock Photo.

operatic machine familiar to music scholars, he also invested in training missionaries to convert those who resided in faraway lands and restored the Gallery of Maps.[41]

As Italy moved away from the center of the European intellectual world, its urban centers became destinations for aristocratic tourists. George Sandys, the English traveler, writer, and translator who was also the first treasurer of Virginia, described in 1615 a voyage he began in 1610 to the far reaches of Italy and the Ottoman Empire. He explicitly located Italy in the Eastern Mediterranean, linked more to the Holy Land and Turkey than to Northern Europe and England.[42]

> The parts I speake of are the most renowned countries and kingdoms: once the seats of most glorious and triumphant Empires; the theaters of valour and heroical actions, the places where Nature hath produced her wonderfull works; where Arts and Sciences have been invented and perfited; where wisedome, vertue and policie and civility have been planted, have flourished. . . . Which countries once so glorious, and famous for their happy estate, are now through vice and ingratitude, become the most deplorable spectacles of extreme miserie: the wilde beasts of mankind having broken in upon them, and rooted out all

> civilitie; and the pride of a sterne and barbarous Tyrant possessing the thrones of ancient and just dominion.[43]

In short, the seventeenth century that witnessed the institutionalization of commercial opera also saw the codification of Western European constructions of Italy as lascivious, decadent, deviant, and weak.

## THE ITALIAN POX

In 1670 Lorenzo Panciatichi, who worked as a librarian for Cardinal Leopoldo de' Medici and was a member of the Academia della Crusca, felt that Italian artistic and intellectual dominance and glory were already over. The French had eclipsed the Italians in his eyes.

> After all, I call it a miracle, that a foreigner should be able to speak in that style, and to write with more appropriateness than we Tuscans. This is one of those things that from time to time makes me angry. We Italians are substandard in almost every genre of literature, seeing through experience that the fine arts (le belle arti) have crossed the mountains, and have come to settle in places that used to be called barbarian, but which are now the most refined: so the sciences, research, and erudition have taken root, and they demonstrate this miraculously in these lands beyond the mountains, while our own land, from whence they have been transplanted, has been exploited almost completely.[44]

Panciatichi was one of many Florentines who spent time in Parisian salons.

This leads back to Mazarin, where this chapter started: Mazarin's productions of *La finta pazza* and *L'Orfeo*. As Frederick Hammond, Margaret Murata, Roger Freitas, and others have discussed at length, he made stunning Barberini productions.[45] Cornelio Bentivoglio, who had ridden in the 1634 Barberini joust, assembled much of the *L'Orfeo* production. The castrato Atto Melani played Vittoria and Orfeo in the operas. The Barberini prized possession, Marc'Antonio Pasqualini, sang Aristeo, a role that came with a magical mad scene, and Gelosia was played by the Roman castrato Pamfilio Miccinello. The final performance concluded with a choreographed ball in honor of Anna Colonna Barberini. The prologue involved sixty performers representing the French army. The production came complete with special effects. Venus descended on a machine designed by the master Torelli. *La finta pazza* was its own travelogue. The Paris ballets choreographed by the Italian Giovanni Battista Balbi were etched into life in engravings by Valerio Spada.

FIG. 7.4 "Four Bear Trainers Sprinkling the Ground," from "Ballet of Bears with Parrots," ballet by Jean Baptiste Balbi for *La finta pazza*, plate by Valerio Spada. Harvard Art Museums/Fogg Museum, Gray Collection of Engravings Fund, by exchange, Photo ©President and Fellows of Harvard College, S7.52.2.

The ballet of eunuchs, bears, parrots, and monkeys was its own emblem of New World alterity (Fig. 7.4). The illustration almost captures the ostriches whose necks extended mechanically, and the Indians with contorted legs and stretched arms embody the grotesque.

But by this time Mazarin was already the subject of a pamphlet war in the form of thousands of tracts leveled at the so-called Sicilian bugger and known as Mazarinades (Fig. 7.5). After the death of Louis XIII in 1643, France fell into the ruling hands of his five-year-old son, the future Sun King, and his Spanish bride, Anne of Austria. Appearing during the Fronde, a series of civil wars in France, these pamphlets and discourses mixed political and sexual misconduct.[46] Thousands of them were distributed in the streets. An urban theater of song and verse created collective responses to the crisis. They portrayed Mazarin as a nasty outsider hell-bent on causing the French people to suffer, and they staged old-school battles between passion and reason, virtue and vice, order and chaos. Mazarin was a devilish monster with Italian vices who sodomized priests and worked against the order of nature. A pamphlet cited by Jeffrey Merrick spelled this out.

The Cardinal [fucks] the Regent;
What's worse, the bugger boasts about it
And steals all her money from her.
To make the offense less grave
He says that he only [fucks] her in the a[ss]:
It's easy enough to believe.[47]

A key part the equation here is Mazarin's association with importing Italian opera in general and castrati in particular as part of the political machine.

LE SALVT
DE LA FRANCE,
DANS LES ARMES DE LA VILLE DE PARIS.

*A* *Le bon Genie de la France, conduisant sa Maiesté en sa flotte Royale.*
*B* *Son Altesse le Prince de Conty, Generalissime de l'armée du Roy, tenant le timon du Vaisseau, accompagné des Ducs d'Elbeuf, & de Beaufort, Generaux de l'armée, & du Prince de Marsillac, Lieutenant general de l'armée.*
*C* *Les Ducs de Boüillon & de la Motte-Haudancour, Generaux, accompagnez du Marquis de Noirmontier, Lieutenant General de l'armée.*
*D* *Le Corps du Parlement, accompagné de Messieurs de Ville.*
*E* *Le Mazarin, accompagné de ses Monopoleurs, s'efforçant de renuerser la Barque Françoise, par des vents contraires à sa prosperité.*
*F* *Le Marquis d'Ancre se noyant, en taschant de couler le Vaisseau à fond, faisant signe au Mazarin de luy prester la main dans sa premiere entreprise.*

LA fatale reuolution de l'Empire des Troyens sembleroit nous rendre hereditaire de son mal-heur, ainsi que cette Ville retient encore le nom d'vn de ses derniers Princes, si dans la consternation publique de cette maladie generale de l'Estat, Paris, le chef de ce grand corps de la Monarchie Françoise, si redoutable à tous ses ennemis, & qui ne peut estre atterré que par sa propre pesanteur. Si Paris ne s'estoit le premier armé pour la deffense de cette Couronne, & la conseruation de son authorité: les armes de Paris, cette Nef plus renommee que celle d'Argos, sous la conduitte d'vn autre Iason, digne sang de nos Roys, assisté des Polux & des Castor, autres illustres Argonottes, dont l'experience & la valeur nous promettent déia vn port asseuré en commançant à desplier les voiles. Inuincibles Herauts, que l'obiet du peril ne peut arrester; vous n'auez à combattre en cette celebre conqueste qu'vn Dragon, gardien de tous les thresors de la France, vne harpie orgueilleuse des despoüilles & des richesses de ce florissant Royaume, vn serpent qui se r'emplit depuis tant d'années du sang des peuples, & que nostre foiblesse laisse laschement sacrifier tous les iours à son insatiable conuoitise, à la honte de l'Estat, au des-auantage de nostre ieune Monarque, & au mespris des Loix & de la Iustice; & qu'apresent tant de sages Nectors s'efforçent de faire reuiure aux despens de leurs propres vies & de tous leurs biens. Mais le mal est si grand & si present, que l'effet du remede consiste à la diligence. Portons donc nos armes vers cét ennemy commun de tous les Estats; & tandis que nostre Prelat assisté de son Clergé porte les bras vers le Ciel comme vn autre Moïse, combattons en vrais Iosuez, & autant armés de foy que de fer, croyons nostre victoire certaine, & que Paris meritera de porter vn iour le nom de deffenseur de L'ESTAT & du salut de la FRANCE.

FIG. 7.5 Mazarinades, *Le salut de la France, dans les armes de la ville de Paris* ([Paris?], 1649). Bibliothèque Mazarine, M 15159.

While castrati did not occupy a central place in French opera, as Julia Prest, Hedy Law, and others have argued, they did appear in the Chapelle Royale, which, like the Italian papal chapel, forbade women from singing.[48] And as Law has argued, French thinkers did not understand the castration procedure in large part because castrati were not made in France. She reads the language of mutilation in the discourse around castrati as part of a long-held distrust of and distaste for Italian music. In 1702, when François Raguenet published his comparison of French and Italian opera, the inherent Italianness of castrati was part of his argument about the superiority of Italian opera. He relished the castrato voice but still pitted it against what he heard as more cultivated drama, plot, instrumental music, haute-contre voice type, and reason of his native France. The castrato was about voice and sound, not reason and intellect.

The castrato phenomenon officially landed in England in 1708 with the arrival of Nicolino Grimaldi (Nicolini), arguably the first British opera mega star. The Neapolitan singer began his career on home turf, sang once in Rome, moved to Venice, and landed in London at the Haymarket Theatre. There he headlined a peculiarly British version of the Neapolitan Alessandro Scarlatti's 1694 *Pirro e Demetrio*, in which Italians sang in Italian and English singers sang in English.[49] Printers quickly put out scores of this massive hit in multiple formats, including a book of piano reductions marketed as lady's entertainments published in 1709 (Fig. 7.6).

Italian opera was by this tine already the stuff of scandal: irrational, erotic, deviant, decadent, extravagant, and feminizing. Outside its geographical boundaries, Britain was engaged in war with the French. Domestically, it staged an internal war with Italian opera that rendered Italy a dangerous, foreign, and feminizing enemy.[50] Before a castrato had even performed in England, critics objected to the alien nature of Italian opera and warned against the southern loss of liberty because of degeneracy and corruption. John Dennis's *Essay on the Opera's after the Italian Manner* situated Italian opera all told as monstrous:

> But yet this must be allow'd, that tho the Opera in *Italy* is a Monster, 'tis a beautiful harmonious Monster, but here in *England* 'tis an ugly howling one. What then must not only Strangers, but we our selves say, with all our Partiality to our selves, when we consider that we not only leave a reasonable Entertainment for a ridiculous one, an artful one for an absurd one, a beneficial one for a destructive one, and a very natural one for one that is very monstrous; but that we forsake a most noble Art, for succeeding in which we are perhaps the best qualify'd of any People in *Europe*, for a very vile one for which Heaven and

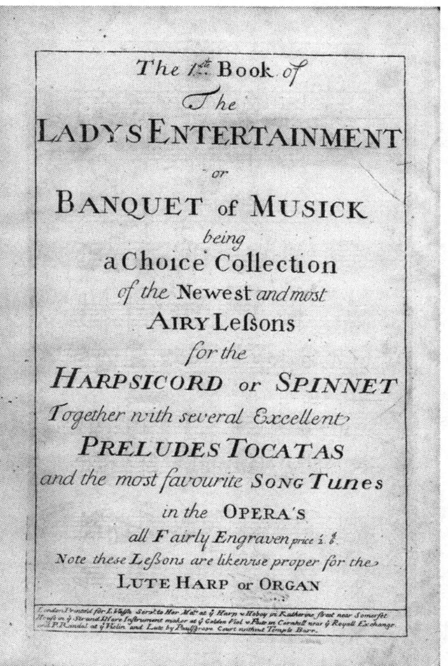


FIG. 7.6 William Babell, *The 1st Book of the Lady's Entertainment or Banquet of Musick* (London: John Walsh, 1709); *a*, frontispiece; *b*, title page. Nederlands Muziek Instituut, The Hague, Netherlands.

> Nature have not at all design'd us, as having given us neither Ears, nor Voices, nor Languages, nor Climate proper to it.[51]

A few years later, Joseph Addison equated taste and nation when he wrote that a composer must

> fit his music to the genius of the people, and consider that the delicacy of hearing, and taste of harmony, has been formed upon those sounds which every country abounds with: In short, that music is of a relative nature, and what is harmony to one ear, may be dissonance to another.[52]

As Jill Campbell argues, castrati stood in England not just at a nexus of gender ambiguity but at the center of ideas about economic instability rooted in nascent capitalism.[53] English texts could, on the one hand, celebrate the potential of the castrato voice in Italy to ravish the listener and, on the other, insist on the absolute threat to English order of castrato singers in England. To describe the castrato voice as the "finest in the world" worked if it was in another part of the world, a part farther removed by the process of translation from French to English. In political terms, while England, France, and eventually Germany became marked as leaders of progress, Italy was marked as an internal Other of Eurocentric continental and English thought.

The 1778 anonymous satire *The Remarkable Trial of the Queen of Quavers* exemplifies almost too well the association of the operatic castrato with the subaltern, monstrous side of the world. The queen of quavers spread the "quavering itch," an epidemic of irrationality and attraction to Italian opera that attacked the body through the ear. The only cure was apparently the mass importation of castrati from other worlds. The castrato was a monster, "an outlaw of nature." An assemblage of animal parts out of a sixteenth-century monster treatise,

> they were more exotic than anything from "the wilds of Africa." They have the look of a crocodile, the grin of an ape, the legs of a peacock, the paunch of a cow, the shape of an elephant, the brains of a goose, the throat of a pig, and the tail of a mouse.[54]

As the eighteenth century progressed, satires against Italian opera generally, and castrati in particular, were metonyms of Italy and leveled labels of vice, deviance, sodomy, and degradation associated with the torrid zone. As Suzanne Aspden has shown, the castrato's body served as a visceral sign of degeneracy.[55] The song of the castrato was imagined as disruptive to society all told, and satires leveled against castrati often mobilized decidedly anti-Catholic and, by extension, anti-Italian rhetoric. In the context of an emergent anxiety about Catholic plots, castrati and other Italian singers signaled the decidedly corrupt vices of Roman Catholicism and papal courts. The following anonymous satire from 1727 illustrates this:

> Who knows but they are sent here to raise Dissensions among true Protestants! There are too many shrewd Causes of Suspicions.
> 1. They come from Rome;
> 2. The Pope lives at Rome;
> 3. So does the Pretender,
> 4. The Pope is a notorious Papist;
> 5. So is the Pretender;
> 6. So is Madam Faustina,
> 7. And so is Madam Cuzzoni.

8. King George (God bless him) is a Protestant;
9. The Pope hates King George;
10. The Pretender can't abide him.
11. But Madam Cuzzoni and Madam Faustina, love the Pope, and in all Probability the Pretender

Ergo * * * * * * * * * * * * * * * * * * * * * * * * * *

From whence I infer, that it is not safe to have Popish Singers tolerated here, in England; but on the contrary, it would be a great Security to the Protestant Interest to have a Clause added to some Act of Parliament, obliging all foreign Singers . . . to abjure the Devil, the Pope, and the Pretender, before they appear in Publick.[56]

At the same time, pornographic literature that critiqued religious and church authorities was on the rise. D'Ancillon's *Eunuchism Display'd*, the already mentioned account that modern scholars often mine for "facts," was a part of a satirical and pornographic tradition. The translator, Robert Samber, who wrote under a fake name, reminded readers that his purpose was to prevent English women from marrying castrati, "to hinder such a Marriage [with a Eunuch] which could not but be attended with dismal Consequences." Samber also translated pornography, collections of "Chinese Tales," and the fairy tales of Charles Perrault, and he partnered with publisher and bookseller Edmond Curll who, Fox News–style, saw the press as a vehicle for creating scandal. The d'Ancillon treatise published by Curll was thus closely associated with pornography, in Curll's own case so much so that he himself was often read as a symbol of British decline. The term *Curllism*, which became a word for literary "indecency," was coined by Daniel Defoe in an attack on *Eunuchism Display'd*. And in 1708 Curll mounted a full advertising campaign for a treatise on venereal disease called *The Charitable Surgeon*. He was censored in 1727 for a treatise on flogging that he called a French anatomy book. And his accounts of hermaphrodites marketed as anatomy sit at the center of the emergent pornographic tradition that Lynn Hunt argues was linked to science and exploration.[57]

## THE CASTRATO IN THE SOUTH

Despite the obvious anxiety about castrati within their borders, English travelers still sought them out and usually looked for them in Naples, often the last stop in Europe on the Grand Tour. A center of castrato opera production, it was also a hotbed of street singing and imagined to be inhabited by "musical people." Naples, a Spanish ruled state, was an internal Other, which does not conflate southern Italy with, for example, the Caribbean. It does locate Italy outside of the hegemonic center. Naples had always had an ambiguous relationship to Rome, was at the border of the torrid zone, and was a gateway to Africa.

Naples was also imagined as particularly lascivious and as a space rife with the dangers of sex. It was associated discursively with dildos, syphilis, and sexual excess. In 1493, when Charles VIII of France invaded Naples, his soldiers left with syphilis, which, by 1520, had spread over much of the world. The French called it the disease of Naples, while others called it the disease of France.[58] The association of syphilis with Naples stuck, and despite intense scholarly debate on its origins, 1495 in Naples gets credit for the fifteenth- and sixteenth-century pandemic. This disease was about more than "public health." In English texts, the dildo was deeply tied to Italy. The earliest reference in English came in Thomas Nashe's "The Choice of Valentines," published in 1595. And in the early eighteenth-century "Monsieur Thing's Origin," the dildo was presented as invented in Italy and named in France.[59]

In the sixteenth century, Jesuits sought to civilize Naples and bring Catholicism to the urban and rural poor, a process that further inscribed the myth of Naples as ungovernable and as the edge of civilization. The first Jesuit missionaries arrived in 1552, shortly after Francis Xavier arrived in Japan. Music was a primary tool in this civilizing process. By the seventeenth century Naples was an important port for the Muslim slave trade, giving the Jesuits foreign and local infidels to convert.[60] The 1631 eruption of Mount Vesuvius, the first since the year 79, created a deeper sense of Naples as unruly, with a people whose lives mirrored the volcanic eruption of their mountain. The volcano was the symbol, even the instantiation of God's wrath. Kircher famously climbed into the belly of Vesuvius in 1638 and found there the potential for an apocalyptic end of the world. He published much of the experience in his *Mundus subterraneus,* whose illustrations look remarkably like stage sets. For him it was an almost theatrical hell that had everything but ghosts and demons.[61]

So, when Burney went to Naples to find castrators and primordial song, he followed a well-trodden path. He was fascinated with street singers whose imagined natural song stood in opposition to reason.[62] To his ear, the music was not of Europe. And Burney, like others, was interested not just in Naples but in the people of Naples. Street and theatrical singers were objects to be observed and found on the Grand Tour, and accounts and guidebooks reveal an interest in customs and morals as well as the people themselves. Until the nineteenth century the categories we understand as composer and performer were not so rigidly defined, so it makes sense that accounts often conflate singers, composers, and "musicians" all told. Writing about the composers of Naples, for instance, Burney wrote,

> There is an energy and fire, not to be met with perhaps elsewhere in the whole universe; it is so ardent as to border upon fury; and from this impetuosity of genius, it is common for Neapolitan composers, in a movement, which begins in a mild and sober manner, to set the orchestra in a blaze before it is finished. Like high-bred horses they are impatient of the rein, and eagerly accelerate their motion to the utmost of their speed; as Dr. Johnson says, that Shakespeare, in tragedy, is always struggling for an occasion to be comic. The pathetic and the graceful are seldom attempted in the conservatorios; and those refined and studied graces, which not only change, but improve passages, and which so few are able to find, are less sought after by the generality of performers at Naples, than in any other part of Italy.[63]

The composers may have musical genius, but they are not refined, rational, or controlled. Not quite human, they work on the automatic, fast-twitch muscles of horses and traffic in unpredictable sounds.

Burney wrote from a long tradition of anxiety and desire around Naples, a city that, in the ancient world, was associated with the mystical power of music, unnatural beauty, and danger. Naples had been since the ancient world associated with sirens who resided somewhere in proximity to the Gulf of Naples. The siren Parthenope, like her sisters, used her enticing song to entrap sailors. Naples, before and after Burney, stood as a nostalgic, originary Eden. In Burney's text, the castrato story aligns with a larger story about the Global South in which Italy and Naples—"a paradise inhabited by Devils"—in particular, were already a problem, seen as excessive, dangerous, and exotic.[64]

> This evening hearing in the street some genuine Neapolitan singing, accompanied by a *calascioncino*, a mandoline, and a violin; I sent for the whole band upstairs, but, like other street music, it was best at a distance; in the room it was coarse, out of tune, and out of harmony; whereas, in the street, it seemed the contrary of all this: however, let it be heard where it will, the modulation and accompaniment are very extraordinary.[65]

Burney's accounts cannot be separated from their sense of the Naples soundscape writ large. His description reeks of Enlightenment judgment:

> The voice part is very slow, a kind of psalmody; the words, of which there are many stanzas to the same air, are in the Neapolitan language, which is as dif-

> ferent from good Italian, as Welsh from English. It is a very singular species of music, as wild in modulation, and as different from that of all the rest of Europe as the Scots, and is, perhaps, as ancient, being among the common people merely traditional. However, the violin player wrote down the melody of the voice part for me, and afterwards brought me something like the accompaniment; but these parts have a strange appearance when seen on paper together.[66]

Here the music of Naples defies the logic of written music and of tonality; it oozes the incomprehensible language that Burney claimed was unintelligible even to other Italians and escapes aesthetic reason. It is sound out of joint.

Burney explained in 1775 in his tour of Germany that

> climate contributes greatly to the forming of customs and manners; and, it is, I believe, certain, that those who inhabit hot climates, are more delighted with music than those of cold ones; perhaps, from the auditory nerves being more irritable in the one than in the other: but I could, by no means, account for climate operating more in favour of music upon the Bohemians, than on their neighbors, the Saxons and Moravians.[67]

And in France the vituperative philosophical battle over the primacy of French or Italian opera used the castrato and Italian operatic tradition to map global hierarchies and to make borders. Rousseau explained in his dictionary of music that "there exist in Italy, some inhuman fathers, who sacrificing nature to fortune, give up their children to this operation, for the amusement of voluptuous and cruel persons, who have the barbarity to require the exertion of voice which the unhappy wretches possess."[68] The castrati and especially their voices stood as sonorous embodiments of racial and global hierarchies.

Rousseau's comments about the castrati sound different when read in the context of his essay "On the Origin of Language," published posthumously in 1781 but written during the time of the Querelle des Bouffons. When Rousseau located the voice and song as particularly human—"man alone sings"—he set up a rubric for marking some humans as nonhuman. Vocal timbre marks a cycle of degeneration, one that arguably goes back to the idea of barbarian, unintelligible languages as existing apart from the language of foreigners. Voice becomes other to language and reason. Castrati sat uncomfortably on the border of music and noise and were inscribed into the process by which the European Enlightenment subject differentiated itself from others based on rational and civilized language as opposed to raw sound and voice. The castrato voice was a part of native otherness in contrast to European order, the opposite of the audibility of reason that Veit Erlmann has written about so extensively.[69]

When Rousseau argued for the supremacy of Italian opera based on an essential, sonorous quality of the Italian language and an essential, musical nature of the Italian people, he positioned Italy and the South as the origins of language, but as arrested by a kind of preliterate passion. He philosophically situated the voice as a natural sound that preceded music and language, with language becoming a motor for the civilizing process. As Gary Tomlinson and others have argued, Western European discourse separated itself from the rest of the world: "Now non-European singing was conceived not as equivalent (in whatever manner) to contemporary European practices but as a survival in far-off places of practices Europe had long since outgrown."[70] The Neapolitan Giambattista Vico, in his *New Science*, argued for a nostalgic mythological world that challenged the proto-Enlightenment domination of reason over passion.[71]

Rousseau's climate politics were in keeping with Enlightenment tendencies. In the 1740s Charles-Louis de Secondat, Baron de La Brède et de Montesquieu, famous in US history classes for his influence on the Founding Fathers, had articulated an Enlightenment theory of climate in the fourteenth book of *The Spirit of the Laws*. He explained that "the empire of the climate is the first, the most powerful of all empires."[72] Montesquieu embraced his climate theory after a voyage to Italy, and it was his trip from Rome to Naples that confirmed his belief in the direct effects of climate on individuals and societies.[73] Montesquieu explained the differences between the delicate, weak, and easily inflamed people of the South and the reasonable, vigorous, heroic, and disciplined peoples of the North: "If it is true that the character of the spirit and the passions of the heart are extremely different in the various climates, laws should be relative to the differences in these passions and the differences in these characters."[74] Montesquieu used music to make his point:

> In cold countries they have very little sensibility for pleasure; in temperate countries, they have more; in warm countries, their sensibility is exquisite. As climates are distinguished by degrees of latitude, we might distinguish them also, in some measure, by those of sensibility. I have seen the operas of England and of Italy; they are the same pieces and the same performers; and yet the same music produces such different effects on the two nations, one is so cold and indifferent, and the other so transported, that it seems almost inconceivable.[75]

And finally, he makes the case for an explicit North-South divide in Europe based on climate.

> In Europe there is a kind of balance between the southern and northern nations. The first have every convenience of life, and few of its wants: the last have many wants, and few conveniences. To one, nature has given much, and demands but little; to the other, she has given but little, and demands a great deal. The equilibrium is maintained by the laziness of the southern nations, and by the industry and activity which she has given to those in the north. The latter are obliged to undergo excessive labour, without which they would want everything, and degenerate into barbarians. This has naturalized slavery to the people of the south: as they can easily dispense with riches, they can more easily dispense with liberty. But the people of the north have need of liberty, which alone can procure them the means of satisfying all those wants which they have received from nature. The people of the north, then, are in a forced state, if they are not either free or barbarians. Almost all the people of the south are in some measure in a state of violence, if they are not slaves.[76]

The cold air of northern countries renders their citizens more equipped for civilization.

## ARCHENHOLZ AND THE BALL-DROPPING ARIA

I want to dwell on a story that is often mentioned in passing by Anglo-American scholars. In 1765 German explorer and traveler Johann Wilhelm von Archenholz visited Naples. His account is dripping with descriptions of natural beauty bequeathed by the gods and Virgil. But he also recorded witnessing a marvel:

> A very particular accident happened last year to a singer of the name of Balami. This man was born without any visible sign of those parts which are taken out in castration; he was, therefore, looked upon as a true-born castrato, an opinion which was even confirmed by his voice. He learned music and sang for several years upon the theater with great applause. One day, he exerted himself so uncommonly in singing an arietta, that all of a sudden, those parts, which had so long been concealed by nature, dropped into their proper place.[77]

This marvel hit the Anglophone presses in 1791 when Joseph Trapp translated the travel narrative.

Archenholz's account has implications for how modern listeners hear the history of the castrato. Archenholz fashioned himself as a rational Enlightenment thinker, committed to objective testimonial. Nevertheless, his account is, it must be stated in twenty-first-century terms, a story: it is not what would count for us as a "true" account. After all, it is hard to imagine that Archenholz saw anyone's testicles pop out, and yet this anecdote appears in much scholarship on the castrato not as truth but not quite as fiction either. He also wrote a four-volume history of Queen Christina that remained a primary text throughout the twentieth century in which he claimed to have acquired a small piece of Descartes's skull while doing research in Sweden.[78] So the ball-dropping aria is an extreme example of the castrato as part of the construction of the South as a contact zone. The castrato—like the dildo, syphilis, hermaphrodites, volcanoes, and monsters—hailed from a faraway place whose climate defied reason.

The Prussian historian and publisher had several axes to grind, and he had already internalized the Grand Tour as a testament to decline and as a voyage to primordial beauty and away from Enlightenment. Naples, for him, was the least enlightened city in Europe. His story of Balami's testicles popping out during an aria appears on the heels of an account of priests walking around with testicles in their pockets: "But as ecclesiastic laws of the church of Rome require, for this purpose a person that has not been mutilated, the Sophists of that persuasion have thought it sufficient for such a priest to have his amputated genitals in his pocket, when he approaches the altar."[79] If any doubt as to Archenholz's assessment of the Neapolitan psyche remains, he explains that "there is no city in Europe where thinking is less in fashion than here. Besides superstition the brains of the Neapolitans are only stuffed with sounds."[80] Sound here symbolizes the opposite of reason and enlightenment, the decadence of the Southern spaces, and the end of Europe.

The point in this historiography lesson, and in providing a different context for well-known texts, is to locate the imaginative construction of the castrato in the imaginative space of the South, what Dipesh Chakrabarty describes as always inadequate and always lacking reason. Southern Italy occupied a liminal space between Europe, Africa, and the Orient. The unification of Italy in 1861 codified, in political terms, the perpetual otherness of the south of Italy. There are around southern Italy overlapping zones of otherness; on the scale of Europe, the circum-Mediterranean was far enough away from the European center of discursive power that it was imagined to be outside of reason. Naples also remained associated with song and vocal excess. In the twentieth century Friedrich Kittler went to the Amalfi coast to stick opera singers on cliffs and impersonate sirens.

This narrative also deliberately inserts the castrato, and in particular the

Enlightenment understanding of him, into the story of music and sound as media that are constitutive of Europe as Self and everyone else as Other. As Gary Tomlinson writes,

> Across the century from 1750 to 1850 music lodged itself at the heart of a discourse that pried Europe and its histories apart from non-European lives and cultures. Perched at the apex of the new aesthetics, it came to function as a kind of limit-case of European uniqueness in world history and an affirmation of the gap, within the cultural formation of modernity, between history and anthropology. It arose, it is not too much to say, in complex alliance with Europe's increasing domination of foreign territories and societies around the world.

Eighteenth-century stories about the castrato are part of a larger historiography in which, as Johannes Fabian wrote in 1983, modernity and timeless cultural value depended on proximity to Europe.[81] The idea of Europe's predominantly written music as superior and distinct was fundamental to the nineteenth-century intellectual world that modern musicologists inherited.

Escaping the Anglophone Enlightenment and using contemporary theoretical ideas allows for a hearing of castrati that both incorporates their early and premodern migrations and leans into the sonic ramifications of the kind of overlapping history Gramsci advocated when he posited that the primary conflict of modernity was between the subaltern and the exercise of hegemony. To insert castrati into this narrative is to think about the ways they were, in their real and imagined incarnations, part of a process of globalization and a human geography that inscribed temperamental differences and, in turn, justified the subjugation and enslavement of southern people.

The southern questions that Gramsci articulated in 1926 formed the roots of much important postcolonial thinking that comes to Anglo-American and European academics through South Asian Marxists: Spivak on the subaltern, Said on Orientalism. Said famously critiqued the West's construction of the Orient, but he also leveled a devastating attack on the very notion of orienting, the essentializing of geographical division. Said took from Gramsci a sense of Western imperial geographies as immanently naturalized. And the notion of the subaltern came from Gramsci's "Notes on Italian History." He wrote,

> it is evident that East and West are arbitrary, conventional (i.e. historical) constructions, because outside real history any point on the earth is East and West at the same time. We can see this more clearly from the fact that these terms have been crystallized not from the point of view of man in general but from

> the point of view of the cultured European classes who, through their world hegemony, have made the terms evolved by themselves accepted everywhere.[82]

The castrato is a part of all of this in Gramsci's zones outside the zone of reason and enlightenment. The voice was, almost as soon as it was popular, fleeting and relational. Naples and southern Italy were the places of Romantic sublime in the nineteenth century and of failed capitalism, and the castrato was always already a sonic marker of this process. This is perhaps a twenty-first-century version of what John Rosselli wrote in 1988:

> The contradictions between the stereotype and what we know of the real thing are best explained by a deep ambiguity in contemporary attitudes, not unlike the ambiguity with which the dominant groups in European society have at various times looked upon alleged inferiors who seemed in some way potent or attractive—Jews, say, or women. As with Jews and women, a good many castrati played into these ambiguous attitudes, no doubt to propitiate hostile neighbors but with the effect of confirming them in their hostile stance.[83]

When Italian opera arrived in the United States in 1825, Walt Whitman responded with an exoticist ear:

> You listen to this music, and the songs, and choruses—all of the highest range of composition known to the world of melody. It is novel, of course, being far, very far different from what you were used to—the church choir, or the songs and playing on the piano, or the [n-word] songs, or any performance of the Ethiopian minstrels, or the concerts of the different "families." A new world—a liquid world—rushes like a torrent through you. If you have the true musical feeling in you, from this night you date a new era in your development, and, for the first time, receive your ideas of what the divine art of music really is.[84]

If my insistence on the technology and techne of the castrato's voice has worked to situate the castrato voice as a mediated sounding instrument, then this historiography explains some of the specific ways that twentieth- and twenty-first-century Anglophone scholars got to the castrato as pure voice and as inherently nonnormative. This stance situates the castrato voice within what Kara Keeling writes about as a relational song and, in the words of Édouard Glissant, a kind of errantry: "Under these circumstances, errantry introduces relatively unpredictable contact between different organisms thereby raising possibilities for new forms, intermixtures, knowledges and sensibilities to be crafted through the pressures, pains and pleasures characteristic of those contacts."[85] The castrato voice functions as what Keeling describes in popular music as the potential to "make audible the social and political dimensions of the spaces it remaps or maps anew and make perceptible the milieus and rhythms from which those territories are formed."[86]

CHAPTER 8

# MORE THAN ONE SEX

At the risk of overreading an anecdote in a travel narrative, let's go back to Johan Wilhelm von Archenholz and his sound bite from "unenlightened Naples." The 1791 English version was marked as "translated from the original German."[1] He reported that in Naples "it was said" that an accident occurred in which a "natural" castrato by the name of Balami sang with such vigor that his balls dropped. For those who comb the castrato literature this metamorphosis may sound familiar; he appears in several studies, at least as a passing footnote. He's a faint trace of Enlightenment fantasies of the castrato's body: a body that exists in song, on the cusp of excess, and on the verge of the freak show.[2] He is a microcosm of the castrato body as a site of disputed real and imaginary boundaries, a body that—like the hermaphrodite, prostitute, courtesan, and widow—exceeds physical and political normative structures. It goes almost without saying that men with mutilated genitals and high voices, who played male and female roles on stage, challenged received binaries of male-female. But the nuance of the challenge is not so obvious. In that dramatic textual moment Balami gained what the current post-Freudian world understands as a phallus, but he lost his vocal prowess. Moreover, even as sterility remains one of the stubborn immutable facts of the castrato, their existence forced debates about fertility, and they emerged as masters of different kinds of reproductions. This does not mean that such singers existed in the utopian third sex that Michel Foucault and Thomas Laqueur imply. But their bodies did unsettle the normative patriarchal order, and they still provoke fundamental questions about bodies, identities, desire, and reproduction.

This chapter uses the figure of the castrato to intervene in musicological engagements with histories of science and medicine. While certain descriptions and narratives about the castrato seem to map onto the so-called one-sex model, the castrato's body calls attention to the broad-spectrum understandings of sex, gender, desire, and reproduction. Manhood was no more monolithic in the era of the castrato than it is now, and to make it even more complicated, the word *man* often meant all of humanity, so it can be especially hard to tease out what applies to men and what applies to humans. And debates and worries about the castrato body emerged within a larger sixteenth- and seventeenth-century crisis of gender. So, the castrato is emblematic of the ways in which gender is always, in Bruno Latour's terms, an assemblage, a series of connectivities.

Building on my arguments about the castrato as a voice of the torrid zone and the Global South, I stress here the castrato as imaginative figure and as a sonic space for anxieties around gender, patrimony, masculinity, reproduction, and marriage. Debates about musical practices that involved castrati were inflected by the essentially errant body of the castrato. This means writing about how stories of the castrato are told and excavating new stories from old texts. To play with the counterpoint between fiction and fact, to engage the history of history helps resist the tendency of music scholars, including myself, to take early modern texts about sex as truths. As the idea of marvel inherited from medieval discourses was slowly written out of early modern narratives, castrati shifted from marvel to deformed and disabled. But they nevertheless circulated in a swoon of desire and pleasure. Then, as now, the business of medical knowledge and knowledge making was deeply political, especially when it involved sex and reproduction, and this discourse was used to build moral, legal, and religious arguments. To take an example that is playing out now, in July 2022, as I review copyedits for this manuscript, the United States is reeling from a Supreme Court decision grounded in premodern medical ideas that is the end result of decades of right-wing political manipulation. The court seems wedded to the Catholic Church's long-standing prescriptions against abortion, which go back to 1588, when Pope Sixtus V issued *Contra abortum*, which made abortion a capital offense.[3] But they seem to have forgotten that it was overturned by 1591 because it was deemed too harsh and not effective.

I begin with my own avoidance of including Balami in scholarship and a discussion of the ways in which the anecdote's own history follows twentieth-century scholarly trends. I add a musicological voice to the critique of Laqueur's one-sex model and argue that the figure of the castrato singer should be understood within a larger crisis in which the issue of reproduction was most at stake.

## AVOIDANCE AND HISTORIOGRAPHY

Balami's testicular effusions made their modern appearance first in Angus Heriot's 1956 *The Castrati in Opera*.[4] By the 1990s, when works on gender, sex, and the body became mainstream in music scholarship, he appeared periodically. More recently, James Q. Davies read the anecdote as symbolic of the death knell or "the looming catastrophe of the castrato," and, for him, this episode that supposedly occurred in Naples is a mechanism to pinpoint the moment when the castrato voice fails, when it loses its "unnaturalness."[5]

While I have mentioned Balami in teaching and oral presentations, I avoided him in print. First, the story cannot be "true" in any way that satisfies a twenty-first-century historian. It reads like an early modern account of a marvel meant more to elicit wonder than to report fact.[6] There is no evidence that anyone's testicles popped out of anything. Second, as a scholar of the sixteenth and seventeenth centuries, I have deliberately decentered eighteenth-century narratives of the castrato as a way of resisting the teleological disciplinary investment in modernism and to avoid temporal hegemonies that are complicit in reinscribing modern sexual hegemonies. Especially with respect to gender, my interest lies more in echoes of medievalisms than in leading tones to the Enlightenment.

Third, I wanted to avoid the trap of a prurient fascination with what exactly the member of a castrato could or could not accomplish sexually. Landing so squarely on the testicles through a pornographic evasion of the actual words—words that can be taken to imply the penis—the story centers the functionality of the male member. Castrati had liaisons with men and with women, and both men and women fantasized about their voices and bodies. François Raguenet is a good example of amorphous desire. As a young French priest in Rome he was enamored of Italian music, and his writing draws attention to the castrato voice as inherently perfect for singing the role of male lover. I quote here from the 1709 English translation that was published in part to offer suggestions for improving English opera.

> These pipes of theirs resemble that of the Nightingale; their long-winded throats draw you in a manner out of your depth and make you lose your breath. . . . Add to this that their soft, their charming voices acquire new charms by being in the mouth of a lover.[7]

He wrote in response to Adriano Morselli's *Temistocle in Bando* of 1685, in which he said that Antonio Ferini playing the part of a Persian Princess was even handsomer than a woman. He noted also that Italians had better instruments, bigger bows, louder theorbos, and so forth.[8]

Finally, the story implicitly amplifies one kind of castrato sound: the heroic, virtuosic, pyrotechnic marvel. To fetishize pyrotechnics is, I have thought, fulfilling a tacit desire to make the castrato into a real man, to hypermasculinize him and to dwell on the sounds that are most familiar and appealing to modern listeners, such as Handel's heroic eighteenth-century roles like Giulio Cesare, Radamisto, Rinaldo, and others.[9] Such hypermasculinizing stands out in descriptions of the nightly contest between Farinelli and a trumpet player that highlighted the singer's vigor and endurance. It also associated the singer with a competitive military tradition; the trumpet was an audible sign of military vigor, and the one-on-one combat mimicked a duel.

> During the run of an opera, there was a struggle every night between him and a famous player on the trumpet. This at first seemed amicable and merely sportive, till the audience began to interest themselves in the contest and to take different sides. After severely swelling out a note in which each manifested the power of his lungs and tried to rival the other in brilliancy and force, they had both a swell and a shake together, by thirds, which was continued so long, while the audience eagerly awaited the event, that both seemed to be exhausted.[10]

Anne Desler has argued that although Farinelli, in his early career, did not usually play heroic roles, his parts were frequently written against relatively dense orchestrations, indicating a remarkably strong voice.[11]

Charles de Brosses commented on castrati in 1739 by describing their voices as clear, piercing, and different from women. He did allow for the fact that Italian women sounded more like castrati than other women:

> You have to become accustumed to the voices of the castrati to savor them. The timbre is as clear and piercing as that of choir boys and much stronger; it seems to me that they sing an octave above the natural voice of women. Their voices have something that is always dry and sharp, very different from the youthful and soft sweetness of female voices; but they are brilliant, light, bright, very strong, and have a very wide range. The voices of Italian women are of a similar type; light and flexible to the nth degree; in a word, of the same character as their music.[12]

The upbeat phrase "one must be accustomed" locates the castrato in a space of alterity. But terms like *dry* and *strong* keep castrati on the male side of the gender continuum.

## RUMORS

It is hard to find the castrato's body in part because generic codes of the early modern period are not intuitive to modern readers.[13] Castrati existed on and off the stage in a fluid, hyperreal space in which the desire to distinguish between sexuality and sex acts—or between discursive and empirical histories—becomes quickly futile. It's like asking a twenty-fourth-century historian of the Trump era to distinguish between the *Onion* and the *Los Angeles Times*, or *Saturday Night Live* and *60 Minutes*. In the sixteenth and seventeenth centuries, medical illustrations were based on erotic prints; satire and so-called factual accounts were often indistinguishable; and literature referenced real events in the language of mythology and ancient history. Raguenet, cited above, loved all things Italian and wrote a guidebook that included among other things dazzlingly erotic descriptions of Roman statues as living, breathing flesh.[14]

The rise of the Italian castrato also coincided with an explosion of medical and fictional literature on hermaphrodites at the end of the sixteenth century that reflected the ways in which masculinity was contested in religious, political, and medical discourses. The fascination with hermaphrodites was also deeply tied to the rise of pornography. Beginning circa 1550, hermaphrodites were associated with sodomy, sexual transformation, and transvestism. André Du Laurens's 1599 *Historia anatomica humani corporis* described in vivid detail every aspect of the sexed body. This was quickly followed by Jacques Duval's *Traité d'hermaphrodites* (1612) and Jean Riolan's *Discours sur les hermaphrodites* (1614). These writers presented the hermaphroditic body as pornographic display.

In the eighteenth century, the interest in humans who defied gender categorization was often explicitly pornographic. A 1718 "treatise" on hermaphrodites by Giles Jacob was packaged as an "innocent entertainment" not intended to incite amorous feelings:

> It is my immediate Business to trace every Particular for an ample Dissertation on the Nature of Hermaphrodites, which obliges me to a frequent Repetition of the Names of the Parts employ'd in the Business of generation; so I hope, I shall not be charg'd with obscenity, since in all Treatises of this kind it is impossible to finish any one Head completely without pursuing the Methods of Anatomical Writings.[15]

The book offers, among other things, descriptions of lustful women who enjoy flogging and masturbating.[16]

Institutionally, castrati were affiliated not with families but with churches, papal courts, and opera, central institutions that inspired suspicion, vitriol, and satirical slams.[17] The satires have been much written about, especially in terms of women who sang. As Courtney Quaintance has argued, Venetian noblemen "consolidated their bonds with one another through the creation, circulation, and consumption of literary fictions of women."[18] Words like *puttana* (whore) and *cortigiana* (courtesan) were as much about what men thought of women as they were about what women did or were. The trope of sexually excessive singers is one that cuts across centuries; the media is still obsessed with the imagined behavior of celebrity singers.[19] Even sources that seem "truer" in modern terms are fraught, mediated, and the written page leaves only a trace. Elizabeth Cohen has argued persuasively for the role of an embodied oral culture even in court records.[20]

To take a notorious seventeenth-century example, Marc'Antonio Pasqualini was caught up in various Barberini scandals, the subject of slander, and the object of desire by men and women. The "truth" of any of it cannot be found in the sources. Amy Brosius and others have written about rumors of a liaison between the castrato and Cardinal Antonio Barberini. The year 1644 saw a surge of street ballads aimed at the Barberini nephews like this one (Fig. 8.1):

FIG. 8.1 Ramona Martinez, *Pasquino Talks*.

What shall we say of Don Antonio
Who has an ill-formed body
has a big nose and little chest
More proud than a demon
he wishes great good to Marc'Antonio
I don't know whether for this music
Or for other inclinations
Take up, O Muse, the *colascione*[21]

The satire demonized the Barberini family and the castrato.

## ONE SEX, TWO SEX

The Balami story does map onto Thomas Laqueur's one-sex model, first posited in the 1980s but popularized in the 1990s via *Making Sex: Body and Gender from the Greeks to Freud*.[22] Despite extensive critique, the model prevails and is often taken by scholars as at least a window into ways the castrato body might have been heard and felt.[23] The book, translated into twelve languages, argues that binary sexuality emerged around 1750. He posits that a one-sex model dominated the ancient world through the Enlightenment and that in essence women were colder, less perfect versions of men. Lacqueur also articulates a teleology in which the Enlightenment and scientific process linked binary sex and gender systems with biology. Using Laqueur's model to understand the castrato would imply that their popularity waned when gender duality emerged. And it might offer a historicized way to understand the castrato outside of modern heterosexual and reproductive imperatives.

According to this model, the qualitative difference between men and women was a matter of temperature; men were hotter; women were colder. Temperament allowed for masculine women and feminine men, with gender metamorphosis as a not-infrequent consequence. Laqueur followed Foucault, who argued that, before the eighteenth century, women were effectively imagined as inverted men; women had the same organs of generation as men except that they were "inside." In this paradigm, which Laqueur illustrated using a now widely circulated illustration by Vesalius, the vagina is an internal penis, the labia a foreskin, the uterus a scrotum, and the ovaries the testicles. The problem is that Lacqueur's illustration is itself something of a mash-up. The page juxtaposes an image from Vesalius and an image by Guido Guidi, also known by his Latinate name Vidus Vidius. And Laqueur elided the fact that in sixteenth-century anatomical drawings, the same skeleton often

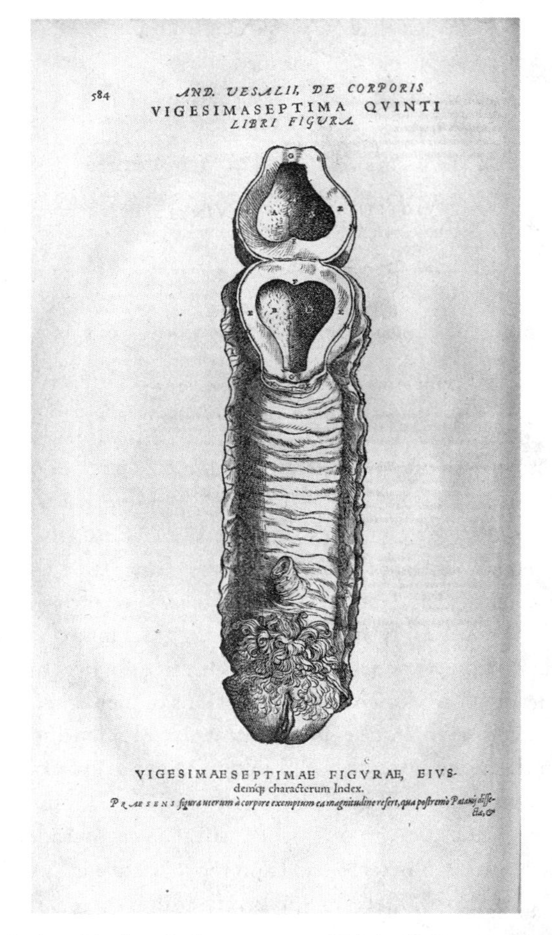


FIG. 8.2 Andreas Vesalius, *De humani corporis fabrica libri septem, liber V* (1555), 584. Courtesy of the Albert and Shirley Small Special Collections Library.

represented male and female bodies, whereas by the eighteenth-century anatomical figures staged secondary sex characteristics, including broader hips and narrower waists for women (Figs. 8.2, 8.3).[24]

For those who have read Lacqueur or who have read about hermaphrodites, the Balami anecdote may ring a bell. His apparent gender transformation works like that of Marie Germain, a story that seemingly also fits very neatly into Laqueur's model. Marie lived as a girl until the age of twenty-two when, while chasing a pig, she jumped over a giant crevice and her male parts popped out. Marie also jumped into the space of reproductive

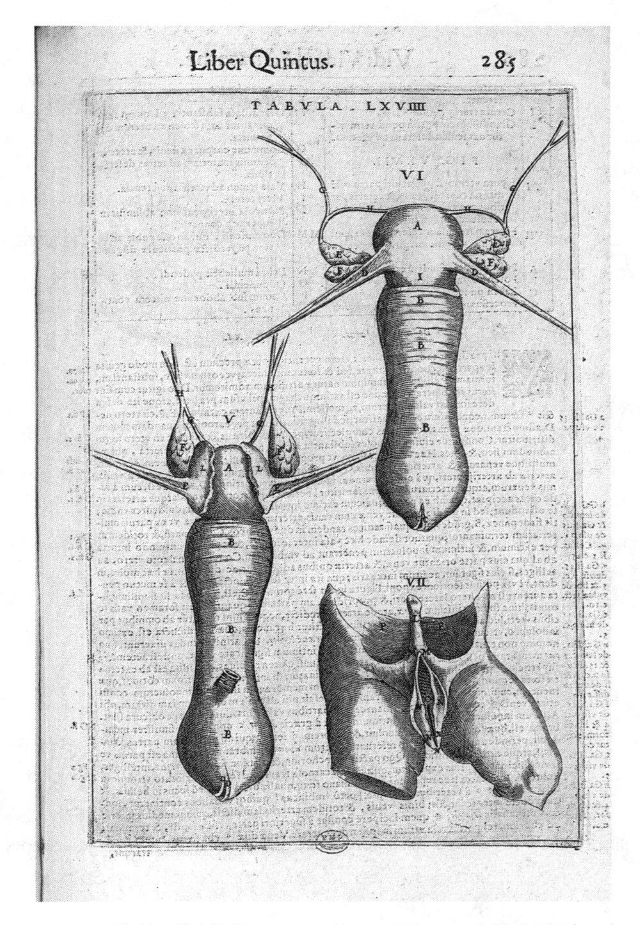


FIG. 8.3 Guido Guidi, *De anatome corporis humani, libri VII* (1611), 285. Courtesy of BIU Santé Médecine—Université de Paris.

heteronormative sexuality. The principle behind these stories and others like them is that male and female categories were determined by heat. So, an activity that generated heat could turn a woman into a man. In cases of spontaneous metamorphosis, women reached perfection when their genitals popped out. Balami leaped into the same zone by singing super high and fast. Balami and Marie experienced an instantaneous gender reorientation. Their stories reflect, depending on whom you ask, either anxieties about boundaries between men and women or an early modern idealized transvestite theater of the world. Valeria Finucci explained that the Balami story "teaches us that nature moves toward perfection, that is, an incomplete man can become complete when the humors in his body are sufficiently hot, like those in a man's body."[25] This despite the fact that he lost his job.

But Marie Germain ought to be a cautionary tale for historians, including me, who have wanted to treat the one-sex model as a truth and who are tempted to read descriptions of bodies that stand outside the gender-sex norm as the equivalent of modern anatomical fact. Marie achieved the most fame among twentieth-century scholars through Montaigne. Montaigne supposedly encountered Marie Germain on his journey from Paris to Rome and first wrote about Marie in his 1580 travel journals, revisiting the case in his 1588 essay on imagination.[26] The 1588 version goes like this.

> [Marie] Germain, of low condition, without any trade or position, who was a girl up to the age of twenty-two, [was] seen and known by all the inhabitants of the town and noticed because she had a little more hair about her chin than the other girls; and they called her Bearded Mary. One day when she made an effort in jumping, her virile instruments came out, and Cardinal de Lenoncourt, then bishop of Châlons, gave her the name Germain. Germain had not married, however; he has a big, very thick beard.

The key for me is the next sentence:

> *We were not able to see him because he was in the village.* In this town there is still a song commonly in the girls' mouths, in which they warn one another not to stretch their legs too wide for fear of becoming males, like Marie Germain. They say that Ambroise Paré has put this story into his book on surgery. The story is very certain and was attested to Monsieur de Montaigne by the most eminent officials of the town.[27]

Even the 1580 travel log clearly stated that the traveler did not actually see Marie: "Passing through Vitry-le-François, I *might have seen* a man whom the bishop of Soissons had named Germain at confirmation."[28]

The resonances between Balami and Marie Germain lie in the representations of their sexual transformation and their existence as marvels described by travelers to faraway places who may or may not have seen said marvel. Importantly, and sometimes elided by modern scholars, Montaigne said explicitly that he and his companions did not actually see Marie. Paré's account of Marie appeared in a chapter called "Memorable stories about Women who Degenerated into Men." The verb *degenerate* is complicated here; in effect women transform to a more ideal body type, but they emerge in a form that is against their nature. Paré wrote with the gusto of a novelist and emphasized

Marie's pain, bursting stomach, and the amazed parents; and he wrote about irregularities of nature. He explained that Marie's stunned parents went to physicians and surgeons to confirm the "metamorphosis" and followed these exams with a new baptism and name: a rebirth.

Marie Germain was a wonder whose flesh and blood self is perhaps less relevant than the surrounding discourse.[29] Paré's account of Marie Germain is in a section on hermaphrodites, which came on the heels of a description of the Ravenna monster, sex between women, and then women who degenerated into men. He drew his readers into a voyeuristic examination of the body parts "la verge virile" and "les testins." He used sexualized and explicit body parts, not innuendo, as well as scientific terms. Montaigne in turn used the story to show the perfection of men: "We therefore never find in any true story that any man ever became a woman, because Nature tends always toward what is most perfect and not, on the contrary, to perform in such a way that what is perfect should become imperfect."[30]

## BEWARE THE ONE-SEX BODY AND THE PERILS OF INTERDISCIPLINARITY

Musicologists, including me, have been drawn to Laqueur's ideas of the one-sex body and to Stephen Greenblatt's connected work on the concept of erotic friction.[31] In an effort to historicize the body and pleasure, Greenblatt read Shakespeare's *Twelfth Night* against legal and medical texts that elucidated early modern understandings of conception. For conception to occur, erotic heat generated by chafing had to arouse and thus heat up both man and woman until they reached their boiling point—the ejaculation of orgasm (both parties ejaculated). Since female bodies naturally maintained a lower temperature than their male counterparts, orgasm and ejaculation required a hotter furnace for women than it did for men. Reading this model of sexual chafing in conjunction with Shakespearean drama, Greenblatt argued that "Shakespeare realized that if sexual chafing could not be presented literally on stage, it could be represented figuratively; friction could be fictionalized, chafing chastened and made fit for the stage by transforming it into witty erotically-charged sparrings."[32] This was a promising model for musicologists. In 1989, for instance, Susan McClary suggested that Greenblatt's depiction of friction shed light on the trio texture that was so popular in the seventeenth century.

Roger Freitas, who provides one of the most nuanced and productive forays into castrato sexuality in life and on stage, is—as he says—deeply indebted to Laqueur, whose vertical hierarchal continuum allows the castrato to occupy a middle ground. The castrato, like the child, does not have the vital heat of a full-fledged man, and the implied effeminacy does not have to mean homosocial or homosexual.[33] My first book, *Monteverdi's Unruly Women,* which began as a conference paper in the mid-1990s, relied heavily on Laqueur and Greenblatt to argue for the subversive potential of the female voice in performances of Monteverdi. Naomi André's work on nineteenth-century singers attributes the decline of the male heroic castrato to a new corporeal regime defined by Laqueur as modern: "With the development of Laqueur's modern sensibility, male and female were no longer connected; they represented opposites."[34] Freya Jarman, writing about the tenor-soprano narrative, makes the case for a shift after the eighteenth century: "Opera in the nineteenth century thus became fundamentally concerned with inscribing the two-sex model and its attendant ideological system."[35] For her, the two-sex system is a model for a narrative framework, not just a medical justification.

The one-sex model and the potential for erotic friction offered a way to understand music and desire in the early modern period. Because this model does not depend on modern gender binaries, it seemed like a more authentic way to hear eroticism in early modern music and a conceptual way out of Freudian castration anxiety. There is a problem, though, with relying too heavily on the one-sex model. To accept it as truth is analogous to a historian of science crafting an argument around the idea that Monteverdi invented opera or that tonality emerged fully formed. As historians of science have argued, writings on sexual difference were no more or less simple and monolithic in the sixteenth and seventeenth centuries than they are today. Taking Laqueur as axiomatic erases the nuance and complexity of discourses of sex before the modern era.[36]

Musing on the tensions in medical knowledge and practice, Nancy Siraisi writes that "some of us who teach in universities today may recognize ourselves in the delays, the self-seeking, the divisional rivalries, the proliferation of professional opinion."[37] Discussions of sex, sexual difference, and gender were more dissonant than consonant and were mobilized to uphold hierarchies and were rooted in contingent categories of natural and unnatural. Furthermore, ways of knowing reflected not just natural philosophy, anatomy, and surgery but also literature, myth, and theology. In other words, the art and science of the human body, like the art and science of musical practice, moved in and out of practical and learned spaces. The Latin learned tradition associated with humanism mattered for sure, but it was not the only truth.

Katharine Park, Lorraine Daston, and others have demonstrated that Laqueur took his argument almost completely from Galen's "On the Usefulness of the Parts of the Body," which circulated in the Western world around

1500. Laqueur, in effect, created a mash-up of contradictory ancient texts, including Aristotle on women as imperfect men and Galen on women as having inverted penises. Park and Daston, writing on hermaphrodites, highlight the differences between Galenic and Aristotelian ideas about sex and gender. Throughout the Latin Middle Ages and early modern period, most authors mined ancient authorities for what they wanted.

In 1600 Lucrezia Marinella, the Venetian thinker who published deliberately polemic texts on the nature of women, made an argument that was exactly the opposite of what Laqueur posits as universal.[38] In her defense of women, she asserted that temperature made women superior to men. Marinella disputes the Aristotle that Laqueur posted as truth. Refuting the idea of women as imperfect, she argued that Aristotle's ideas came from envy of his wife and rage against women all told. She refuted the idea of women as imperfect:

> Some others say, as did the good Aristotle, that women are less hot than men, and therefore are more imperfect, and less noble, than them: what unimpeachable and all-powerful reason! I believe now that Aristotle did not consider with a mature mind the operations of heat, what it means to be more or less hot, and what good and ill effects derive from this [heat]. If he had considered properly how many very bad actions [men's] heat (which is greater than women's) produces, he would not have said a word about it. But the naughty fellow went along blindly, and so committed a thousand errors. There is no doubt, as Plutarch wrote, that heat is an instrument of the soul; but it can be good, or [it can be] ill-suited to its tasks, as one must seek in it a sort of compromise between too much and too little. Too little and inadequate [heat], as in old men, is the least capable of action. Too much and excessive [heat] makes people hasty and wild. Therefore not every [quantity] of heat is good and acts to serve the tasks of the soul, as Marsilio Ficino says. But [it is] good in a certain amount and in a suitable proportion, as that of woman. Therefore Aristotle's explanation that men are nobler than Women because hotter is invalid.[39]

She says that women are superior to men because of their temperament. Marguerite Deslauriers has argued that Marinella's defense matters in part because it shows the fungibility of early modern interpretations of ancient sources and the complicated relationship between political and bodily hierarchies.[40]

To use Laqueur to talk about castrati or, more broadly, musical practice in the sixteenth and seventeenth centuries also presents a temporal problem. He sees a vacuum of knowledge between the ancient world and around 1600. It is true that, as I have been arguing, ancient texts remained deeply influential, but interpretations changed. Joan Cadden articulated the complexity of understandings of sex in Latin Europe that resulted from, among other things, the twelfth-century translations of Arabic texts. Ibn Sīnā's (Avicenna's) *Canon of Medicine* was first translated around 1200. Debates around Islamic medicine revealed tensions between investments in Greek medicine and new discoveries.[41] Just to take one example in the thirteenth century, Albertus Magnus wrote extensively on sexual difference. He was steeped in Aristotelian scholasticism but inflected his writings with his own political and moral convictions and conveniently didn't quote passages that did not enhance his arguments.

Moreover, historians of science stress that anatomical discoveries, especially in the realm of gynecology, of the late sixteenth century called into question a wide range of received assumptions. By 1493 Leonardo was interested in reproduction and in the inside of the female body.[42] The sixteenth century saw not just translations of crucial Latin and Greek texts but the emergence of anatomical practices rooted in sensory experience.[43] According to Nancy Siraisi, over sixty translations of Ibn Sīnā were published between 1500 and 1675.

In the late sixteenth century gender was contested. Universities, churches, courts, and artisan guilds were hotbeds of debate about categories we now call sex, gender, and sexuality, and what it meant to be male and female was far from clear. Meanwhile, Marco Fabio Calvo published the first Latin translations of Hippocrates in Rome in 1525. They included "On Sterile Women" and "The Nature of Women," both of which challenged the then-reigning Galenic ideas.[44] The works of thinkers like Ulisse Aldrovandi, Gabriele Falloppio, and Realdo Colombo reveal the ways that new translations of Hippocrates challenged Galen and Aristotle.[45] Vesalius and Charles Estienne presented new anatomies of testicles.[46] Mandated special treatment of women's bodies, and the so-called discoveries of the clitoris and fallopian tubes in the 1560s, worked against any kind of neat homology between male and female bodies. Realdo Colombo's "discovery" of the clitoris is a case in point. Vesalius, of the inverted penis illustration, never drew a clitoris and argued that only hermaphrodites had them. After Falloppio published a commentary on Vesalius, Vesalius responded in kind. His answer to the clitoris problem was that

> it is unreasonable to blame others for incompetence on the basis of some trick of nature and you cannot ascribe this new and useless part, as if it were an organ, to healthy women. I believe that a structure like this appears in hermaphrodites as Paul of Aegina describes, but I have never once seen in any woman a penis (which Avicenna called *albathara* and the Greeks called an enlarged *nympha* and classed as an illness) or even the rudiments of a tiny phallus.[47]

In short, there was no universal agreement on the nature of the so-called imperfection of women and the teleological materialization of masculinity.

## IT'S ALL ABOUT THE TESTICLES

Castrato bodies troubled received understandings about gender in the sixteenth and seventeenth centuries. Their bodies didn't quite fit into a system of power and capital rooted in patrilineal descent, but as Martha Feldman has persuasively argued, many castrati occupied places very close to absolutist rulers. For ancient Rome, Shaun Tougher makes the case that the position of eunuchs was more about sex and power than about gender:

> Castrates could signify lack in ways that highlighted the fertility or fecundity of emperors and could indicate the literally transformative power of the ruler to change (male) bodies into something else. On these terms, castrates came to symbolize royalty, and once associated with courts and rulers, were difficult to dislodge.[48]

In Catholic Europe these issues of fertility came up aggressively in legal battles over marriage that centered on what semen could and could not do and on what the testicles did.[49] Legal and moral questions around marriage related to power and authority and to what counted as male and what counted as female. Questions of castrato impotence, in other words, were vectors to much larger political, religious, and cultural worries about generation, marriage, and gender. In the thirteenth century, ecclesiastic courts could end marriages if one partner was deemed impotent. Leah DeVun stresses that medieval jurisprudence was reinvigorated by the rediscovery of Roman law and that by the early fourteenth century, male medical practitioners testified frequently in ecclesiastical court cases about impotence and marriage.[50]

A 1586 letter, discovered by Richard Sherr, involves the recruitment of castrati by Guglielmo Gonzaga for the court of Mantua, where Cardinal Ippolito Olivo Gonzaga ran into trouble over a castrato's marriage:

> I spoke immediately with the little castrato who told me that at present he cannot make any plans for himself because of a great trouble he has at the moment which is this; having taken a wife a few month ago, and having slept together for some time, having gotten permission to do so from a parish priest, who is now in prison because of this and is being prosecuted by the pope—and he also is being prosecuted, it being said that he could not take a wife being a castrato—it

> appears that until this negotiation is finished he cannot make any resolution about himself or promise himself to anyone.[51]

Marriage and virility were especially vexed in Mantua. In 1585 Vincenzo Gonzaga's marriage to the thirteen-year-old Margherita Farnese was annulled because of the bride's "unbreachable gate." Vincenzo was accused of natural and functional impotence and had to prove that he was capable of sexual relations with a virgin. His member was examined repeatedly to assess its functionality. In 1608 when Vincenzo was producing the famous wedding festivities that elicited Monteverdi's *L'Arianna*, he was also sending explorers to the new world in search of what can best be described as natural Viagra.[52]

The exact functioning of the testicles had been a matter of debate for centuries. And early modern writers did indeed draw heavily especially on Aristotle and Galen in their controversies. Aristotle thought that the testicles played no role in semen production. Instead, blood that spent too long in the testicles turned into semen. Galen, on the other hand, understood the testicles to be key in the generation of what sixteenth- and seventeenth-century commentators called *verum semen*, which meant fertile semen. Aristotle gave a particularly materialist description of the role of semen, one that turned semen into a tool set in motion by a natural process. For Aristotle, semen was effectively both matter and form. Semen, like air, carried the force of life; it was both an immaterial thing and a transmitter of material things.

> Nothing passes from the carpenter into the pieces of timber, which are his material, and there is no part of the art of carpentry present in the object which is being fashioned; it is the shape and the form [*morphe kai eidos*] which pass from the carpenter, and they come into being by means of the movement of the material [*hyle*]. It is his soul, wherein is the "form," and his knowledge, which causes his hands (or some other part of his body) to move in a particular way (different ways for different products, and always the same way for any one product); his hands move the tools and his tools move the material. In a similar way to this, Nature [*physis*] acting in the male of semen-emitting animals, uses the semen as a tool, as something that has movement in actuality; just as when objects are being produced by any art the tools are in movement, because the movement belongs to art is, in a way, situated in them.[53]

Aristotle's materialist sense of semen was mobilized in premodern discussions of sex difference.[54] He also located the soul in the semen, hence the idea that masturbation is a sin; in effect, you give away your soul.

The politicization of reproductive discussions is clear in the writings of

Juan Huarte de San Juan's *Examen de ingenios*, which was first published in 1575. Music scholars, including me, tend to cite his passages on temperament, intelligence, and inverted genitals. Huarte wrote from a Galenist point of view, and his descriptions of genital reproduction seem to endorse the kind of one-sex picture that Laqueur via Galen painted. But Huarte's medical-political rhetoric reveals a particularly early modern formation of anxieties around masculinity and power, not just an objective description of reproduction. He uses the inward genitals of the woman to prove that men are the goal of nature's perfection.

> And in this way it is true that if Nature had finished making a perfect man, and she wanted to transform him into a woman, she had no other job but to turn the instruments of generation inwards; and, if she was made as a woman, and she wanted to turn her into a man, by throwing her uterus and testicles outside, there would be nothing more to do.[55]

The Spanish text uses the words *útero y los testículos*, which translate to "uterus" and "testicles." Huarte went on to say that if such a transformation occurred in utero, the man would be inclined to have female traits later in life.

For Huarte the changes in body and voice that accompanied testicle alteration stemmed from an alteration in temperament; castrated men turned cold and moist, a natural state for women but the result of an external process for men. Huarte's humoral vocabulary presents an especially mechanical process as if the humors were gears and castration was a process that changed their arrangement. Mechanical changes also directly affected the more ephemeral categories of imagination and intelligence:

> How much harm it does to man to deprive him of these parts, although small, will not need many reasons to prove it. We can see from experience that then his hair and beard fall out; and his thick and big voice becomes thin; and with this he loses his strength and natural warmth, and is left in a worse condition and more miserable than if he were a woman.[56]

Huarte located imagination and temperament in the testicles.

> But it is important to note that, if before the man was castrated [*capasen*], he had a lot of ingenuity and skill, after cutting the testicles he loses it as if he had received an injury in the brain itself. This obviously proves that the testicles give and take away the temperament to all parts of the body. If it does not, let us consider, as I have done many times, that out of a thousand castrated men that

> are attracted to letters, none comes the skill; and in music, which is their ordinary profession, it is clear that they are very crude; and this is because music is the work of the imaginative, and this power demands a lot of heat, and they are cold and humid. Then it is true that by ingenuity and skill we can tell the temperament of the testicles. And, therefore, the man who is keen in the works of the imaginative will have heat and dryness of the third degree; and if man does not know much, it is a sign that with the heat humidity has been added, which always spoils the reason; and this confirms that if you have more heat you have a lot of memory.[57]

Note that he used the word *capasen*, which is usually translated as "castrated." And he uses the word *capones*, which translates to "capons" but was the word usually used for *castrato*. It usually referred to castrated singers and in this case was linked directly to the profession of singing castrato.

But Huarte's book is no objective medical text. It was deeply political, motivated by a nationalist urge to create a more robust Spain and tied up in the climate politics that I discussed in the previous chapter. In the guise of collecting ailments explaining psychology, Huarte proposed a method so that every man found the perfect occupation, which in effect categorized men by their intelligence. He linked complexion to the Protestant dominance of the north and the Catholic dominance of the South. The Spanish had a better sense of religious feeling but poor memories; northern Europeans had good memories but lacked spirit.

The book begins with a dedication to King Philip II and a commitment to building a stronger Spain.

> So that the works of the artisans would have the perfection that suited the use of the republic, it seemed to me, Catholic Royal Majesty, that a law must be established: that the carpenter should not do work related to the trade of the farmer, nor the weaver the work of the architect, the lawyer should not cure, the doctor should not be an advocate; rather, each one should exercise only that art for which he had a natural talent, and leave the others alone. Because, considering how short and limited the ingenuity of man is for one thing and no more, I always understood that no one could know two arts perfectly without failing at one. And, because he cannot succeed in choosing the one that is better for his nature, there must be deputies in the republic, men of great prudence and knowledge, who would discover the ingenuity of each one at a tender age, forcing him to study the science that he was taught. It is convenient to not leave individual choice. From this would result in your estates and dominions losing the possession of the greatest craftsmen in the world and the most perfected works; this happens by joining art with nature.[58]

One of the most widely circulated early modern Spanish books of natural philosophy, it was translated into Italian, French, English, Dutch, German, and Latin, and it saw over eighty-two reprints during the seventeenth and eighteenth centuries. Huarte was a one-hit wonder; he published only this, and in 1583 the book landed on the Index of Forbidden Books. Huarte revised the book to accommodate the Inquisition, and it came out after he died in 1594. Huarte understood wit and ingenuity as tied directly to climate and origin. He used climate to argue that Spanish men are better at learning languages and insisted that English and Dutch men resemble drunkards.[59] And he tied the worth of men to vital heat as well as the most innate human characteristic. Even "verse and song" arise from different climates.[60]

Meanwhile the English translation inflected the book with its own climate politics. The English translator aimed to prove that English was a more perfect language. He cut passages that leveled criticism of people from northern climates.[61] The book in Spanish is called *Examen de ingenios*, and Carew's 1582 translation was titled *The Examination of Men's Wits, in Which by Discovering the Varietie of Natures is showed for what Profession Each One Is Apt and How Far He Shall Profit Therein*. The word *ingenious* is not straightforward. It is translated in English as "wit" and in Italian as *ingegno*, which better translates to "ingenuity." Huarte himself links the word to the power of generation. Carew worked from the 1582 translation by Camillo Camilli. Huarte did not even need the word *men* in his title, as the assumption is that only men have wit; indeed, he explained that parents wanting wise children should try to produce males. In 1698 Edward Bellamy did a translation called *The Tryal of Wits* that was based on the edition modified after the book was censored by the Inquisition. The book was on two indexes of forbidden books.

### *CUM FREQUENTER*

While busily banning books the Papal court also tried hard to regulate castrati marriages. On June 12, 1587, Pope Sixtus V issued the papal bull *Cum frequenter*.[62] Sixtus V made the marriage of castrated men a violation of canon law and ordered that marriages involving castrated men be dissolved. The bull is frequently referenced in studies of castrati and it arguably dictated the Catholic Church's concept of marriage, impotence, and infertility for four hundred years. Like Huarte's treatise, or the Supreme Court in 2022, it is in the end more about legitimizing masculinity and politicizing marriage than it is about a sex act or a body part. Sixtus wrote in response to Cesare Speciano, bishop of Novara, the Spanish nuncio. The nuncio sent two letters about the problem of the "numero infinito" of marriages of castrated men and had written about it the year prior. This bull is mentioned in countless discussions of castrati and was used as late as the twentieth century to render the marriage of someone with a vasectomy sinful.[63] Sixtus V often appears in musicological literature because, in 1588, he created the Congregation of Rites, a group that legislated ecclesiastical practices including singing in church.

It's worth repeating that the brief was exactly contemporaneous with Huarte's text and that it was issued during a tense period in Spanish-Papal relations. Spain was ruled by Philip II, who was also the king of Portugal, Sicily, and Naples. Philip and Sixtus did not work well together. Sixtus was mad at the Spanish Armada and its leaders, whom he thought were incompetent. Consequently, he frequently threatened the ambassador to Rome, Enrique de Guzmán Oliveras, who served from 1581 to 1592, with expulsion. Meanwhile there was a raging conflict between Jesuits and mendicant theologians. Sixtus is often euphemistically called an austere pope. In Rome he imposed heavy taxes, locked up vagabonds, and had no problem with capital punishment. He cracked down on women, Jews, and infidels and tightened his grip on papal outposts. When he died, a mere seven years into his brutal reign, a crowd of two thousand nobles and commoners stormed the capital demanding the statue of Sixtus so they could cut off his head.[64]

The bull itself made potency, virility, and marriage dependent on *verum semen*, something that eunuchs, castrates, and spadones were understood not to produce. It brings into sharp focus the castrato as a failed man and writes them out of the community. They are both there and not there; written into a gender binary world only to exclude them. Sixtus did not center the male-female continuum but focused on eligibility for marriage and, by extension, patrilineal descent and authority. The bull further provided a mode of criminalizing sex without generation. Canon law had previously understood a marriage as consummated if semen made it into the vagina. Sixtus changed it so that the man had to also produce viable semen. The brief essentially held until May 13, 1977. The castrato, in other words, forced the church's hand on matters of impotence.[65]

As was conventional, Sixtus began by restating the question that the Spanish nuncio asked. He asserted that eunuchs who lacked both testicles and who were thus "fridigae naturae et impotentes" (of frigid nature and impotent) could not contract a valid marriage. This assertion already set him up as a Galenist.

> Since frequently in these regions there are certain eunuchs and spadones [*eunuchi et spadones*], who lack both testicles, and are thus certainly and manifestly not able to emit true semen [*verum semen*]; because they as a result of impure lust for flesh and with unclean embraces have relations with women, and emit

> a certain liquid that is similar to semen, although not by any means useful/suitable for generation or the consummation of marriage, and are presuming to contract marriage with women who are very much aware of that very defect of theirs, and stubbornly persist that they act licitly, and various trials and controversies concerning this matter are being brought before your and other ecclesiastical courts, your fraternity has asked us what should be established pertaining to marriages of this kind should be made.[66]

The pope ruled that those who were "frigid and impotent" could not marry. Beyond the content of semen, he revealed a deep concern for the validity of already-existing marriages:

> We thus take into account that, according to canonical sanctions and nature's rationale, those who are of frigid nature and impotent are not in any way considered fit for contracting marriage, and that the previously mentioned eunuchs, or Spadones, do not wish to have as sisters those whom they cannot have as wives, because experience demonstrates that such men as long as they insist that they are capable of intercourse, likewise the women who marry them, under the illusion and structure of marriage, strive after/aspire to sinful relations of this kind, such that they cannot live virtuously but that they are united carnally to each other with a depraved and libidinous intention; since such unions cause opportunities for scandal and sin and cause damnation of souls, they must be completely exterminated from the Church of God.[67]

The pope took two positions. First, he forbade the marriage of eunuchs, and second, he rendered both parties involved in these marriages sinners and condemned their souls to damnation. The sin was false pretense: feigning capability of procreation while being incapable not only of procreation but of consummation. This made marrying for pleasure a sin. His final paragraph reemphasized procreation as the purpose of Catholic marriage.

> And most importantly/above all considering that there is no utility to the marriages of eunuchs and Spadones of this kind, but rather enticements of temptation and incentives for lust arise, we likewise entrust this business to your Fraternity with these present documents document and give you the power to prohibit the contracting of marriage by those mentioned and by any other eunuchs or spaded men, lacking both testes, with any women, whether aware or unaware of the aforementioned defect, and to declare them, by our authority, unfit for contracting marriage in any way, and to forbid local Ordinaries from allowing unions of this kind to happen henceforth in any way, and to make sure

> that those who have already entered unions of this kind are separated, and to declare such marriages already contracted as null, void, and not valid.[68]

Sixtus implied an essential lascivious and sinful nature in eunuchs and the women who married them.

Medical practitioners were increasingly asked to testify in courts in order to prove masculinity or femininity. Toward the turn of the seventeenth century, experts in canon law took up the question of sanctioned marriage, and for the most part they followed Sixtus's investment in semen. Tomás Sánchez, a Spanish Jesuit legal expert who wrote on the moral and legal aspects of marriage, glossed the brief by arguing that even if eunuchs emitted watery semen it was not "true."[69]

Paolo Zacchia added medical evidence to Sánchez's legal theories. His seventeenth-century commentaries on *Cum frequenter* are often cited by scholars interested in castrati.[70] He argued that marriage required virile semen. A product of the Jesuit Collegio Romano and the Sapienza, he worked for both the Barberini and the Pamphili pope as a physician. In seventeenth-century Rome, medical practitioners and theorists played important roles in secular and papal courts, especially around the causes of death and paternity. Physicians were effectively custodians of the papal body and important text makers in legal matters involving the body. They were also frequently consulted on miracles.

Between 1621 and 1651, Zacchia published a nine-volume magnum opus *Quaestiones medico-legales*. The book established legal medicine as a scholarly field and powerful regulatory instrument. Following scholarly norms of his day, Zacchia cited authoritative texts from Ovid and Lucretius to contemporary thinkers.[71] The book, an exemplar of the intersection between early modern medicine and Counter-Reformation theology, is a series of case studies. Zacchia outlined medical problems in terms of legal questions. The section on impotence, sexuality, and reproduction is followed by one on simulated diseases. The book concludes with a volume of eighty-five "consilia," case studies used for teaching. Topics include things like the soul of the human fetus, divorce, chocolate, and diseases that prevent communion. The cases are diverse; a woman who gave birth to monsters twice and was accused of bestiality, and a murder that was disguised as plague.[72] He also compiled one hundred deliberations and decisions made between 1569 and 1657 by the Roman Rota, the highest appeals court of the Catholic Church. It's an overwhelming list in print (Fig. 8.4). He discussed, for example, the birth of a cyclops in Sicily as a provocation for the question of whether the semen of humans and animals could reproduce. Zacchia addressed other questions around reproduction,

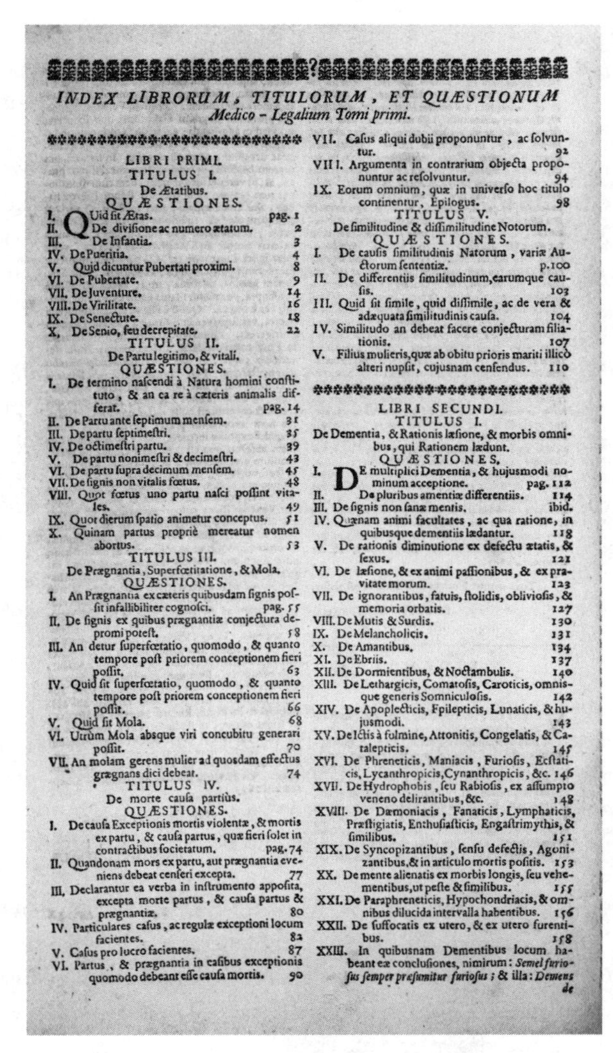

*INDEX LIBRORUM, TITULORUM, ET QUÆSTIONUM*
*Medico – Legalium Tomi primi.*



FIG. 8.4 Paulo Zacchia, index page of *Quaestiones medico-legales* (Lugdunum, 1726). Bayerische Staatsbibliothek München.

including the case of a woman who apparently gave birth to a dog, which for him related to the same human-animal reproduction question as the cyclops.

Zacchia claimed not to establish laws but to prove medical truths, and he devoted extensive attention to sex, marriage, adultery, birth, and pregnancy, with an emphasis on varieties of sexual impotence.[73] Legal questions tended to involve capital and marriage; paternity, for instance, determined inheritance. He wrote on an encyclopedic array of legal issues, relating to sex, reproduction, and sexuality including hermaphrodites, eunuchs, pregnancy, rape, and abortion.[74] He offered extensive discussion of the legal status of hermaphrodites; the short answer was that they broke laws if they went against their dominant nature. And he weighed in on cases that aimed to determine whether a woman had lost her baby by natural causes or by abortion.

For Zacchia, a potent—and thus eligible for marriage—man had to have an erect penis, put it inside a female vessel, and produce semen. The issue of marital impotence came up in legal situations frequently in the early modern period, not just around castrati. Zacchia came out much more as a Galenist and saw the testicles as key to *verum semen*. To him, semen came from the blood, and the testicles transmitted it: "Testicles had the distinct purpose of transmitting perfect semen [*perfectissimam seminis*]."[75] Testicles also did the work of making a vigorous sexual response, which he called "spiritus turgens." The debate he summarized was around exactly what the semen had to do to be marriage-worthy. It turns out that there was quite a bit more to the business of ejaculation than is apparent from Laqueur's model.

## MAGIC

And now for a pivot. I could keep spinning arguments around the castrato voice by excavating the medical, liturgical, political, and musical discourses around the castrato and his body. Regardless of testicles, semen, one-sex, two-sex, or hidden bodies, what the castrato voice did was to put into air words, sounds, and timbres of love, spirituality, and power, and it did this in sacred and secular spaces. No matter how you slice it, the voice was a vehicle for transformation, and commentators from antiquity until now have tried to control it. There is something magically ephemeral about the voice, and there was a powerfully erotic potential to it that reverberated through secular and sacred performances.[76] The documents discussed here made futile attempts to regulate pleasure, desire, and affect.

In hunting for the truth of sex and bodies, it is easy to forget that the mess of voice, desire, conversation, and ecstasy makes any one objective truth impossible. Music, sound, and voice are about pleasure; that's partially why they were so contested in the sixteenth and seventeenth centuries. Susan McClary has shown again and again the ways in which pleasure derived from sound lies at the center of the emergence of European modernity and of the tonal vocabularies that began to crystallize around 1600. Musical desire, for her, plays out sexually but also spiritually in the Catholic mystical drives for divine union.[77] Suzanne Cusick has also revealed the swirl of desire in early modern discourses many times over from the undertone of desire in debates around the seconda prattica to the *eros* of singing the close dissonances of baroque trio textures. For her, early modern elites "constructed music as an acceptable medium for sublimating sexual energies of all kinds."[78]

My questions do not center on whether the castrato was trans, intersex, third sex, or queer, nor do I wish to make claims about their identities. They were, and they were not; the categories are anachronistic but also heuristically useful. Kathryn M. Ringrose suggests that Byzantine castrati functioned as a third gender, as perfect servants. By the twelfth century, they were integrated into Byzantine society. In Ringrose's reading, eunuchs retained boyish looks, a penchant for servitude, and a lack of developed male aggression.[79] Katherine Crawford argues that castrati aligned in some ways with the medicalization of transgender and intersex bodies, given the removal of otherwise "healthy tissue" that also entailed, in cis-normative terms, a loss of masculinity.[80] The castration procedure was a disabling, involuntary transgendering; demolishing male testicles made castrati more like women: "Medically, then, castrates were transsexuals in that surgical intervention altered their bodies and changed its fundamental orientation in the world from male to (partly) female."[81] Emily Wilbourne posits a kind of queerness:

> The voluntary castrato is quintessentially queer, for he slips between the cracks of heteronormativity: he modifies his body to suit his desire, and he abdicates the procreative responsibilities of the heterosexual. At the heart of this identity is a phenomenal, superhuman voice—and this marks a crucial factor for queer musicology.[82]

These arguments can be extended by noting the radical nature of using a surgical procedure to alter a body that was not broken and thus go against the Hippocratic Oath. The procedure used to make Italian castrati was not uncommon. But, as I will focus on in chapter 10, the decision to intervene when there was no obvious medical problem was radical. Premodern surgical practices aimed to render things normal, not to make them different. The castrato voice reached an operatic climax at precisely the moment when European thinkers began to imagine abnormal physical morphology as a disability, not a miracle, portent, or other divine marvel. Modernism rendered strange bodies as embodiments of error, not of wonder.

To conclude in theoretical terms, there is an opacity to the matrix of sex, desire, and gender within which the castrato existed and exists and which, as C. Riley Snorton articulates in *Black on Both Sides*, must exist in the tension between the past and present.[83] He means that the past is always being reenacted and encountered through the present, that there is always a process of recovery, and that the past has an aesthetic presence in the now. The castrato's body was and is a site for debates around vocal and sonic pleasure. Almost obsolete at the height of their prevalence, they served as instruments in musical performances of recovery again and again: recovery of ancient drama, recovery of saints' lives, recovery of liturgical rituals.

Vibrations of desire resound through the bodies of castrati, an overtone of what Carla Freccero describes as a haunting effect of the past; the past is not quite in the present, but the traces resound.[84] Such an asynchrony works especially well for castrati. This is what Valerie Traub calls thinking sex: embracing the opacity of sex and considering sex itself as an epistemological problem.[85] The castrato allows for a ghostly meditation on the notion of sex as a way of thinking, not a practice to be captured in knowledge. Eroticism and desire are always a set of performances and practices. This was especially the case for castrati who were, in every aspect of their lives, embedded in performative theatrical cultures: church, court, opera, salon, and academy. And they did this most potently with their voices as instruments of affective and spiritual transformation, moving between body and mind, spiritual and corporeal, psychic and material.

# OUT OF SYNCH

CHAPTER 9

# TIME TRAVEL / LIQUID ECSTATICS

In 1637, at the age of thirty-seven, Loreto Vittori was accused of kidnapping Plautilla Azzolini, the wife a Florentine painter, Francesco Borbone. Banished from Rome, the singer spent his two-year exile in his hometown of Spoleto. In 1640 he was allowed to sing for the pope again. As the story goes, he threw himself at the pontiff's feet, Mary Magdalene style, "begging his pardon for the public error that he had committed now two years ago; our lord heard him willingly and then gave him absolution, for everyone heard the priest's dismissal after confession: Go in peace; sin no more."[1] Three years later in a ceremony at the Chiesa Nuova, he was ordained a priest with music by "150 singers, divided into six choirs, [singing] music of the best Roman choirmasters of the time, such as Stefano Fabbri, Virgilio Mazzocchi, Francesco Foggia and Giacomo Carissimi."[2]

The most detailed accounts of the incident appear in the report of Governor Giovanni Battista Spada and the Roman satirist Gian Vittorio Rossi, also known as Giano Nicio Eritreo and Janus Nicius Erythraeus. It is presented as a "letter" from "Olertus" (Vittori) about his "fall from grace," which explained that the valorous castrato merely meant to help the young woman and her mother.[3] According to Rossi a.k.a. Eritreo's account in February of 1645, Vittori gave a performance of Domenico Mazzocchi's *Lagrime amare*.

> Once in a while in winter, he could be heard in Rome in the chapel of the Fathers of the Congregazione dell'Oratorio S. Maria in Vallicella, where I heard him sing one evening a lament of the Magdalene, who weeps for her crimes/sins and throws herself at the feet of Christ: he who in our eyes was endowing the Magdalene with such ardor of inspiration, with such force of voice, with such soft and delicate inflections in his song, that if she had lived again, she would have perfectly recognized and admired the true suffering and grief in that imitation of his repentance.[4]

The ecstatic account reveals a deep investment in the affective experience of Vittori's voice. Captivated listeners were freed from their own sin; tears welled up in their own eyes when singer reproduced the groans of the Magdalene. Vittori manipulated his own voice and thus the minds and bodies of his listeners: "He was showing that he could twist and bend it like the softest wax in whichever way he wanted."[5] His voice could do anything: "(he could make) a graceful and splendid voice, and one most suited to bringing about any variation and modulation, such that in reliance of chord it can react to any solicitation, whether acute, grave, fast, slow, strong, [or] weak."[6]

Athanasius Kircher riffed on Rossi and positioned Mazzocchi's lament as the quintessential example of the wonders of a kind of music that he called the metabolic style. No one seems to doubt that Vittori performed something on that cold February day or that he did so with state-of-the-art melodic sighs, jarring chromatic inflections, and disorienting rhythmic variation that seventeenth-century elite listeners were programmed to hear as spirit and body altering. Whether Vittori sang Mazzocchi's lament remains debatable, but it's plausible.[7] One of the most gifted singers of the day, he was certainly equipped to perform a kind of music meant to entice listeners to real tears and ecstasy.

The performance that Rossi described occurred in Santa Maria in Vallicella during Lent, a season in which listeners were primed for spiritual ecstasy or, perhaps, sacred eroticism. Roman churches were theatrical spaces, and Santa Maria in Vallicella, now called Chiesa Nuova, is an important site for the history of sound, music, and the oratorio.[8] Located less than a quarter mile from Piazza Navona, the building was designed to maximize acoustic potential and was one of the first churches to have two organs. By 1600, spiritual plays and massive polychoral extravaganzas that enveloped listeners in surround sound occurred there frequently. That same year Girolamo Rossini, the first Italian castrato to be employed by the Sistine Chapel, was installed as the maestro di cappella.

The figure of Loreto Vittori, the penitent performance, and Mazzocchi's lament of the Magdalene create a mess of desire and a convergence of the spiritual and the erotic. Centering in part on a performance whose details remain fuzzy is deliberate. The castrato voice could entice an ecstasy that superseded imaginings of the body as monstrous, deviant, abnormal. Perhaps this is what Elisabeth Le Guin calls carnal musicology, an acknowledgment that—even in an obstinately and ostinato-like, historically driven project—there must still be place for pleasure, pain, sonic sensuality, and imaginative leaps.[9] As for the sounds, the music, it's time for a sonic imaginary: think whatever screams carnal spirituality, surround sound, ecstatic listening. A double choir in a Roman church was enveloping; a solo voice was piercing. For me it might be the choral singing in the Brahms requiem, or it might be conjuring up a Beyoncé and Jay-Z concert I saw in Rome with my then fourteen-year-olds, a dizzying sonic and visual technological spectacle that climaxed in an acoustic duo.

Swirling around one body and its residue allows for an exploration of ecstatic listening and desire that deliberately eschews sexual practice, identity, and the matrix of discursive systems enumerated in the previous chapter. Moving between the castrato's body and the absence of the castrato's body purposefully implies that it might not be wrong to consider the role of fantasy of the contemporaneous and contemporary kinds. After all, it is almost impossible to know what someone you can talk to in real life thinks about desire, sex, and eroticism, so why should scholarship be limited by attempting to figure out what it all meant to people who have been dead for four hundred years?

Vittori's voice and performance exist in what Carla Freccero calls queer time: a temporal dislocation that does not center on "what really happened."[10] She builds on Carolyn Dinshaw, who, in exploring the Middle Ages, wrote,

> Queerness works by contiguity and displacement, knocking signifiers loose, ungrounding bodies, making them strange. . . . Queerness articulates not a determinate thing but a relation to existent structures of power . . . and it provokes inquiry into the ways that the "natural" has been produced by particular discursive matrices of heteronormativity.[11]

I am indebted to two theoretical interventions. First, Virginia Burrus, in *The Sex Lives of Saints*, pushes for a "sublime art of eroticism rather than a repressive morality of sexuality."[12] And second, Emily Wilbourne, describes castrati as inducing *melophilia*, a term that describes

> patrons with an excessive love for vocal music—often but not always castrato song. . . . The word is tied specifically to the melody-dominated genres of the seventeenth and eighteenth centuries, both the "new" (and newly professionalized) music of the early baroque and the virtuosic, focused musical content of the da capo aria as it flourished in operatic and chamber music from the later seventeenth century onwards.[13]

## HYPERREAL

Like many castrati, Vittori's life existed and exists as an assemblage of stories. The stories were inflected with and gave weight to the complicated debates around sex, body, and gender that preoccupied writers quoted in the last chapter. A celebrity in seventeenth-century Rome, the charismatic, successful, and talented castrato sang male and female roles in churches, theaters, and streets. Consequently, he was the subject of literary celebrations and derogations such as the Cesarini poem that calls him a new Orpheus (Fig. 9.1).

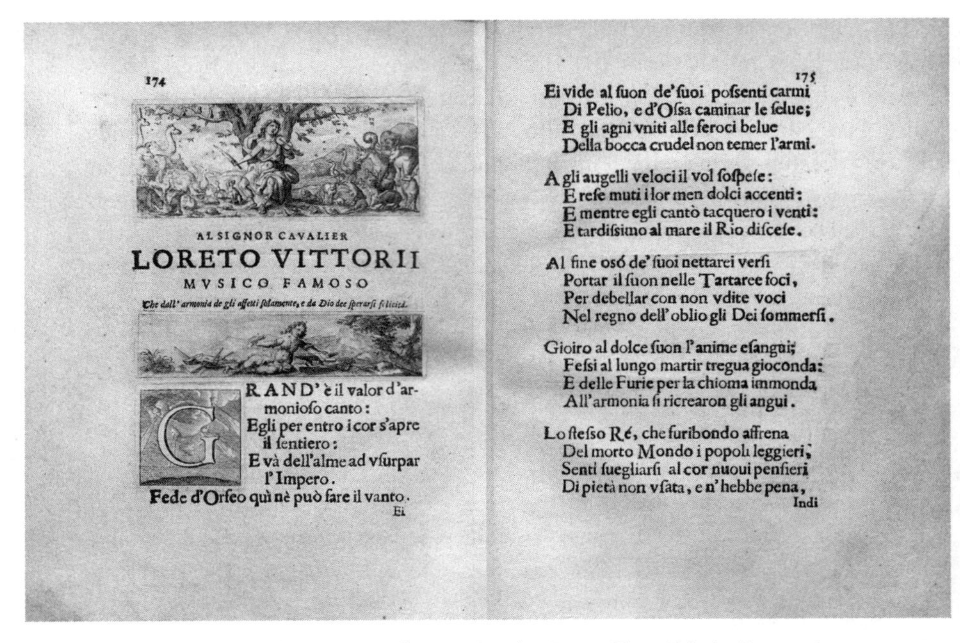

174

AL SIGNOR CAVALIER

LORETO VITTORII

MVSICO FAMOSO

*Che dall' armonia de gli affetti solamente, e da Dio dee sperarsi felicità.*

GRAND' è il valor d'ar-
moniofo canto:
Egli per entro i cor s'apre
il fentiero:
E và dell'alme ad vfurpar
l' Impero.
Fede d'Orfeo quì nè può fare il vanto.
Ei

175

Ei vide al fuon de' fuoi pofsenti carmi
Di Pelio, e d'Ofsa caminar le felue;
E gli agni vniti alle feroci belue
Della bocca crudel non temer l'armi.

A gli augelli veloci il vol fofpefe:
E refe muti i lor men dolci accenti:
E mentre egli cantò tacquero i venti:
E tardifsimo al mare il Rio difcefe.

Al fine osó de' fuoi nettarei verfi
Portar il fuon nelle Tartaree foci,
Per debellar con non vdite voci
Nel regno dell' oblio gli Dei fommerfi.

Gioiro al dolce fuon l' anime efangui;
Fefsi al lungo martir tregua gioconda:
E delle Furie per la chioma immonda
All' armonia fi ricrearon gli angui.

Lo ftefso Ré, che furibondo affrena
Del morto Mondo i popoli leggieri,
Senti fuegliarfi al cor nuoui penfieri
Di pietà non vfata, e n' hebbe pena,
Indi

FIG. 9.1 Virgilio Cesarini, *Al Signor Cavalier Loreto Vittori Musico Famoso*, in *Virginii Caesarini Carmina* (1664). Biblioteca Nazionale Centrale, Rome.

Maffeo Barberini "discovered" Vittori in 1617 when the singer was seventeen. Vittori trained in Florence and sang in the productions of Peri and Gagliano until Maffeo Barberini discovered him and brought him to Rome. Then at age seventeen, he entered the service of Cardinal Ludovico Ludovisi. When his patron died in 1622, he entered the Sistine Chapel choir. Sometime between 1622 and 1623, Urban VIII (the Barberini pope formerly known as Maffeo) made him Cavaliere della Milizia di Gesù Cristo.[14] In 1628 he returned to Florence to sing as Zephyr in Marco da Gagliano's *La flora*, composed for the wedding of Princess Margherita de' Medici and Duke Odoardo Farnese. The still young Vittori already had all the chops; the role demanded extreme virtuosity, breath control, and affective nuance. He sang a lament in act 4 that is Arianna-like in its falling sighs and irregular rests. And a duet with Clori at the beginning of act 2 starts off simply but quickly begins an escapade of melismata (Ex. 9.1).

Thanks to fragments of official records, satires, and musical scores, Vittori's life is by the standards of seventeenth-century virtuosi quite legible. And yet his life is a series of opacities and displacements, perhaps the most dramatic of which was his role in the abduction of wealthy Plautilla Azzolini, who lived on the Corso. In 1637 Plautilla's family reported her abduction to Giovanni Battista Spada, the governor of Rome, who unleashed a full-scale investigation, including interviewing witnesses. According to the report she was on the way home from a late-night mass with her mother and brothers when armed men snatched her and took her to an undisclosed location. The brothers reported the event two days later. Vittori, meanwhile, hiding in Rome and feeling his reputation deteriorate by the day, claimed he could not in good conscience remain silent and reported the event to his confessor, Father Gabrielli, who in turn told his protector/patron Antonio Barberini, who ran interference. The confessor tried to call in favors, but Spada insisted on an investigation. During that investigation Plautilla's painter husband agreed to take her back. According to Spada, the cardinal, however, wished that she would be returned "not into the hands of her husband but rather that she should be placed in a convent to ensure that she not lead a bad life, since accounts of her depraved inclination and the consent she gave before the abduction persuaded him that she was not able to live chastely."[15]

Urban was nevertheless furious and did not want the castrato in his chapel or, indeed, in the Papal States. The castrato retired to his hometown of Spoleto for a year. Not unlike sexual misconduct sanctions today, the official records in the papal chapel reveal nothing: Vittori was supposedly on a medical

EX. 9.1 Gagliano, *La flora*, act 2, mm. 65–84, Zeffiro

leave, but no records of an illness exist. Antonio Barberini, who was at the time embroiled in a scandal involving Marc'Antonio Pasqualini and Leonora Baroni, paid him the entire time.[16]

This vignette is not as far fetched as Balami's ball-dropping aria, but it is wrapped in the trappings of masculinity and in creating the equivalent of a social media profile. Valeria Finucci uses this incident as evidence for the heterosexual attachments of castrati and as an assertion of masculinist power, especially in contrast to the female roles that Vittori played in operas and the sacred contexts in which he sang.[17] To extend her argument, one could focus on the carriage as the site of abduction. Carriages were liminal sites of performance and display, in which matters of honor were hotly contested. The tension between display of wealth in carriage ornamentation and the invisibility of the interior of the carriage itself created an erotic tension. As John Hunt has argued, carriages existed within the theatrical culture and the conspicuous consumption of early modern Rome. And while they started as female modes of transit, they were, by the mid-seventeenth century, deeply entrenched in matters of honor.[18] Ancient Greek and Roman gods, infamous for their interest in mortal women, did a lot of kidnapping, and Renaissance and baroque artists replayed these scenes repeatedly, often with the main action happening in carriages (Fig. 9.2).[19]

It's almost impossible to reconstruct the purported abduction. Rome was a city of violent rivalries and street violence, and Spada's official record is itself mired in politics and may say more about the way an abduction could have occurred than about an actual event. In early modern Rome the office of governor was part of the criminal and civil justice system and often served as stop on the path to cardinalhood. Urban VIII made Spada a cardinal as soon as he finished his eighteen year term as governor in 1645. Governors tended to have a vast network of spies, which meant they knew about crimes first. The governor also adjudicated matters of honor, crimes like the abduction of a painter's wife by a famous castrato that implicated the papal court.

Reconstructing the abduction is further complicated by the fact that the most detailed account comes from Rossi/Erythraeus, who told it as a story within a story.[20] And anything from Rossi must be taken with many grains of salt, as he was deeply invested in satirizing the papal court. A product of Jesuit education at the Collegio Romano, he joined the Accademia degli Umoristi when it was founded in 1603. The academy was dedicated to humor and to the bodily humors. His *Pinacotheca Altera Imaginum, Illustrium* included three hundred short biographies of artists, including castrati; many of these biographies lampooned the papal court and the literati. It reads like a combination of Vasari and the catalogs of courtesans that flowed through Venice. The modern version would be a tabloid.[21] In the realm of satire, the castrato

FIG. 9.2 *La rencontre et combat des ambassadeurs d'Espagne et de Portugal arrivé à Romme l'an 1642*. Engraving. Paris, Bibliothèque Nationale de France.

served as a prompt for social critique, usually of marriage, sacred celibacy, papal courts, and literati.

In his version of the events, Erythraeus invited his friends to converse, and they were sad that "Olertus" was not there to amuse them with song. According to one speaker the castrato had been exiled because he kidnapped the woman he was having an adulterous relationship with and was condemned to pay her ten scudi for each month she was in captivity. Erythraeus claimed to have a letter telling the "real story," which he read aloud as part of the events. Olertus painted a picture of an abduction that was staged.

Olertus's letter explains that the whole incident started with a song:

> Last June, when I was walking around the city at night after dinner in order to avoid the heat, I, in front of the palazzo of a distinguished woman, under whose care I was, (and) out of respect for her I sung a little song. But when this [song] was completed, I was warned by the servants/attendants of the house that a certain servant had been gravely offended by my song who was living under that roof along with his wife and children and was especially eager for a good reputation, because he had a daughter of exceptional beauty in his house who had not long ago been abandoned by her husband.[22]

The servants then informed Olertus that should he sing more he would have rocks thrown at him. Already here the voice of the singer projects the potential indecency of singing to a woman, the damage even a castrato could do with his voice to the honor of a woman. Vittori himself evidently referenced the incident in his comic epic *La troia rapita* (The kidnapped pig) calling it a youthful error. Martha Feldman points out that *troia* can also be translated as "whore."[23] The poem seems modeled on Alessandro Tassoni's satirical *La secchia rapita* (The kidnapped bucket), published in 1622.[24]

In 1639 the exiled singer wrote the first version of his opera *La Galatea*. Five years later it was staged at the Prince of Cariati's palace and was first opera performed in Naples.[25] Vittori dedicated it to Cardinal Antonio Barberini to, as he put it, "bring light on my name." In Vittori's rendition, featuring the ill-fated lovers Acis and the sea nymph Galatea, music works powerfully. In a rage, the jealous Polyphemus (a cyclops) kills Acis, and Galatea makes the cyclops turn him into a river spirit. The opera is full of laments and the fire of desire. The end of act 2 is downright creepy: Cupid as a disembodied echo manipulates Galatea's speech, which makes her doubt her lover's commitment. Next there is a gruesome scene in which satyrs announce that they will chase and capture Cupid. One says, "Ché com'un tordo spennachiarlo vo'" (because I want to pluck him like a thrush). Another says, "Ch'io vo' castrarlo" (for I plan to castrate him). The tune is three verses of one nasty satyr singing a strophic song with sprightly instrumental ritornello followed by a chorus of satyrs that is built on jarring rhythmic baselines that can be sung in a "Spanish manner." This dance alternates between duple and triple meter.

## HIDDEN BEAUTY

The fascination with the erotic lives and scandals of performers is unremarkable, especially from the vantage point of the twenty-first century, where social media has made the performance of the personal ubiquitous. But in early modern Rome it mattered in some very particular ways. Thus Vittori, who may or may not have abducted the young Plautilla and who may or may not have sung as the repentant Magdalene, was part of the sacred and political soundscape of the papal court. What Vittori did on and off the stage and the pulpit intersected, and his voice was in some fundamental way an instrument of the pope.

Vittori and Plautilla were characters in a long story about sound and church and about the contested place of the erotic in sacred contexts. Throughout the sixteenth and seventeenth centuries, intertwining worries about what kinds of music were appropriate for church and about the castrato's presence in church reached an almost feverish pitch, so that by the eighteenth century the tensions were deeply embedded in the castrato's voice and body. Pietro della Valle, a Roman nobleman, traveler, musician, and thinker who wrote an essay in 1640 defending the music of his time, also wrote about the castrato voice. Della Valle described being drawn into church by the voice of the castrato. He wondered whether he was a "uomo troppo sensuale" (a man too sensual) but explained that he was most drawn to churches with excellent music and that his family's church, with its lack of musicality, was his least favorite church in Rome:

> Their song, as a matter of fact, does not attract me. In sum, I go so much more willingly where I can hear good singing, and more than once, I remember that good music has excited within me feelings of devotion, of compunction, and even of a desire for the afterlife and celestial things.[26]

The problems with women singing the kinds of music that Vittori sang have been much discussed. To some extent the female voice, not the voice of the castrato, was the exotic Other barred for the most part from Roman stages and churches; women were barred from holiness all told. The canonization of women, for instance, presented particular problems. Saint Teresa of Avila, the subject of Bernini's *Ecstasy of Saint Teresa* (discussed below), for instance, somehow transcended her gender through ecstasy. The burning fire of divine love canceled the wet and cold temperament of women.[27]

Moreover, the links between female virtuosity and sexuality are clear. As I and others have argued, thanks to early modern mouth anxiety, the direct link of these activities to sexual reproduction turned singing into an erotic activity that compromised the reputations of even the most chaste singers. The body parts that make speaking and singing possible directly affected those that make sex and reproduction possible. Articulated in the throat, *gorgia* induced the rapid closing and opening of the glottis, an action that paralleled the opening of the uterus imagined to accompany intercourse. These were not the issues for castrati, but they still caused something of an epistemological pickle.

Meanwhile, Christian writing approached music and song with suspicion. In their view, music so viscerally imitated and stimulated the passions that it could corrupt the soul. St. Augustine's classic account of the inner conflict between his love of music and his Christian consciousness is a well-known example. Augustine wavered between denigrating the dangers of songs that merely gratify the senses and praising the potential of musical liturgy to instill spirituality in the masses.[28] Following this tradition, Christian fidelity and its incumbent distrust of song tempered late Renaissance explorations of the power of ancient music by exposing the dangers of a force so intimately connected to the spirit and capable of so profoundly moving the emotions.[29]

In opposition to Renaissance writings that praised the power of music, thinkers—fueled by Christian piety and skepticism—approached song with suspicion and denounced its lascivious effects, focusing on its pernicious potential to wrench the body out of rational control. Such worries stood behind debates about the appropriateness of music in church and about its place in instructing comportment. It is not a coincidence that Pope Paul IV, who authorized the hiring of the Spanish castrati Francisco and Hernando Bustamante, also set in motion the Index of Forbidden Books.[30] He made this list a year after he requested that the papal choir audition the two Spanish sopranos.[31]

Gioseffo Zarlino, in a Ciceronian and Bembist argument, warned against excess in composition: "if it does not corrupt the sense, it at least corrupts the instrument [the singer]."[32] Similarly, in describing song's affective power, Giovanni de' Bardi reminded his readers that music could embody lascivious forces: "among Moors and Spanish women one may see shameless and wanton customs represented in music and dancing."[33] Kircher, true to form, wrote this into a modal taxonomy and found the tenth mode especially troubling: "The tenth tone/mode[:] lamentable/tearful, amorous, and soft; if it is relaxed, [it] moves a man inclined to soft conversation toward spiritual things and pious works; if it is intensified, it easily carries him into lasciviousness, a soft mind, and worldly loves/desires."[34]

Music also came under fire by the Council of Trent in 1562, with prescriptions leveled against sounds that obscured the words, took too long, or generally felt lascivious. As Craig Monson has demonstrated, the council aimed to regulate "impious sounds." The resolution, which was never passed, stated,

> If something from the divine service is sung with the organ while the service proceeds, let it first be recited in a simple, clear voice, lest the reading of the sacred words be imperceptible. But the entire manner of singing in musical modes should be calculated, not to afford vain delight to the ear, but so that the words may be comprehensible to all; and thus may the hearts of the listeners be caught up into the desire for celestial harmonies and contemplation of the joys of the blessed.[35]

In the same decade, Girolamo Cardano, in a quirky text called the *Proxeneta,* slammed the vices of castrato singers at Bologna. The book, apparently written in the 1560s and 1570s, is a kind of conduct book for managers of households and for university life. Of singers in Bologna, he wrote: "The boys from Bologna are beautiful, tasteful and for the most part excellent musicians."

They were also prone to lasciviousness and intellectual failure and were left out of inheritance structures because they had parents who "usually rent these boys exactly as if they were houses." And finally, their strength was all used up from making music: "Their strength already exhausted because of the musical studies, the boys cannot endure to accomplish anything admirable to the attainment of virtue. Therefore, they cannot get into the master's good graces and keep their close familiarity with him."[36]

In the middle of the seventeenth century, the critiques that began with the Council of Trent cracking down on lascivious music or lascivious singers intersected with multiple discourses that pitted the body against the soul and that dug into the tension between spiritual uplift and carnal ecstasy. The dueling Neoplatonic and ecclesiastical anxieties about the potential of song led both to extreme celebration of the music of Roman churches and to serious regulation. Because listening could lead to a spiritual transformation, it required careful management.

These debates played out on the bodies of castrati. The Jesuits, who were deeply invested in the sound of speech and music, engaged in rigorous debate about the place of the castrato voice and body. On the one hand, they used music to engage listeners; on the other, they worried about its ill effects. More specifically, as Eric Bianchi has elucidated, the castrato's singing body—with the powerful ability to transform the soul—put on display precisely the tensions around reproduction and testicles.[37] The castrato could lead a listener to spiritual transformation, but he also was a kind of excessive mutilation of the commitment to celibacy discussed in the last chapter.

In 1648, Melchior Inchofer, best known for his role in convicting Galileo of heresy, wrote a diatribe called *De eunuchismo* (On eunuchism), which condemned the castrato:

> Indeed so to speak, if it were of such great concern to the melodiously harmonized praises of the creator, then wouldn't he himself, who rendered everything complete, also have foreseen this deficit/indigence: or perhaps do we mere mortals find fault with the providence of God and presume to amend those things that he himself contrived on purpose? But if desire for the softness of honeyed voices is so enticing, why don't we just allow women to sing rather than making men effeminate, causing them to abandon the core of their nature, and sing like women just as Nero made Sporos (a male Eunuch) his wife.[38]

Kircher articulated an inner conflict that sounds a lot like Augustine. At points, he praised the voices of the castrati and the virtue of those who heard them:

> Finally, to say nothing of the great/illustrious singing teachers of the highest voice who are found in the German College, they are most excellent both for charm/sweetness of voice and skill of their art. Therefore, I am now convinced that with these or other most excellent musicians, any skilled and industrious choirmaster—especially with the addition of a mixture of elegant and ingenious system—could excite in the souls of listeners the same marvelous effects and affects that the ancient historians used to praise extensively and commended to posterity.[39]

But he also attacked the unseemly bodies and vocal excesses of the castrati. They sang like animals, with no regard for decorum: a not-so-subtle accusation of a lack of humanity.

> They wish only their own voice to be heard; hence they are wont quite deliberately to drown out those of the rest with indiscriminate noise straining their voices so disgustingly that you think you are listening to the music we hear when the bray of asses accompanies bleating sheep, which in all ways offends against the laws of decorum. I shall say nothing here of the absurd bodily posture that singers display while singing; you can see a number of them purporting to beat the time by moving their whole bodies in the unseemliest manner. You may see some now raising their heads at each interval, now bowing them, now shaking and twisting them from side to side—you would say they were actors.[40]

He ultimately wanted the castrati hidden from view:

> Some people are right to say that musicians should be shut away and not seen by anyone, lest by the unseemly movement of their bodies they should break the power of harmony. It is therefore of great importance for arousing the emotion that performers should take care to combine a seemly deportment of the body and controlled movement with seemly delivery and seemly suppleness of voice.[41]

Kircher used the Latin word *musicos*, which in Italian would likely be *musici*, which often meant castrati. To hide the unseemly body was to make the body into a disembodied voice.

Finally, in 1665 and 1667, Pope Alexander VII issued proclamations designed to hide castrati and limit ornamentation. This pope, previously known as Fabio Chigi, was an important musical patron, and the Chigi family continued to be important opera patrons.[42] On July 30, 1665, he laid down the law for choir directors when he said that to promote piety they had to make sure that solo voices did not sing the entire song, and if solo voices must be used, they had to alternate high and low voices.

> We, occupied in looking after the decorum and reverence of the churches destined for divine praises and prayer, and of the oratories of our gracious city (from which examples of good works go forth into all parts of the world), are compelled, by the desire of pious solicitude, to keep far away from them anything ostentatious, and especially choirs of music and symphonies in which anything indecorous or divorced from ecclesiastical rite is mixed in, with offense of the Divine Majesty, scandal of the faithful, and impediment of the elevation of hearts and devotion to things that are above.[43]

The new legislation punished musical offenses by excommunication and the forfeit of wages and explained that within twenty days churches would have to put grates and stalls in place that would hide the singers.

## BLURRED LINES

Despite these debates and prescriptions, voices always pushed boundaries. Pietro della Valle's writings reveal some of the ways this happened. His writings are often read for what they reveal about the seconda prattica, polyphony versus monody, virtuosity, and instrumental music. But they also document the experiences of elite Romans, and particularly relevant here is his emphasis on the ecstatic potential of the castrato's voice, which he characterizes as "natural," as opposed to falsettos and boy sopranos. The voices he wrote about crashed through boundaries between sacred and secular, spiritual and corporeal:

> While they [boy sopranos] did have a voice, as people who had no judgment because of their age, they sang without taste and without grace, as if they had learned by rote, such that sometimes hearing them made me feel like I was being beat mercilessly with a rope. The sopranos of today, being people of judgment, maturity, feeling, and exquisite skill in the art, sing their music with grace, with taste, and with true elegance; clothing themselves in emotions they are ravishing to hear.[44]

The word *sopranos* referred to castrato singers. Comparing music of his contemporary moment to the past, della Valle explained that previous eras lacked a Loreto Vittori or a Marc'Antonio Pasqualini. He was talking here about music in and out of the church.

Accounts of castrato voices, and voices more generally, sounded the same whether or not the music occurred in church, which makes sense given that music and other arts, sacred and profane, flowed seamlessly into one another. To take a well-known example, Bernini's *Ecstasy of Saint Teresa* (Fig. 9.3) was installed in Rome in 1652 around the time that Kircher, della Valle, Melchior, and others were debating the position and role of castrati in church. This was seven years after Vittori's performance as the Magdalene.

Bernini's statue—an icon of baroque ecstasy loved by tourists and scholars alike—writhes and moans in a way that makes her look precariously alive. The real Saint Teresa of Avila, who wrote massively sexualized descriptions of her relationship to God, arguably rekindled a fascination with spiritual ecstasy. In the eighteenth century Charles de Brosses, who was so moved by the castrato's vibrant voice, wrote of the statue, "There is a marvelous expression, but frankly too lifelike for a church. If this be divine love, I recognize it."[45] Saint Teresa herself described spiritual ecstasy in words that could just as easily be used for erotic ecstasy: "While the soul is seeking God in this way, it feels with the most marvelous and gentlest delight that everything is almost fading away through a kind of swoon in which breathing and all the bodily energies gradually fail."[46]

The practice of music drama rendered the boundaries between secular and sacred fluid from the very start. Emilio de' Cavalieri's *Rappresentatione di anima et di corpo* showed by example the potential of sonic delight to incite religious ecstasy and directly connected Roman Catholic reform with Florentine academies. It is not worth relitigating debates about where monody, recitative, and expressive song speech began, but it is worth remembering that the project of music drama was one of recovering an affective and spiritual process grounded in experience. Cavalieri's was in the Jesuit tradition of using pleasure to teach and of using devotional music meant to excite human passion for pious ends. This drama, one of the originary ones, centered on a sonic conflict between the body and soul, pleasure and reason.

Cavalieri's oratorio, with libretto by Agostino Manno, evidently heard by all of the cardinals in Rome, was staged in 1600, the year that Loreto Vittori was born. The performance occurred in the church where Vittori sang (or did not sing) the Magdalene's lament. Cavalieri's oratorio, a sacred theater production, was published in a virtuosic commemorative volume by Alessandro Guidotti and printed by Nicolò Muti in the same year.[47] The entire creative team had their hands in court production and spiritual lamentations. Cavalieri had been responsible for many Medici productions and excelled at realizing the affective potential of dramatic musical compositions.

FIG. 9.3 Gian Lorenzo Bernini, *Ecstasy of Saint Teresa*, 1647–1652, marble. Rome, Santa Maria della Vittoria. Paolo Romiti / Alamy Stock Photo.

Rossi, a.k.a. Eritreo, who published Vittori's side of the abduction story and who later wrote an encomium to Vittori's voice, also described the experience of Cavalieri's piece, writing decades after the actual performance through the mouth of his associate Giulio Cesare Bottifango, whom he said was an ear witness. At the end of the description Rossi explained that Bottifango had demonstrated its power by playing parts of it for Rossi from the printed score.

> So that you yourself can see that what I say is true, he led me to the cembalo and sang some pieces from that *Rappresentatione*, and in particular the part of the Corpo which moved him so much, and I liked it so much that I begged him to give me a copy, which he very courteously did, copying it out for me in his own hand, and I learned it by heart and would often go to his house to hear it sung by him.[48]

Bottifango held up Cavalieri's piece as the epitome of music that moved the passions. Part of the reason for the drama's staying power was the remarkable printed score, itself a celebratory commemoration and virtuosic production. It was the first complete edition of a music drama and included careful instructions on performance that served as an almost phonographic repetition. The description from Rossi describes in vivid detail the affective response of the listener to the singers' voices.

> He was there the day that it was performed three times and never was able to satiate himself. He told me that in particular, when he heard the section sung by Tempo [Time], he felt himself overcome by great fear and trembling. At the speech of Corpo [Body], performed by the same [boy] who played Tempo, when he doubted what he should do, namely to follow God or follow World, and then resolved to follow God, tears in great abundance fell from his eyes. He felt stirring in his heart great repentance and pain for his sins. Nor did this happen only at the moment, but every time he sang it, since every time he wanted to take communion, in order to arouse devotion in himself, he sang that section and burst out in a river of tears. He highly praised the speech of Anima [Soul]. Besides being performed divinely by that little boy, musically it was of incomparable artifice, that expressed the feelings of pain and sweetness with certain false sixths moving to a seventh, that ravished the heart. In a word, he concluded that in that genre it was not possible to do anything more beautiful or more perfect.[49]

The speech Rossi referenced was the prologue sung by the scary old man Time just before the day of judgment. The description is remarkably detailed; the depth of the bass voice terrifies. It is the high voice, Anima, that has the false sixths, and sevenths are gritty, close dissonances.

In seventeenth-century Rome well-known castrati performed in both sacred and secular works and composers wrote for sacred and secular contexts. Ecclesiastical leaders understood the power of song to convert, move, and convince followers of power. Popes and princes alike regularly hired castrati from the Sistine Chapel to perform their operas. As opera codified as a genre, the papal court frequently staged operas based on the lives of saints. Through operatic preaching, they performed the moral codes of the Catholic Church and upheld the authority of the church and of the spirit. For instance, Virgilio Mazzocchi, Domenico's younger brother and Francesco Barberini's maestro di cappella, was an important trainer of castrato voices who also composed the music for *San Bonifatio* in 1638. The sacred opera, performed at the Palazzo della Cancelleria, featured ten or eleven boys from Mazzocchi's singing school. This was a year after he wrote the comic piece *Chi soffre speri* with Giulio Rospigliosi; the latter was the librettist for both and would become Pope Clement IX. Bernini designed sets for Marco Marazzoli's intermedio *La fiera di Farfa*.

When Rossi identified Vittori's Lenten performance as Domenico Mazzocchi's Magdalene, the singer was already a figure of vocal ecstasy and one who perhaps embodied the vituperative conflict between sacred and profane. Nineteen years earlier Vittori had starred in Mazzocchi and Ottavio Tronsarelli's *La catena d'Adone* (The chains of Adonis). The opera libretto was based on canto 13 ("La prigione") of Giovanni Battista Marino's *L'Adone*, published in 1623, and was quickly listed in the Index of Forbidden Books. The poem led to a heresy trial under Urban VIII. Vittori played the sorceress Falsirena, who turns the forest she rules into a pleasure palace when she falls in love with Adonis. Mazzocchi's music moves easily between recitative and aria, and the sounds embody what the appendix explains as a character whose insides battled reason and sensuality. The opera stages sensuality with Falsirena as its persona.

> FALSIRENA, counseled by Arsete to do good, but persuaded by Idonia to do evil, is the soul counseled by Reason but persuaded by Lust. And just as Falsirena easily yields to Idonia, it is thus demonstrated that each *Affetto* is easily overpowered by the Senses. And if in the end, as the wicked Falsirena is bound to a hard rock, one must again understand that Punishment is the final consequence of Sin.[50]

The opera also relishes the fine line between art and nature; it's hard for Adonis to tell whether the pleasure garden is artificial or natural. Falsirena uses the chain of sensuality to ensnare Adonis, and the singer Vittori embodied the power of music and, more specifically, song to bind. Falsirena causes tempests, thunder, and lightning, moving the earth's spirit with her voice. But the character is ultimately damned. And the jury is still out on whether or not the opera was originally written to stage a musical battle between Margherita Costa and Cecca da Padule, both courtesan singers.

## JESUIT TEARS

The performance Rossi described occurred during the height of the Jesuit spiritual year. Jesuit spiritual practice mobilized aural ecstasy, a sonic experience of conversion, repentance, ecstasy, and loss of self through suffering. They used as a template the spiritual exercises of Ignatius of Loyola, a meditative practice that ties individual suffering and conversion to collective spirituality. The Jesuits mobilized a range of creative practice aimed explicitly at channeling the physical and emotional experience of conversion. They aimed for tears. During the reign of Clement VIII (1592–1605), when castrati occupied more and more sonic space in papal Rome, the lament was increasingly present.[51]

Vittori, Domenico Mazzocchi, and Pope Urban VIII—to whom the lament was dedicated—were all deeply tied to Jesuit spirituality. In 1640 Vittori wrote and composed an oratorio based on the conversion of Saint Ignatius for the centennial of the Jesuit order. According to an *avviso* from July 14 of that year, it was "recitata in musica" (recited in music). The description sounds very much like a Roman opera, complete with a marvelous scene in which Ignatius entered atop a carriage with pages and knights, and Christ descended from heaven like an opera star:

> Christ descended from heaven to the earth, seeing which, the King threw himself on his knees with great fury at the feet of Christ, who after having spoken, returned to the heavens, and the scene closed, and the aforementioned King remounted his chariot, and exited by way of the stage.[52]

The music is not extant, but it probably sounded a lot like *Galatea* and any number of Barberini sacred operas.

Using contemporaneous sources, Andrew Dell'Antonio has written about the ways that listeners were cultivated to experience this kind of ecstasy. He uses the writings of Grazioso Uberti, a Roman jurist and amateur musician, as key texts in the promotion of ecstatic sound. Neither lawyers nor judges, jurists were legal scholars who aimed to understand the law. Uberti, in his

*Contrasto musico* of 1630, posited the voice as a vehicle for devotion, as a force that exceeds description, and as uniquely capable of creating sensual delight. While he contended that words alone could not explain heaven, it was possible to convey its pleasures through music (which, of course, was not an original idea). He claimed that you could not explain the glory of heaven with mere words:

> It is too little to speak of that jubilation, that joy, that sweetness, that happiness of the Blessed. People do not understand it enough, especially the very simple. For to make men love it even more, in order to move their souls, we paint choirs of angels in the act of singing and playing various music instruments. But since painting is mute and does not give the ears any melody, we add the singing of voices and the sound of instruments, to make a more vivid aural representation of sweet harmony.[53]

Musical thinkers articulated spiritual music though the language of eroticism and bodily processes. For the Jesuits especially, oratorios were tools for persuasion, and drama was a full-bodied mode of conversion in and out of liturgical contexts; they believed in using pleasurable media to transmit spiritual truth.

## WATERWORKS

The text of *Lagrime amare* is attributed to Cardinal Ubaldini with the subtitle *La Maddalena ricorre alle lagrime* (The Magdalene has recourse to her tears). The dialogues and sonnets both contain laments, notably including a Latin setting of Dido's lament. *Lagrime amare* is quintessentially wet and bloody: tears flow from eyes to wound and back. They pour tears, blood, and other fluids. Humors transform into one another, and the text connects the blood of Christ to the listener's tears:

> Lagrime amare all'anima che langue
> Soccorrete pietose; il dente rio
> Già v'impresse d'inferno il crudel angue
> E mortifera piaga ohimè, v'aprio.
> Ben vuol sanarla il Redentore esangue
> Ma indarno sparso il pretioso rio
> Sarà per lei di quel beato sangue
> Senza il doglioso humor del pianto mio.
> Sù dunque, amare lagrime correte

A gl'occhi ogn'or da questo cor pentito,
Versate pur, che di voi sole hò sete.
Se tanto il liquor vostro, è in Ciel gradito,
Dirò di voi, che voi quell'acque sete,
Ch'uscir col sangue da Giesù ferito.

Bitter tears, mercifully comfort
My languishing soul: the evil tooth
Of the cruel Hellish serpent bit you
And, alas, has opened a fatal wound.
The bloodless Redeemer wishes to heal it,
But the precious river of that blessed Blood
Will be shed in vain for it
Without the sorrowful liquid of my weeping.
Therefore make haste, bitter tears, and run
To my eyes from this repentant heart;
Pour forth, for I thirst only for you.
If your liquor is so pleasing in Heaven
I will say that you are the waters
That flowed together with blood from Christ's side.[54]

The voice sounds dramatically from the get-go (see Fig. 9.4): a descending vocal and bass line with chromatic alteration and messa di voce on the word *amare* (bitter). The rhythmic flow is irregular, giving the impression of halting words. A melodic descent on "mortifera piaga" slides after a dramatic pause into a not quite repetition on two cries of *oimei*. The final poetic line descends an octave with more sighs interrupting.

The piece exudes falling intervals, long melodic descents, and chromatic inflection. The most explicitly sung melodic melisma is at the end of the third section, when she thirsts only for him, and the piece concludes with a definitive cadence including a falling fifth in the bass line.

It's easy to hear why the Jesuits found laments like this such a powerful tool. Scholars from multiple fields have explored the affective potential of the lament as it oozed from the mouths of abandoned human women, nymphs, saints, lovers, sometimes mourning men, and more. Pain, suffering, weeping, and wailing had long inspired harmonic creativity, and the similitude of sounds in sacred and secular contexts made any distinction between spiritual and corporeal hard to believe. This sounds clearly in Monteverdi's 1641 *Pianto della Madonna sopra il lamento d'Arianna*, a rescored contrafactum of his iconic *Lament of Arianna*. Instead of Arianna lamenting the loss of Theseus,

FIG. 9.4 Mazzocchi's *Lagrime amare* (*La Maddalene riccore alle lagrime*, 1638), from *Dialoghi e sonetti*. Museo internazionale e biblioteca della musica di Bologna.

Mary laments the crucifixion (Weeping of the Madonna). Both pieces capture agony in what sounds almost like an improvised performance accompanied by violins and violas. The rests, the drawn out second syllable, the chromatic inflection, and the falling vocal line embody grief. As Emily Wilbourne has argued, Virginia Andreini, who envoiced and made famous Monteverdi's lament, was a lament specialist in the commedia dell'arte tradition and frequently played the role of the Magdalene; her voice is part of what gave the lament its potency in the original performance.[55]

There are so many seventeenth-century laments that take a listener's breath away. In them composers and singers tended pull out all the stops, especially during this time period. They were performed invocations to weep collectively. The descending tetrachord that Ellen Rosand offered a sonic archaeology of in the 1970s was a sonorous embodiment of emotion.[56] Seventeenth-century theorists often used the term *catabasis*, ancient Greek for "descending somewhere," for descending bass lines that reflected melancholy and penance. The style of vocal music with instrumental accompaniment allowed for vocal play with harmonic structures, the harmonic creation of desire.

Like the Magdalene herself, singers who animated her words sang backward and forward in time and became vessels for embodied affect. When Vittori sang, laments in sacred and secular contexts had been stylized as extreme versions of an affective sound system that pervaded emergent baroque aesthetics. Incantations that came from the mouths of penitents, lovers, nymphs, saints, and deities mobilized the same sonic effects in sacred and secular contexts and projected seventeenth-century musical versions of ancient theatrical catharsis. The goal was conversion and ecstatic release, and it was part of a larger project of experimenting with affections.

Laments did such visceral affective work because of the ways that sensation, sound, and spiritual ecstasy were imagined and experienced in the early modern world. Pain, despair, and sorrow enticed specific musical gestures, which in turn took the listener through an experience. For the humoral body, which functioned via a complicated set of internal processes, good health depended on maintaining the proper corporeal fluidity. All fluids could be continuously transformed into one another so that sweat, blood, breast milk, semen, saliva, and tears comprised, at their base, the same materials. Unbalanced humors resulted in bad fluids, which in turn caused illness and pain. A healthy body required a particular balance of the four qualities that described all matter—heat, cold, moisture, and dryness. It was made up of four elements—fire, earth, water, and air—and was physically continuous with the rest of the material and immaterial world. In this model, the parallel substances of voice, tears, vomit, and sweat turned into one another and flowed in and out of the porous body through open orifices, purifying, nourishing, and flushing it. Kircher wrote that the air of sound "excited a natural moisture altogether equivalent to the object and to the harmonic motions."[57]

While the worlds of natural philosophy, anatomy, and medical technology were far from monolithic, there was a dominant blend of Galenic principles around temperament, Christian theology, and embodied sensuality that linked not just bodily and spiritual well-being but what might be called physical and mental health. The gradually increasing emphasis on physicality in religious devotion, which Caroline Walker Bynum has traced for the Middle Ages, continued through the seventeenth century.[58] The lament worked so viscerally in part because of the humoral linking of tears, breath, and, by extension, song.

Even as medical practitioners and natural philosophers vituperatively debated the workings of the body, spiritual writers remained committed to a liquid body, one where the soul melted and where the milk of the Virgin's breasts and the blood of Christ loomed large. Scholars have written extensively about the ways that music was uniquely equipped to represent and entice an experience of the spiritual erotic. Melting souls made it possible for liquids to merge: the breast milk of the Virgin Mary, the blood of Christ.

The lament fused song and tears and modeled a physiological process. Most famously, court chronicler Federico Follino made the totally conventional comment about Virginia Andreini's performance of Ariadne's lament at the wedding of Margherita di Savoia and Francesco Gonzaga that "there was not one lady who failed to shed a tear."[59] In 1638, while Vittori was exiled, the then nineteen-year-old Barbara Strozzi read both sides of a debate written by Giovanni Francesco Loredano and Matteo Dandolo about whether tears or song worked more powerfully as weapons in love. The debate presumed a similitude between singing and weeping and assumed that song could embody tears and even become tears. In Strozzi's mouth, song triumphed: "I do not question your decision, gentlemen, in favor of song, for well I know that I would not have received the honor of your presence at our last session had I invited you to see me cry and not to hear me sing."[60]

In some cases, the lament was a metonymic stand-in for theatrical sound all told. Indeed, Kircher's description of the Roman theater sounds very much like Follino. Following Jesuit theatrical tradition, he celebrated music that made listeners into "viri perculsi" (men who were moved). The Roman theater, to him, had miraculous and indescribable effects that put audiences into a physical state of excess and almost inner battle. He wrote of "audiences who often were unable to be restrained, breaking out into screams, wails, sighs, and strange/foreign movements of their bodies clearly imitating with outward signs the degree to which they were being incited by the ardor/fervor of internal passion."[61] Sung performance was sensory overload that radically affected the listener's connected mind.

## MODULATORY JUSTIFICATIONS

*Lagrime amare all'anima che langue* keynoted an expressive compositional program. The piece appeared at the very end of the *Dialoghi e sonetti*, a volume that featured several laments, including a Latin Dido. Mazzocchi followed the lament with an explanation of notational symbols titled "Avvertimento sopra il precedente sonetto." In fact, the "Note on the previous sonnet" is almost four pages, and it articulates an expressive practice that he called the enharmonic style.

The "Avvertimento" instructed that the *Lagrime amare* be sung "scritto à rigore" (strictly as written) and that no alterations be made unless they are explicitly notated, an instruction that betrays his deep investment in a style and practice that used compositional and vocal devices to create sensuality and ecstasy. Following the trend toward chromaticism, Mazzocchi leaned into devices that privileged dissonance and jarring key relations, and he wrote in vocal ornaments and techniques that highlighted textual excess that, as he put it, made music less boring. A priest, musician, and member of a noble family, he embraced Jesuit rhetorical practices and classical laments and throughout his compositions put cathartic penance into sound. This was in keeping with the agenda of the Accademia dei Virtuosi, the musical academy that circled Cardinal Francesco Barberini and performed and discussed musical performances. In addition to the piece that Vittori made famous, Mazzocchi also set the *Dialogo della Maddalena,* published posthumously in the *Sacrae concertationes* of 1664, and *Maddalena errante,* published in the 1638 *Dialoghi e sonetti.* So, the liquid and vociferous conversion of the Magdalene worked well for Mazzocchi, whose music displayed a fascination with chromaticism, dramatic modulation, and musical rhetoric.

Mazzocchi described in detail the messa di voce, extensive use of flats, and the enharmonic divisions called enharmonic dieses, what we would call quarter tones. He gives a kind of answer key or chart for his enharmonics in which he showed readers what his notational symbols meant (Ex. 9.2). He referred specifically to the last stanza of the lament which enumerates the bloody wounds of Christ; the basso continuo moves through jarring chords, and he adds melodic inflections in the melody.

*Lagrime amare* appeared in the same year that Mazzocchi published an entire book of Pope Urban's texts called *Poemata* and within a few months of Monteverdi's *Madrigali guerrieri et amorosi.* The music and the justifications are indebted to Gesualdo, Marenzio, and Luzzaschi, and they entered the contentious fray of Roman musical debates around the affective potential of sound. Kircher used Mazzocchi as a key exemplar of what he called the metabolic style, which followed Gesualdo and Monteverdi in extreme chromaticism, enharmonicism, and expressive dissonance and which was described in terms of intense effects on listeners. Since the seventeenth century, commentators have focused on this as a modal mutation. And as Jeffrey Levenberg argues, it was also tied to the invention of enharmonic instruments like

EX. 9.2 Mazzocchi's enharmonic division of the whole tone

Vicentino's chromatic keyboard.[62] Kircher explained the style in terms of "mutation" (motions): "Furthermore, each mutation [*mutatio*] has great force, and causes noticeable changes in listeners, and can be varied in infinite ways, and is the most appropriate for expressing any affect whatsoever."[63] Kircher quoted the whole second stanza of Mazzocchi's Magdalene lament celebrating this style as a modern incarnation of the "metabolism" of ancient Greek music. The style, a favorite of Barberini academies, has been extensively discussed by scholars of sixteenth- and seventeenth-century music, and Patrizio Barbieri traces the style as an experimental practice of ancient recovery that began with Doni and was enacted by della Valle, Mazzocchi, and Carissimi. Barbieri also links it to a mode of modulation begun by Gesualdo.[64] The compositional practice, like the seconda prattica, focused on using modal shifts and variation for maximum affect and effect.

## MESSA DI VOCE

In Mazzocchi's setting of the first nine lines of *Lagrime amare*, the notation "V," a sign for the messa di voce, appears twice: first, on the second word of the poem (*amare*; Fig. 9.4), and then on the word *lagrime*. The technique required great strength and control. It also allowed for shifts in dynamics, pitch, and register. Breath support must be adjusted to the change in intensity to produce a constant pitch and compensate for the natural tendency to strain or push. Essentially, the singer tends to go sharp during the crescendo and flat during the decrescendo, but breath control allows for a continuous pitch. Think about the way vocal pitch rises when people yell, and then think about what it would feel like in the throat and stomach to try and control that. The messa di voce is inextricably intertwined with the castrato voice. By the eighteenth century, the messa di voce, as Martha Feldman has argued extensively, was not just a technique but a hallmark of the castrato voice, a means of practicing vocal technique and the height of breath control.[65] It's a vocal activity that is all about the throat and breath.

Mazzocchi was likely the first to use the term *messa di voce* itself and to explicitly write it into a composition. But he did not invent the ornament. Giulio Caccini, in 1602, published *Le nuove musiche*, which included a lengthy preface on the techniques used in the new type of singing and a method for learning them. He tied the messa di voce explicitly to pedagogy by calling it the primary means to mastering tone and called it "crescere e scemare la voce" (crescendo and decrescendo of the voice).[66] For him, the messa di voce was as much a practiced technique as it was an ornament, and it was about sonorous expression of an idea or feeling.

Mazzocchi described the messa di voce in both places he wrote about the lament: in the *Avvertimento* and in his preface to the madrigals published the same year. In each case the ostensible reason for the explication had to do with notation. He described the notation "V" as raising the pitch by a quarter step while simultaneously increasing the volume. In the madrigals, he offered more details: "One must raise the voice, or as is commonly said messa di voce, which means to raise little by little the volume and the pitch together in particular by half of the aforementioned, as one does with the enharmonics."[67] He went on to say

> But whenever it is necessary to raise the voice only in volume and spirit, and not in pitch, it will be designated by the letter C, as has been done in some madrigals, and then it will be observed that just as the voice must first be sweetly raised, afterwards it must be reduced little by little, until it is so faint that it can barely be heard, or until it disappears into nothing, like an echo from a cistern which responds in that way to certain voices. But this and other advice can be followed by a good choirmaster and with the discretion of a good singer.[68]

### IT'S ALL ABOUT THE VOICE

Mazzocchi said that chromatic tones can be sung by the knowledgeable but enharmonic music requires the most knowledgeable. Needless to say, Vittori's facility as a writer and singer of laments hovered between the real and the imaginary. His *Lamento del re di Tunisi* makes a classic lament out of the story of the conversion of a Muslim princess by Jesuits at Palermo. His opera *La Galatea* had two showcase laments. To execute this music required a combination of book learning and musical practice. And it was all to be used sparingly. Ironically, Mazzocchi extracts a Latin phrase to make this point about judicious use of such inflections: "following the advice of Juvenal, *Voluptates commendat varior usus*."[69] This is from the end of the eleventh satire, "An Invitation to Dinner," and is embedded in this line:

> You can head for the baths at once with a clear conscience, although there's still a full hour till midday. This is something you'd not be able to do for five days in a row, because even this kind of life is enormously tedious. Pleasures are enhanced by rare indulgence.[70]

Kircher explained that the metabolic style required the ingenious talents and voice of a Vittori, Pasqualini, or Argenti, castrati known for affective and virtuosic power. Later, Doni attributed this kind of singing specifically to Vittori.

> The judicious singer must also try to pronounce distinctly and emphatically, so if it is possible not even one note of what he sings is lost. In this Sig. Cavaliere Loreto is outstanding, since beyond his natural talent, he has been brought up in Florence, where a particular profession is made of good pronunciation, as was so in Athens, whether in singing or speaking.[71]

And here is where Gian Vittorio Rossi, who was so mired in prurient details of the church, becomes a key source. I agree with Gallico, who concludes that in effect the performance transcends the piece itself and that Eritreo's musical judgment is what we need to listen for.

It is far from surprising that when Rossi rhapsodized and riffed on Vittori's voice he serenaded the affective power of the voice at large. Rossi explained that when Vittori first arrived in Rome, his patron Ludovisi controlled his singing, allowing him only to sing for noble people who could bestow him with gifts. Rossi highlighted the sound and the mysterious effect of that sound, a kind of conversion that can be secular or sacred and is physical and spiritual. Vittori's great skill meant he could modulate his voice to any emotion and could do anything with his voice.[72]

Rossi stressed Vittori's ability to act and embody the motions of the moment and to rapidly shift mood on and off the stage. Vittori, he writes, was a true artist whose rhetorical skills were so advanced that he could offer clear articulation of words and syllables with no affectation whatsoever.

> But he [Vittorio] was endowed by nature and the precepts of the best teachers. So if he has to represent the voice and the speech of a man agitated by anger, he uses a piercing, excited, often precipitous type of voice; if he is to show compassion and sadness, he uses a flexible, interrupted, full voice with lamentation; if he is to express dread, he uses a slow, hesitating, humbled quality; if the force of a truculent man has to be rendered, he uses a fervent, urgent, threatening voice; if the pleasure of a delighted mind has to be put forth, he uses a voice that is profuse/extravagant, light, tender, cheerful, and relaxed; if the pain of a man crushed by some sickness has to be portrayed, he applies a heavy voice without commiseration and expressed with a single tone and articulation. All these things are taught by Lucius Crassus, and the best teachers are seen performed more felicitously in the measured voice of that fellow as he sings than in the books of these teachers. The ability to so knowingly form and shape himself in song to whatever affection of the soul he desires is a natural gift.[73]

Rossi responded to the feeling of the singing and to the power of the sensual expression.

About the voice all told Rossi explained that Vittori had more skill than others and that in particular he avoided "limpid articulation" and obscuring the final syllables of words. With most singers, listeners dozed or did something else while listening.

> However, as soon as the song [sung by Vittori] begins, everyone stands upright and attentive, stops their conversation, which they had begun, and holds their breath so they do not make any noise at all by breathing frequently. Then, wonderfully softened by the song of the same fellow, they are alienated/estranged from it [the song] with no contempt/dislike and finally, far from walking away exhausted with sated/satisfied ears, they are held back as if bound and imprisoned by him.[74]

## THE MAGDALENE

The fantasy, not the actuality, of Vittori as Magdalene was and is an assemblage of historic events filtered through a kaleidoscope of sound, sexuality, and desire. A New Testament moment made into narrative in the Middle Ages, it was filtered through the Counter-Reformation, set to text by a cardinal with music by a learned composer, and performed by a man-made voice. The sounds were the excess of suffering and angst that singers—in the roles of lovers, penitents, saints, and warriors—sent into the vulnerable ears of listeners. It was, perhaps, a musical embodiment of relic culture, a sonic imbuing of a thing with sacred power that transcended materiality and time.[75] And it was a sonic embodiment of crying meant to entice tears.

The event can be heard not as Vittori impersonating the Magdalene but, rather, as Vittori bringing the sense of her to life at a particular moment. Leaving us modern readers with a sonic residue, this performance of what Elizabeth Freeman calls temporal drag embodies "a kind of historicist *jouissance*, a friction of dead bodies upon live ones, obsolete constructions upon emergent ones."[76] In theoretical terms, the saints embody queer time, reaching backward and forward always. They are what Freeman articulates when she encourages a temporal complexity: "To reduce all embodied performances to the status of copies without originals is to ignore the interesting threat that the genuine *past*-ness of the past sometimes makes to the political present."[77]

The performance sounded and sounds a displacement of affect that is intertextual and cross-temporal. Mazzocchi's Magdalene sounds a lot like Titian's penitent Magdalene looks. And Titian's penitent Magdalene looks uncannily like Titian's Venuses, who were never all that repentant (Fig. 9.5). Federico Gonzaga, who commissioned Titian's Magdalene as a devotional

FIG. 9.5 Titian, *Santa Maria Maddalena*, 1531–1535. Oil on canvas. Gallerie Degli Uffizi.

work for the poet Vittoria Colonna, asked that the artist make her as lachrymose as possible. Titian made the Magdalene naked and left her barely covering herself with her unruly golden hair; it is a painting that forces the viewer to confront the flesh. Unabashedly and voluptuously sensuous, it turns a sacred image into a vehicle for delight.

In a classic seventeenth-century Roman Counter-Reformation response to the painting, the mannerist poet Marino returned to the Magdalene as a luscious exemplar of sensuality and spiritual transformation. Marino's 1630 poem on the painting, *La Maddalena di Tiziano*, focuses on her body parts and on her tears: mouth with nectar between vivid pearls and lovely rubies, odorous roses, lips that know chaste kisses. It is a fiery blazon of a sacred goddess who sings. Placing the reader at his side admiring the painting, Marino first invites the reader to experience the painting with him:

> Ecco come con lui si lagna, e come
> Del volto irriga il pallidetto Aprile,
> E, deposte del cor l'antiche some,
> Geme in sembiante languido & humile;
> E fanno inculte le cadenti chiome
> Agl'ingnudi alabastri aureo monile;
> Le chiome, ond'altrui già, se stessa or lega
> Già col mondo, hor col ciel, e piagne, e prega.[78]

> Look how she cries out to him and how
> She floods the pallid April of her face
> And, the ancient burdens gone from her heart,
> She sighs with a languid and humble countenance;
> And her wild, trailing tresses make
> A golden necklace on her alabaster nakedness;
> Locks that once bound others, now bind her,
> Once to the world, and now to the heavens, she weeps and prays.

The painting was so real that he could hear cries. The visual ecstasy was the virtual sensuality he imagined, and he perpetuated the idea that you could fall in love with a statue, feel its tears, hear its sighs:

> Oh come dolce in flebil voce e rotta
> A ragionar col sommo Amor ti stai!
> Si vivi espressi son gli atti e i lamenti,
> Ch'io vi scorgo i pensier, n'odo gli accenti[79]

> In a gentle and broken voice,
> Ah, how sweetly you engage with the highest Love!
> Your laments and gestures are so vividly expressed
> That I see your thoughts and hear your sounds

And in the final stanza, Titian's Magdalene embodied moving the passions with even more vigor than nature, invoking the seventeenth-century sense of art as moving the body and soul:

Ma ceda la natura e ceda il vero
A quel che dotto artefice ne finse,
Che, qual l'havea nel'alma e nel pensiero,
Tal bella e viva ancor quì la dipinse.
Oh celeste sembianza, o magistero,
Ove nel'opra sua se stesso ei vinse,
Fregio eterno de' lini e dele carte,
Meraviglia del mondo, onor del'Arte.[80]

But let Nature cede and let Truth cede
To that which learned artifice has feigned
For he painted her here, as alive and beautiful
As he had conceived her in his soul and thought
Oh, celestial semblance, Oh masterly craft,
In which he surpassed himself in his own work,
Eternal ornament of cloth and paper
Marvel of the world, honor of art.

Marino also discussed the Magdalene in the fourth part of *La musica* from his *Dicerie sacre*, which positioned music as a sonic signifier of Christ's suffering:

> Let's become tuned once more to this piteous music; if not with our mouths, then at least with our eyes; if not with song, then with the lament; if not with voices, then with tears. Let's imitate with these waters, the waters of those melodious founts which respond naturally to the sound. This, this was your music, oh Magdalene.

Magdalene was pardoned because of her tears: "Notice my tears with your ears." She begged only that he hear her: "The lord has heard the sound of my weeping."[81] Rossi's reaction to Vittori's performance sounds a lot like Marino's reaction to looking at Titian. The act of listening to Vittori as the Magdalene made him cry his own tears and enact his own repentance. When Vittori embodied the Magdalene weeping for her crimes and throwing herself at the feet of Christ, no man in the audience was immune:

> None of those present were of such a placid and indifferent mind as not to feel led to those motions of the soul, which the artist wanted to arouse: that is to say, in truth, to tears, to anger, to hatred for sins. The witness does not know about the others. For his part, when that perfect interpreter of the character of

> Magdalene cursed the crimes of her past life, he flared up vigorously against his own sins; when the virtuoso, with a miserable sound of his voice, reproduced the groans of the weeping sinner, he felt the tears gush copiously from his eyes, and he felt enraptured in unparalleled admiration when he showed it, bringing his voice little by little from very serious to very high pitch, and then again from very high to very grave, and leading it for several turns with incredible volubility, to be able to twist and flex it as he wanted, as if it were very soft wax.[82]

Rossi was moved by the magic of Vittori's voice, a force manipulated by human effort and a force that, like a relic, enacted an intrusion of the past into the present.

The voice is, as Kara Keeling writes, a connection that is never fixed, resists description, and is a medium for queer temporalities. Here Vittori, Mazzocchi, and all of the commentators were an embodied link to the affective and transhistorical work of seventeenth-century monodic song and, particularly, an embodied moment of the kind of erotohistory Elizabeth Freeman writes about, where a sensory experience is an assemblage of events, experiences, and text. Writing about the monster, she suggests that the body is not ever just a body but is a figure for corporeal relationships between past and present: She explains further that "in this sense, the monster's body is not a 'body' at all but a figure for relations between bodies past and present, for the insistent return of a corporealized historiography and future making of the sort to which queers might lay claim."[83] Another way to think of it might be an archive of feeling: feeling about music, body, space. For literary descriptions of castrati performances, this may mean that rather than relegate them to the realm of conventional description, modern readers can resurrect them for what they say about affective, embodied experiences. Rossi's description of the voice presents a sound world that elucidates the voice of Vittori as the carnal moment, a performance of displacement and an assemblage of historic events filtered through a kaleidoscope of sound, sexuality, and desire.

From a twenty-first-century perspective, the Vittori/Magdalene performance is both hyperspecific and radically transhistorical. It reflected a particular Roman moment for sure. Mary Magdalene was the quintessential weeper, a favorite of the Barberini popes and—as mentioned above—the Jesuits. Pope Urban VIII featured her in much of his poetry and engaged a rather public campaign to acquire a Magdalene relic from France. And in 1620 Francesco Barberini tried to get a little finger for the chapel of Sant'Andrea della Valle. The Barberini palace had a piece of Mary Magdalene stored in an enamel gold ring inside a golden embroidered bag and entombed in a reliquary.[84] More broadly, by the middle of the sixteenth century, her conversion was almost always one that remained wrapped up in an enjoyment of pleasure, which rendered her an ideal vehicle for sensual performances of conversion and a powerful incarnation of the impossibility of separating sacred and secular.

In the end, I wonder, via the words of Heather Love, "Are we sure we are right to resist the siren song of the past?"[85] She asks this provocative question about queer histories, arguing that in the interest of progress there is always an attempt to avoid looking back but that perhaps Orpheus's survival offers a metaphor: listen, but don't be destroyed. This chapter has been an attempt to listen to the affective responses of the seventeenth century and to take seriously the work of the voice as a feeling, an affect.

CHAPTER 10

# CYBORG ECHOES

In 1565 Guglielmo Gonzaga sent his ambassador, the humanist scholar Girolamo Negri, on a fruitless mission to look for castrati. After one died and others were too expensive, a frustrated Negri came up with a new plan. "I would think it would be less difficult to try to recruit them as young boys, and that all they need is the ability to make themselves [i.e., to be trained], and I believe that this would be better done there [Mantua] than here."[1] Richard Sherr, who discovered the letter, suggested that Negri meant for Gonzaga to "make his own." Given the learned doctors in the Gonzaga employ, Negri's suggestion was not unreasonable. Except that it was, and they didn't.

Castrati were not made by learned surgeons. They were mostly made by *norcini*, itinerant specialists from the region of Norcia. The making of eunuchs—a practice that, by the eighteenth century, was a symbol of abject alterity—was known to premodern Europe through ancient sources, the Arab world, and the Christian Bible. In each case a surgical procedure remade a prepubescent boy into a figure whose physical and psychological properties were different from those of a man who went through puberty. For singers, vocal training enhanced the process initiated by surgery. No matter how you sliced it, the castration procedure changed a boy's or man's body, reconstituted his body, removed him from his normative social situation, and made him into something new. In contemporary theoretical terms the castrato's voice exemplifies the notion of cyborg as Donna Haraway famously defined it: "a fusion of the organic and the technical forged in particular, historical, cultural practices."[2] The castrato voice merges cultural and natural artifacts and connects the organic body to the technologically produced object.

This chapter links the practice of surgical alteration and vocal training to an ontology of body remaking and tunes in to structures of knowledge that, between the 1560s and the 1660s, led to increased professionalization of both surgery and singing: new techniques for vocal training, and new techniques for working on the body. Surgeons emerged as what Sandra Cavallo calls artisans of the body, artisans who worked with human flesh as their material.[3] There was a growing relationship between anatomists, surgeons, and learned physicians. As John Gagné puts it, "manual production (techne, Aristotle's lowest form of industry) formulated claims to equality with practice (praxis) and theory (episteme). Knowledge built in bodies, could, in turn rebuild bodies."[4]

Music making and surgery both remake the body, and in the era of the castrato, both were enmeshed in struggles between art and nature and with new ways for the human hand to match or perhaps exceed nature's ingenuity. So, Julius Casserius could write in 1601 that "when the larynx is formed as nature intended, voice too is produced in a natural way; but if the larynx is not shaped naturally, voice also is uttered in corrupted fashion; the voice that is not natural is corrupt and against nature."[5] And surgical alteration, like vocal training, intersected with discussions of virtue, prescriptions against ornamentation, and the dangers of beauty. Music historians associate a paradigm shift with the rise of instrumental music and with a decreased investment in ancient texts. Historians of surgery associate a paradigm shift in part with the invention of gun powder and the ensuing severe and novel injuries as well as increasing investments in cosmetics, prosthetics, and armor.

## HEARING BACKWARD

The early history of castration and of eunuchs has been told in many contexts and it might seem redundant to rehearse it. But I link it to the practice of technologically extending the human body, bringing to the fore the premodern links between the making of castrati and the making of slaves. The era of the castrato was the era of the continued rediscovery and re-entrenchment of classic texts that inflected ideas of what it meant to be a human, a man, a woman, a eunuch. Spanish conquerors drew on ancient climatic hierarchies (described in chap. 7) to delineate essential differences between Spanish and Indian bodies. Aristotle's theory of natural slavery was used to justify the enslavement of Indigenous peoples by the Spanish in 1512 via the Laws of Burgos (Leyes de Burgos), the first text to legislate Spanish behavior toward the Indigenous peoples.[6] Tomás Ortiz declared in that year that Indians were incapable of learning, ate other humans, and engaged in bestiality. And ancient texts also served as evidence in theological and medical discussions of castrati in ways that implicated voice in the reification of naturalized racial and gender hierarchies. This is not to posit a continuity between the ancient and the early modern period but is, rather, to contemplate the ways that the castrato challenged the philosophical and theological limits of humanity that, in the sixteenth and seventeenth centuries, still reflected ancient texts and their early Christian inflections.

In the ancient world the making of castrati was intertwined with the fashioning of the human body to meet specific needs and with an intervention that prolonged the physical markers of youth in men. Claudian, the fourth-century Latin poet whose writings on the hydraulic organ I cited earlier, connected the surgical making of eunuchs to the arresting of a developmental process and construction of a new kind of body. His writings were part of a cold war between the West and the East and represent what Shaun Tougher has called an orientalist projection of an Eastern practice.[7] Claudian accused the Parthians of castrating boys for the pleasure of their women: "Parthians employed the knife to stop the growth of the first down of manhood and forced their boys, kept boys by artifice, to serve their lusts by thus lengthening the years of youthful charm."[8] The process increased the value of a body on the slave market but decreased the humanity by cutting away the ability to procreate:

> Up hastens the Armenian, skilled by operating with unerring knife to make males womanish and to increase their loathly value by such loss. He drains the body's life-giving fluid from its double source and with one blow deprives his victim of a father's function and the name of husband.[9]

Aristotle articulated the connection between testicles and voice through a discussion of animal castration. Aristotle understood that the practice of castration by removing the testicles changed morphology and especially the voice. In *Generation of Animals* he wrote,

> When the testes are removed the tautness of the passage is slackened . . . and the source or principle which sets the voice in movement is correspondingly loosened. This then is the cause on account of which castrated animals change over to the female condition both as regards the voice and the rest of the form.[10]

In the second century, Galen suggested removing the testicles of Olympic athletes to make them more perfect and more useful as athletic tools.[11] He also advised dietary modification and an exercise regimen. And he suggested celibacy as appropriate for athletes and singers, though in general he found intercourse important for maintaining good health. Juvenal's satires from the same century used eunuchs to denigrate women and to warn against their excessive desire, making clear that a process of using surgery to modify a human body was well established:

> There are girls who adore unmanly eunuchs—so smooth, so beardless to kiss, and no worry about abortions! But the biggest thrill is one who was fully-grown, a lusty black-quilled male, before the surgeons went to work on his groin. Let the testicles ripen and drop, fill out till they hang like two-pound weights; then what the surgeon chops will hurt nobody's trade but the barber's.

> Slave-dealers' boys are different: pathetically weak, ashamed of the empty bag, their lost little chickpeas.[12]

Juvenal implied two kinds of eunuchs: one essentially made as an adult for the pleasure of women and one made early for the purpose of enslavement.

Around the same time, Quintilian tied the business of making eunuch slaves to the business of rhetorical excess:

> Declaimers are guilty of exactly the same offense as slave-dealers who castrate boys in order to increase the attractions of their beauty. For just as the slave-dealer regards strength and muscle, and above all, the beard and other natural characteristics of manhood as blemishes, and softens down all that would be sturdy if allowed to grow, on the ground that it is harsh and hard, even so we conceal the manly form of eloquence and power of speaking closely and forcibly by giving it a delicate complexion of style and, so long as what we say is smooth and polished, are absolutely indifferent as to whether our words have any power or no.[13]

Joy Connolly argues that in Quintilian's model, the eunuch occupies an anxious space between man and woman, slave and free, and that it is a space filled with oratorical excess, vice, and corruption.[14] The artificiality of the eunuch body made to appeal on a slave market directly matches the empty oratory that, by the seventeenth century, would be associated directly with virtuosity:

> But I take Nature for my guide and regard any man whatsoever as fairer to view than a eunuch, nor can I believe that Providence is ever so indifferent to what itself has created as to allow weakness to be an excellence, nor again can I think that the knife can render beautiful that which, if produced in the natural course of birth, would be regarded as a monster. . . . Consequently, although this debauched eloquence (for I intend to speak with the utmost frankness) may please modern audiences by its effeminate and voluptuous charms, I absolutely refuse to regard it as eloquence at all: for it retains not the slightest trace of purity and virility in itself, not to say of these qualities in the speaker.[15]

Empty eloquence, like the eunuch, lacks virility, authenticity, and reason.

## THE MAKING

The earliest mention of what we understand as a surgical procedure for making eunuchs occurs in the Byzantine medical encyclopedia of Paul of Aegina, produced in the seventh century, which detailed several procedures for correcting bodies that appeared abnormally gendered. The book draws heavily on Galen and reflects contemporary practices. He moves through removing excessive breast tissue, corrective cosmetic surgery for damaged foreskin, and adding a tube to a damaged penis tip to retain urethra function. In all cases, the surgery fixes something damaged and works to recover normal function. As mentioned in the prologue he treats the procedure in a section dedicated to elective surgery. In the text the procedure for castration sits between descriptions of "correcting" a scrotum and "correcting" hermaphrodites. He begins his description by explaining that this operation goes against his practice and is performed only at the request of powerful men:

> The object of our art being to restore those parts which are in a preternatural state to be their natural state, the operation of castration professes the reverse. But since we are sometimes compelled against our will by persons of high rank to perform the operation, we shall briefly describe the mode of doing it.[16]

In the thirteenth century, when surgeons, who had been trained largely through an oral tradition, and their apprentices began writing textbooks in Latin, the books prominently featured techniques for bilateral orchiectomy. Part of a larger effort to legitimize the profession, these texts focused on defining the natural body and on infusing bodily alteration with aesthetics, beauty, and embellishment. By the fourteenth century the well-established procedure existed within what Leah DeVun describes as the emergence of a medieval science of sex. *Chirurgia magna* by Guy de Chauliac, for example, was written in 1363 and was circulated widely in manuscript before its first printing in 1478.[17] Chauliac, physician and surgeon to Pope Clement VI, was a clerk and master of medicine, not a barber. His widely translated how-to manual came complete with rigorous Latin and Greek citation. Chauliac pioneered the "external inguinal ring" and associated hernia repair with castration.

He also associated the castration procedure with a surgical working out of sexual difference, including the "correction" of hermaphrodites who threatened the natural order. This had serious social consequences. Medieval texts, for instance, associated Jews and Muslims with hermaphrodites, animals, and cannibals, which in turn classified them as nonhuman. As DeVun explains, "hermaphrodites perhaps uniquely traversed the sexes and confounded sexual taxonomies — making them pivotal figures in a new set of arguments that proposed sexual difference as a basic component of humanity."[18] Surgeons excised physical defects and returned their patients to normative physical standards.

Almost a century later the Venetian surgeon Giovanni Andrea della Croce aimed to capture in one massive book everything a surgeon would need. His

*Chirugiae universalis opus absolutum* (Complete universal surgery) was published in Latin in 1573. A year later he put out a vernacular Italian edition designed for a broader readership and more invested in the surgeons themselves. It was organized differently, had more pictures, and further stressed the living body as the surgeon's purview. Croce cared deeply about practical skill and wrote about filling the surgeon's tool kit. The book reads a bit like Kircher's *Musurgia universalis* with a long history of surgical practice, articulations of current debates, descriptions of how to make and use instruments, and more. As Cynthia Klestinec argues, especially in the vernacular Italian translation, Croce emphasized that surgeons were aligned with artisans who made things with their hands.[19]

> I say in the human body (to show the differences between the art of surgery of *medici* and that art of *marescalchi*, who work on bodies that are inhuman and animal); I say living to make it understood that surgery is very different from the anatomical activities, which are done solely on dead bodies . . . the anatomical art is different from surgery, since surgery works on the living human body and anatomy on the dead body; and because of the goal, since surgery works to unite the parts that are separate or divided in the human body while anatomy seeks to separate and divide the parts that are continuous and united.[20]

At the end of the Italian edition, Croce added a section called "Officina della Chirurgia," which means the surgeon's workshop.[21] In this virtual workshop, a "castrator" (Figs. 10.1 and 10.2) is described like this:

> The flesh grows also between the *tunichae* (layers of tissue), which the Greeks call *oscheon*, the Arabs *barichem*, and the Latins *scrotum*, from which originates a certain fleshy hardness, called *sarcocelum* by the Greeks, *burum* by the Arabs and *hernia carnosa* by the Latins; it is absolutely not natural, therefore it needs to be removed: nevertheless, if it is attached to the testicle, or if it is inside out, one must perform a castration, for which procedure we need to have four things, the castrator, the razor, the thread, and the cauterizer.[22]

Meanwhile, in the second half of the sixteenth century the church began to employ medical practitioners to assist in the process of sanctification. These experts used "contemporary" understandings of anatomy to prove the holy in a dead body or to prove or disprove a miracle.

This brings us back to the Gonzaga court in the late sixteenth century. The gradual infiltration of castrati into papal chapels and Italian court culture in the late sixteenth century is now well documented. And thanks to the

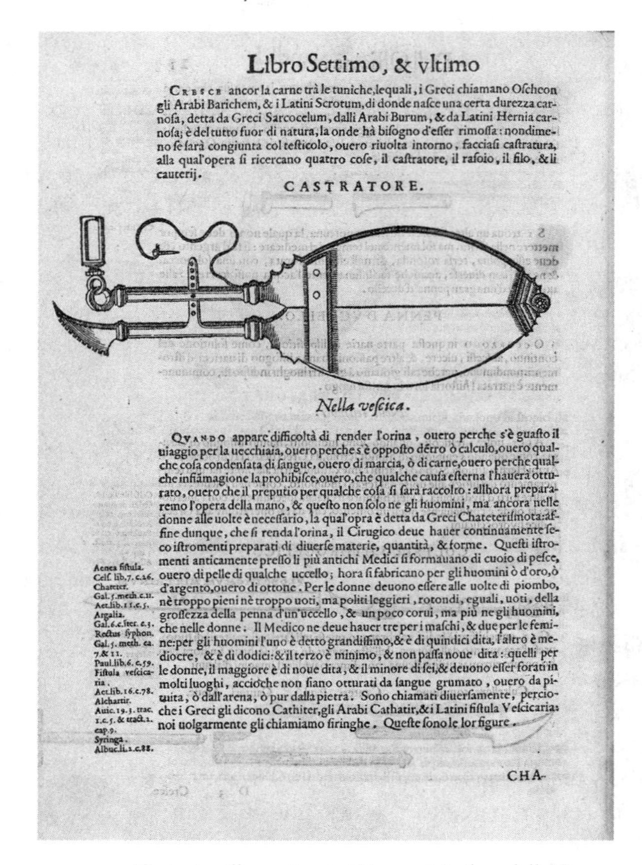

Libro Settimo, & vltimo

Cresce ancor la carne trà le tuniche, lequali, i Greci chiamano Oscheon gli Arabi Barichem, & i Latini Scrotum, di donde nasce una certa durezza carnosa, detta da Greci Sarcocelum, dalli Arabi Burum, & da Latini Hernia carnosa; è del tutto fuor di natura, la onde hà bisogno d'esser rimossa: nondimeno se sarà congiunta col testicolo, ouero riuolta intorno, facciasi castratura, alla qual'opera si ricercano quattro cose, il castratore, il rasoio, il filo, & li cauterij.

CASTRATORE.

Nella vescica.

Qvando appare difficoltà di render l'orina, ouero perche s'è guasto il uiaggio per la uecchiaia, ouero perche s'è opposto dẽtro ò calculo, ouero qualche cosa condensata di sangue, ouero di marcia, ò di carne, ouero perche qualche infiãmagione la prohibisce, ouero, che qualche causa esterna l'hauerà otturato, ouero che il preputio per qualche cosa si sarà raccolto: allhora prepararemo l'opera della mano, & questo non solo ne gli huomini, ma ancora nelle donne alle uolte è necessario, la qual'opra è detta da Greci Chateterismota: affine dunque, che si renda l'orina, il Cirugico deue hauer continuamente seco istromenti preparati di diuerse materie, quantità, & forme. Questi istromenti anticamente presso li più antichi Medici si formauano di cuoio di pesce, ouero di pelle di qualche uccello; hora si fabricano per gli huomini ò d'oro, ò d'argento, ouero di ottone. Per le donne deuono essere alle uolte di piombo, nè troppo pieni nè troppo uoti, ma politi leggieri, rotondi, eguali, uoti, della grossezza della penna d'un'uccello, & un poco corui, ma più ne gli huomini, che nelle donne. Il Medico ne deue hauer tre per i maschi, & due per le femine: per gli huomini l'uno è detto grandissimo, & è di quindici dita, l'altro è mediocre, & è di dodici: & il terzo è minimo, & non passa noue dita: quelli per le donne, il maggiore è di noue dita, & il minore di sei, & deuono esser forati in molti luoghi, acciòche non siano otturati da sangue grumato, ouero da pituita, ò dall'arena, ò pur dalla pietra. Sono chiamati diuersamente, perciòche i Greci gli dicono Cathiter, gli Arabi Cathatir, & i Latini fistula Vescicaria: noi uolgarmente gli chiamiamo siringhe. Queste sono le lor figure.

Aenea fistula. Cels. lib. 7. c. 26. Chateter. Gal. 5. meth. c. 11. Aet. lib. 11. c. 5. Argalia. Gal. 6. c. iter. c. 3. Rectus syphon. Gal. 5. meth. ca. 7. & 11. Paul. lib. 6. c. 59. Fistula vescicaria. Aet. lib. 16. c. 78. Alchartir. Auic. 19. 3. trac. 1. c. 5. & tract. 2. cap. 9. Syringa. Albuc. li. 2. c. 88.

CHA-

FIG. 10.1 Castrator illustration in Giovanni Andrea della Croce, *Cirugia universale e perfetta di tutte le parti pertinente all'ottimo Chirurgo* (Venice: Ziletti, 1574). Wellcome Collection.

research of Richard Sherr, Iain Fenlon, and Giuseppe Gerbino, there is no doubt that Guglielmo Gonzaga, the Duke of Mantua and an intense musical patron, was an early adopter of the castrato.[23] He built his church in large part as a vessel for music making, and though he worked to recruit composers and performers, he remained disdainful of the more festive atmosphere of Ferrara. To acquire singers, he used the Gonzaga diplomatic machine. Vincenzo, Guglielmo's heir, assumed his position as duke in 1587 and established a *concerto delle donne* the same year, continuing the tradition of acquiring musicians and musical instruments. The Mantuan recruiting operation was massive.[24]

The Gonzaga court was a hub of medical knowledge and practice and a place where the human body was a site for the vexed relationship between

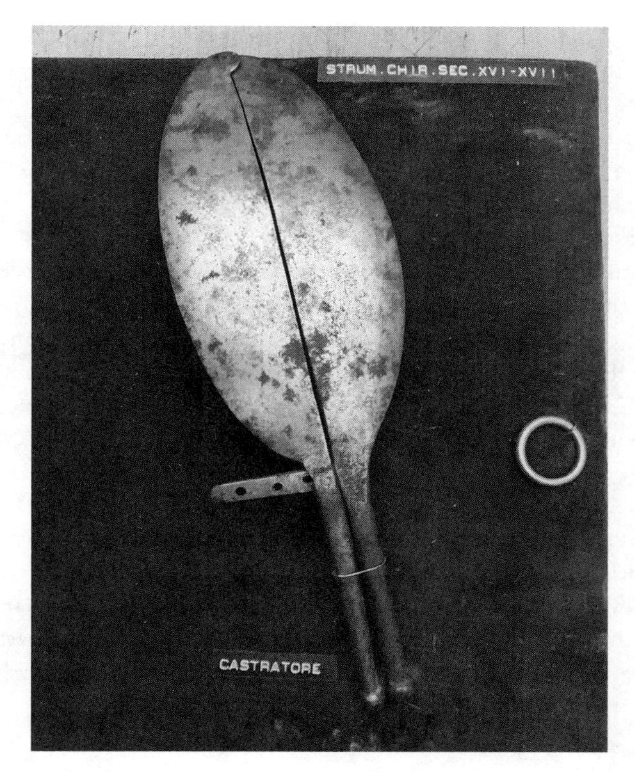


FIG. 10.2 Castrator, Castrato Rizzoli Orthopedic Institute Bologna historical collection.

art and nature. The dynastic family had a host of medical problems and themselves inhabited nonnormative bodies. In efforts to manage their own conditions they kept close tabs on medical developments. Guglielmo was a hunchback who suffered from bone tuberculosis. Vincenzo Gonzaga's failed marriage mentioned in chapter nine led to a lifelong search for "cures."

The duke had as his personal physician Gaspare Tagliacozzi, who was one of the earliest learned surgeons. Born into the artisanal class, he excelled at reconstructive plastic surgery, a practice utilized primarily by nobles. Tagliacozzi's pioneering nose repair surgery played out in early modern reconfigurations of art and nature and challenged the Aristotelian idea that an artist either completes nature's work or refines it. The coexistence of castrato singers and a pioneer of plastic surgery in Mantua is important. When, in 1595, Vincenzo became afflicted with a disfiguring skin condition that sent him home from the battle field, Tagliacozzi came to work on him. In 1597 the surgeon and anatomist dedicated his book on reconstructive surgeries of the ears, lips, and nose to the ailing duke. Valeria Finucci argues that Tagliacozzi dedicated the book to Vincenzo not because of the duke's related ailments but because of the duke's investment in the arts and with the search for beauty. As the artist corrected and extended nature in clay and paint, so, too, did the surgeon correct imperfections with his knife.[25] Tagliacozzi articulated his nose correction surgery not as a new or artificial nose but, rather, as a procedure that made a natural nose more perfect. He claimed to restore beauty and restore the face to its natural state. His notion of plastic surgery emerged from the idea of shaping and molding a form and was distinguished from cosmetic surgery.

Tagliacozzi intentionally distinguished himself from barbers and castrators like those described in Tommaso Garzoni's 1585 *La piazza universale di tutte le professioni del mondo*. An encyclopedic, ethnographic, performative, and literary rendition of a piazza, the text presents a satirical picture of late sixteenth-century professions. Garzoni portrays surgery as an artisanal practice, depending on ingenuity and natural ability. In Discourse 132 Garzoni describes the castrator, locating him among the professions that counted as mechanical, along with plumbers and masons. They learned their trade in the south, perhaps even traveling to Alexandria, Cairo, or Aleppo. Garzoni explained that in antiquity the castrators were a noble profession because eunuchs were honored in ancient history. But in the current moment they were highly suspect.

> As for [the castrator's] relationship to medicine, in and of itself it does retain some honor, but because of the medical subject, it is easily more base and more undesirable than anything else, because in the end a castrator is nothing but a doctor of testicles, or rather, a barber, who, filled with harshness, doesn't know how to cure the wound without first creating it. This profession is often taken up by norsini, as from Norsia [i.e., Norcia] often come those who fix broken arms, and those who heal broken arms, and those who put *brachieri*, which in Latin are called *fasciae*, or bandages, on the virile parts of another kind of medicine that is very different. And because this profession comes down to a small thing, that is to say, the mere cutting of a sack while the man is tied down and held in the manner of a beast.[26]

Garzoni was for sure deliberately debasing those who castrated. Neither Garzoni nor any of his figures would have had a seat at any papal tables, and by the middle of the seventeenth century, medical professionals laid some claim to at least theorizing the procedure.

The term *norcini* refers to a class of practitioners who came from the village of Preci in the region known as Norcia, a town in the south of Umbria, and who specialized in removing kidney stones, cataracts, and inguinal hernias. They also frequently removed growths on the urethra associated with venereal disease. The town of Norcia was and still is known for raising pigs to make sausage, pigs that were castrated to fatten them up.[27] This was a hereditary profession passed down from father to son that defied categorization. They were not barbers, a profession associated with the street and the unclean business of altering humans, but they were frequently associated with them. And although they were not physicians trained in the learned Latin tradition, *norcini* were frequently hired by hospitals and some served as personal physicians to popes. Preci had an important surgical school attached to the Benedictine monastery. Surgical texts produced there used examples of procedures performed by *norcini*. For instance, stones were removed by crushing them first. They used what were at the time cutting-edge instruments and anesthesia, and they put their instruments in a hot fire before using them, an early practice of sterilization. In 1615 Pope Paul V founded a confraternity of *norcini*.

Making castrati prompted legal debates. The team of men who, in 1633, presented reports on Galileo's misdoings to the Holy Offices—Zaccaria Pasqualigo, Melchior Inchofer, and Théophile Reynaud—also debated the making of eunuchs with, among others, Athanasius Kircher.[28] Reynaud called eunuchs a crime against God and nature. Pasqualigo justified the act of castration for the making of a voice by positing the removal of a body part for the good of the whole body as morally good:

> Hence it happens, that in turn much ought to be made of an excellent singing voice both since perfection is for the good of the body; which [perfection] is preferred to many members; and because by reason of the same [excellence of voice] a person is rendered honorable even among princes; finally since it is sufficient, that thenceforth noble sustenance/maintenance of life might be attained, and more comfort is perceived from it [excellence of voice], than from that thing that is fitting for begetting children. Whence it may be judged there is a greater good to excellence of voice than to not being castrated. Therefore, excellence of voice will be able rationally and according to natural law to be preferred over virile things, and accordingly castration is to be employed, lest excellence of voice may perish.[29]

Pasqualigo, in other words, argued that to extend the human voice was to make the body more perfect.

As debates about the castrato's body emerged again during the seventeenth century, there was also an increasingly codified practice examining and evaluating bodies for the places they occupied on a sliding scale of humanity. The idea was that holiness could be proven through human dissection and anatomical investigation. The materials of the body as they reflected the soul and the relationship to the divine were more and more explained by science. When practitioners like Paolo Zacchia and the writers on hermaphrodites, whom I discussed in earlier chapters, described gender abnormality as an aberration made by God or demons, by extension they implied that the medically proven holiness of a dead saint could solidify the boundaries of gender and of the human. Zacchia was consulted on miracles, and his book was a prime subject for physicians called to testify before the courts. Much of his writing showed how to distinguish between the preternatural and the natural, which was effectively the evaluation of whether an event or human body part was part of nature's work or excessive to it. The supernatural, as expressed in bodies scrutinized by the church whether dead or alive, were part of a dangerous game.[30]

## MAKING THE VOICE

The surgical procedure was just the first part of the bodily transformation that made a castrato singer, with the postsurgical training more rigorous, intense, and extensive than what other singers received. The interest in molding the bodily materials that gave rise to the voice bespoke a desire to change human processes and human anatomy. It also mirrored the artisan's desire to mold his materials. To be sure, many of the training techniques were practiced by female singers as well as tenors and basses, but the extent and scope of what castrati received reflected the "investment" already made in the alterations of their bodies for the service of their singing.

The new styles of singing that were codified in print around 1600 depended on a mastery of the body and on the appropriate and expressive use of techniques and voice that earlier musics did not require. Vocal training altered the body by increasing breath control and vocal agility; in practice and performance, singers animated words and feelings through their breath. Most singing teachers argued that the best production was by bodies that were naturally suited to making the proper sounds but also that practice could improve and alter the instrument.

The 1560s are an important decade for these changes. Camillo Maffei, whose account of vocal training was first published in 1562 as a letter to his patron, is a good example. The letter was included in a volume that also dealt with comets, and described the earthquake of 1561.[31] Maffei's document, which has most often been read for its instructions about learning ornaments, names the practice of vocal ornamentation *gorgia* in reference to the relevant body part. Maffei, who was a physician and not a surgeon, also inaugurated a learned discourse around modifying the voice. He asserted that the human voice could be modified by art. Maffei said he wrote the letter because his patron Giovanni di Capua asked whether it was possible to learn to sing without a teacher. Steeped in Aristotle and Galen, the document is deeply indebted to the then-current philosophical and anatomical understandings of the voice and respiratory system.

Maffei's treatment of the voice is very similar to Tagliacozzi's treatment of the human body all told. He presented the voice as an object that, like any piece of art, could be constructed. Following Aristotle, he compared making a voice to making a brass vase. The master of the voice was the soul, he argued, and the instrument was what we would now call the trachea. The materials were air or breath. The voice was a malleable raw material that a singer could work and mold. A good singer, he wrote, could "gain the proper disposition of the throat, with industry, using the method I have described."[32] His instrumental vocabulary prevailed well into the eighteenth century.

For Maffei, nature provided the materials of voice, but training made it complete: "Nature has made, from the sinews descending from the head to the stomach, a branch there, together with its muscles."[33] His description of voice rested in the materials of the body, and he said that neither ancients nor moderns had a good method for making a throat more effective for singing. Maffei then claimed that he was the one who could accomplish this task because it took a medical doctor to understand the voice: "For anyone who tries reasonably to talk about it must be not only a musician but also a learned physician and philosopher."[34] His instructions for vocal training emerged out of his anatomical and philosophical descriptions; detailed instructions for practicing included asserting that it was best to practice four or five hours after eating because if the stomach was full, "the pipe cannot be as polished and clean as is necessary."[35] Maffei concluded with the assertion that nature provided the bare materials, but art made the method. After describing modes of practice, he offered cures for an unhealthy voice, including placing a leaden plate on the stomach and various herbal recipes.

A century later, the castrato voice infused every aspect of Roman sound culture. The Perugian castrato, composer, and historian Giovanni Andrea Angelini Bontempi described the training method used in the school of Virgilio Mazzocchi. Like Maffei a century before, he described the voice in materialist terms. His work was, in part, a defense of musicians' practice:

> The Roman schools oblige their pupils every day to employ an hour in singing difficult and uncomfortable things, to acquire experience, another [hour] in practicing the trill, another in *passaggi*; another in the study of letters, and

> another in instruction and practice of singing, both with the teacher listening and before a mirror, to accustom themselves to make no unbecoming motion; neither of the waist, nor of the brow nor of the eyes, nor of the mouth. And all these were the occupation of the morning.[36]

After lunch, the vigorous protocol continued with half an hour each on theory and counterpoint, followed by an hour of written counterpoint and the study of letters. Students spent the rest of the day practicing the harpsichord or composition. And their business was not yet done, for they completed their regimen with outside exercises in which they

> were to go often to sing and to listen to the response of an echo outside Porta Angelica near Monte Mario, to develop judgement of their own intonation [*accenti*], going to sing in almost all the music that was performed in the churches of Rome, and observing the styles of singing of the many illustrious singers who flourished in the pontificate of Urban VII; imitating these styles and justifying it [the imitation] to the teacher, upon returning home. [The teacher], then, in order to imprint these [lessons] more firmly in the minds of the students, gave necessary explanations and admonitions.[37]

Bontempi's program sounds very similar to the programs that, in the next century, Porpora prescribed for his students (Farinelli among them). The point is that the voice was a technical instrument modified by extreme training, and this view arose out of premodern conceptions of both voice and body as nature and art.

By the end of the eighteenth century, the training regimens described in Maffei's and Bontempi's treatises were well rehearsed. In 1774 the castrato Giovanni Battista Mancini wrote,

> It used to be said in the past that if the trill was not given by Nature that one could not acquire it by art. I do not deny this but let me say that the teachers in the old days, did not leave any voice, agile by nature, imperfect, as we see done in our own times; on the contrary, we have seen on many occasions that they had recourse (whatever obstacles they encountered) to industry, and endless and tirelessly persistent patience in using the most valuable aids in helping their students attain this end. . . . Certainly, I do not wish to reform Nature, even when she is avaricious and miserly with her gifts but whatever she has given some disposition no matter how slight it seems to me to be advantageous and should not be overlooked, and not without fault could I neglect to teach that from which the profession reaps such luster and success.[38]

Mancini made it clear that a good vocal instrument depended as much on training and practice as it did on natural materials. Craftsmanship could alter the materials of the voice.

## EXTRA HUMANS

To conclude I will dwell on two men whose bodies pushed the limits of the human in the middle of the seventeenth century. Jean Royer, a professional puker from France, and Baldassare Ferri, a singer from Perugia, were part of the spectacular (as in both spectacle and wonder) fabric of early modern Europe and especially Rome, performing alongside fireworks, stagecraft, and simulated monsters. Both were cosmopolitan performers who earned themselves prime places in critical writings; they were, in essence, instruments in multimodal productions.

The archives don't have many traces of Royer.[39] In 1653 he arrived in Piazza Navona and treated audiences to feats that called into question his status as a human. Royeur performed his tricks in the shadows of ingenious hydraulic machinery pushing water through an opening between large man-made rocks. He ingested water and, through an invisible force, turned it into something else. Like a charlatan in the piazza, Royer charged money for his deed, and Kircher and his assistant Gaspar Schott were sent out to see whether he was practicing dangerous dark magic. According to a description by Schott, Royer first consumed large quantities of water and then quickly began to vomit:

> he who from his stomach pushed out twelve to fourteen perfumed waters of different colors, most exquisite liquors, wine alight with flame, and oil burning without wick, many kinds of herbs and flowers with full and fresh petals. He also made a fountain by projecting water out of his mouth into the air for the duration of two *Misereres*.[40]

Blurring the relationship between the organic and the man-made, he did the work of a machine and, in effect, became a work of art. Schott made this connection clear with his direct parallel to the stone fountain.

Royer performed his trick in the Piazza Navona, the site of so many of the spectacles described earlier in this monograph and at the center of a city replete with spitting and vomiting fountains. The Jesuit investigators brought the puking Frenchman to Kircher's museum, where he revealed his methods for puking up flowers and wine and exposed the workings of his hydraulic instrument. He extended his body with artifice that remained invisible to those around him. Schott and Kircher worked in performance and, later, in print to show that Royer's talent involved the manipulation of natural causes rather than the work of some intervention, demonic or otherwise. This conceptually reversed the process of sanctification; rather than proving that the human made a miracle and was connected to the divine, the investigation proved a mechanical, artisanal process. It was a human version of Michele Todini taking apart his elaborately grand automatic combinatory keyboard and then putting it back together so that the visiting nobles would see it as a feat of technical virtuosity and not dark magic. The puking Frenchman eventually appeared in Schott's 1657 massive encyclopedic machine book, *Mechanica hydraulico-pneumatica* as Machina VII alongside other hydraulic machines, such as a two-headed Hapsburg eagle that vomited water and had also been in Kircher's museum. The machines celebrated human innovation and revealed secrets of nature and God. In the book Royer became an artifact, and Schott put on the page what had been demonstrated in performance; think printed versions of improvised performances.

In contrast to Royer we know a lot about Baldassare Ferri, who was by all accounts a rock star castrato. Born in Perugia in 1610, the family legend claimed a wound that mandated castration surgery. He studied in Orvieto, Naples, and Rome. In 1626 Cardinal Francesco Barberini sent sixteen-year-old Ferri to the Polish Vasa court as part of a ploy to recruit the Polish court into a coalition against Turkey.[41]

Baldassare Ferri, like Royer and like many of the castrati mentioned in this book, was also a natural wonder, a specimen whose voice was anatomized, justified, and recorded for posterity. The most complete account of Ferri's voice was penned by Giovanni Andrea Angelini Bontempi, and it became a model text for descriptions of the castrato voice, for capturing in words these marvels of sound. I'll come back to that in the epilogue.

Bontempi's encomium anatomized the invisible interiors of Ferri's voice, a narrative version of the pictures shown in anatomy textbooks and elaborate machine books. He explained Ferri's anatomy to prove that the description was "real." The crucial sentence to me in Bontempi's much-quoted description is "The reader [i.e., student] of music should not think that this is a fable because it includes things that seem impossible to execute. For Ferri they were not only possible but they were easy."[42] Bontempi thus assured his readers that while Ferri's voice might seem to do impossible things, it in fact was completely possible because the invisible interior of his body was special and extreme. Bontempi used words to penetrate the surface of Ferri's body and explore the viscera—the cavities, pores, and insides—

> because he did not have a diaphragm like others, but his viscera were much more vigorous, natural, and had more spirit. And because he had a whole and

> large breast, one can also be persuaded that he also had rare and spongy lungs and a more frigid temperament; hence it was that he could hold more spirit than others and make miracles of art with which he made it seem like he was a greater marvel and almost exceeded the humanly possible.[43]

The very large chest of the castrato was filled with humoral spirits that made more sound.

The description of what Ferri's voice could do is grounded in a description of the invisible interiors of the body. It also works to make the singer superhuman or, in modern terms, a kind of cyborg who is already posthuman. Bontempi came back to the spiritual power of the castrato voice in a later discussion of modes. This small segment of Bontempi's theoretical discourse has been used to equate semen and voice.[44] Corollario XXI follows a long discussion of modern modes:

> The effectiveness of singing, governed by the true rules of either modern mode, is vast; for just as *virtù*, or natural spirit, when it is very potent, immediately softens and liquefies the hardest elements, and makes the harsh sweet, and generates propagation outside itself with the production of seminal spirit, thus vital and animal *virtù*, where it is most efficacious with its own spirit, stirs by means of song its own body with its own movement, and through effusion moves neighboring bodies, and with a certain stellar property, which it conceives both from its own form and through the chosen opportunity of time, orders both its own and the alien body. This agitation is born only from the force and virtue of the harmonic number, so widespread in the diversity of the intervals and modes, which are so powerful, that if by observing the modes used in both this region and the other, and by which stars each is dominated, one directed at them the appropriate melodies, composed according to the expediency of those dominant stars, the result would be an almost shared form, and in that form yet another celestial virtue would be created. But given that such knowledge, without the aid of divine fortune, could not be achieved through human diligence, one can in the doctrine of the modes leave aside consideration of the stars and take up instead that of the affects. Their movement arises from the diversity of the harmonic numbers, constituted by the different modes, and from the differing characters of the listeners who hear them.[45]

In some ways this passage is unremarkable. To make the modes work requires a special kind of voice; song affects the body's innards, and doing it with special virtue makes divine harmony. Moreover, it is one paragraph in a two-hundred-seventy-six-page book that begins with the invention of music and covers what was then modern music theory. But read against Bontempi's celebration of Ferri's voice and against his detailed description of Mazzocchi's training program for castrati, it reinforces the argument I am making: that the castrato was a posthuman figure whose singing voice radically transformed his body inside and out.

Song is, as others have insinuated, equated with semen here. But it's not about sexual reproduction. Vocal productivity equals life productivity. Bontempi, throughout the volume, also explained vocal production and anatomy in classic Neoplatonic terms. Voice was animated by breath, pushed from the chest through the throat and into the mouth where tongue, lips, and teeth gave it articulation as speech or song. The voice, song, spirit—whatever it was called—messed with the listener's and singer's souls and bodies; it moved the affections and changed the body. The vocal control of a singer like Ferri was also control of spirit, and it was an animating process. The questions surrounding Ferri's voice, the ardor Pasqualini incited, and Bontempi's celebration of the castrato become questions about what constitutes a human. These descriptions highlight the flesh and blood turned instrument of vocality.

## CONTRA ENLIGHTENMENT

To write a book that situates the voice of any human—castrato or not—as a musical instrument and as a product of techne necessarily questions the limits of the human body and points to the dehumanizing potential of aesthetic practice. Posthumanism has been almost inextricably intertwined with cyberspace, computers, and technological assemblages of bodies in parts. To posit the castrato voice as a kind of machine is to argue that the posthuman turn is not new, that the act of critiquing the Enlightenment notion of human happened before the Enlightenment. It is to argue that humans have always been malleable and remade, and it is maybe even in agreement with Bruno Latour that we are closer to the sixteenth century than the twentieth "precisely because the agreement that created the bifurcation in the first place now lies in ruin and has to be entirely recomposed."[46]

The castrato is most often read through the lens of the Enlightenment. But as I have been arguing, the castrato was a product of the early modern, medieval, and ancient world and exists now in the historical imaginary between fiction and reality. That he was posthuman before humanism suggest that the roots of posthumanism lie within humanism itself. Posthumanist thought is grounded in critiques of the Enlightenment project and especially in dismantling modernist renditions of rationalism and liberty. But the fantasy of technological conquest of nature assumes a particular model of humanism that was never there. Humanism and its discontents were never as monolithic as they seem to contemporary theorists. Rosi Braidotti, for instance, writes against the Vitruvian man as

> the standard of both perfection and perfectibility [that] was literally pulled down from his pedestal and deconstructed. It turned out that this Man, far from being the canon of perfect proportions, spelling out a universalistic ideal that by now had reached the status of a natural law, was in fact a historical construct and as such contingent as to values and locations. Individualism is not an intrinsic part of "human nature," as liberal thinkers are prone to believe, but rather is a historically and culturally specific discursive formation—one which, moreover, is becoming increasingly problematic.[47]

Posthumanism at its root suggests that humans may not be as exceptional as they think; that we share much with animals and with machines; and that humans are inextricably intertwined with animals and technologically created beings. Following Donna Haraway, scholarship on technological extensions and mediations of the human body tend toward an interest in the machinic, and this is the direction I mostly take in this book.[48] But posthumanism also provides a way to reorganize human relationships with other species, and Haraway's more recent work does just that, offering ways to conceptualize the way humans share the planet with other species.[49] Such species-centered scholarship tends to approach a limited historical period and take as universal Enlightenment ideas about individualism. If my project is, at root, an effort to show the machinelike iteration of the castrato voice as premodern posthumanism, then I would be remiss in not also considering the very vexed borders between animals and humans, many of which were enacted in the anatomy theater.

That animals and humans had so much in common anatomically caused philosophical problems for many thinkers. The Gonzagas, who collected castrati, also trafficked extensively in curated and manufactured animals. Over the course of the sixteenth century, the connected Gonzaga and Este courts, for instance, aimed to breed and train especially perfect horses. If humans were created in God's image and animals were not, why did their innards sometimes look the same? Why could Claudius Auberius, a student of Galileo, use the enormous testicles of pigs to show how human testicles worked?[50] These thinkers were already turned toward the nonhuman, and along the way they relied on sympathetic connections. They did not just practice vivisection because of prescriptions against human dissection; they did it because it was not all about exceptionalism. Pythagoras, who was so important to music thinkers, believed that souls could migrate from one species to another. Leonardo's sketches of human body parts, which are easily available these days, notably compare a human leg with a horse's leg, a monkey's arm and a man's arm, in ways that reflect more than page design. He saw them as embodying parallel kinds of motion.

The castrato is a sonic reminder that there was never an actual Vitruvian man; he emerged as an abstract ideal. But it's not just him. Even while Leonardo was drawing his abstract man, the figure was always precariously positioned between multiple materialities. In the century after Leonardo made his man, countless treatises and books of recipes and secrets explained to both men and women how to perfect their bodies through cosmetics, hair removal, and other remedies, making the body already an abstract assemblage of parts.[51] And in the sixteenth century, as the practice of castrato singing became institutionalized, debates about the physiology of sex centered on the capacities of men and women, which sex was suited to rule, and which sex was suited to be ruled.

EPILOGUE

# CADENTIAL HAUNTINGS

When you play certain kinds of improvised music, the hardest part is ending. Even a cadence doesn't offer a prescription. Maybe you start a whole new idea, maybe you go back to the opening melody, or maybe you fade out on a slight dissonance. It turns out that the cadence to a book project that lasted twenty years is equally vexing; the writing becomes more and more a practice of improvisation, of tropes and rhythms that fit together until they don't, of starts and stops. To get out of the ending problem, I'll take a cue from sixteenth- and seventeenth-century writers who often wrote more than one epilogue.

Both endings evade the cadence. The elusiveness of the castrato signals the unknowable nature of sound, desire, and eroticism. Valerie Traub writes that "sexual knowledge is difficult because sex as a category of human thought, volition, behavior, and representation, is for a variety of reasons, opaque, often inscrutable, and resistant to understanding."[1] So as I implied earlier, scholars can use the histories of science, medicine, and sound to think through some aspects of singers with crushed testicles and stunning voices. But the feeling isn't in the texts. Especially with a figure as captivating as the castrato, there is inevitably a taste of what Jacques Derrida called "archive fever," an ecstatic relationship to discovery and a failed consummation. The castrato, as many of my colleagues have written provocatively about, exists in a state of constant disappearance. Their vocality requires excavation.[2] On the sole recordings of the castrato Alessandro Moreschi, Martha Feldman says, "Why not imagine them instead as aural palimpsest, the scraped and funneled parchments of an acoustic past?" Jessica Peritz writes of Gaspare Pacchierotti as a living monument to the recent past.[3]

My first ending resembles a reharmonization of the opening phrase, an explication of a well-discussed text: Giovanni Andrea Angelini Bontempi's description of Baldassare Ferri. It's a quiet restatement of the recurring themes, a reading of a "real" castrato story, an account of a castrato voice by a castrato that encapsulates much of what I've been arguing. The description gives a vivid sense of the vocal tricks and reiterates the value of attending to sources that predate the more familiar Enlightenment ones. Encapsulating the historiography and the philosophical traditions that have been the backbeats to this book, it is an "in addition" kind of epilogue. It keeps sixteenth- and seventeenth-century Italy at the center and rereads a frequently studied text through an Anglophone academic lens.

The second ending is a riff in a related key. Vernon Lee, my favorite queer Victorian—writer of ghost stories, histories, aesthetics, and imaginary portraits—wrote of the castrato as a voice machine. But also, her desire to capture the lost voice of the castrato ghosts that entranced her is a case of archive fever. Her writings invite thoughts on telling stories and the sensory nature of doing history. The study of the castrato almost requires the kind of play that Derrida incited by embracing the complicated relationship between truth and language and an ontological frustration with both.

## FIRST ENDING: WHERE THE BODIES ARE BURIED

Evidently Bontempi was acquainted with a volume of poetry published by Ferri's nephew Giovanni Angelo Guidarelli shortly after the castrato's death called *Il pianto de' cigni in morte della fenice de' musici* (The cry of the swans on the death of the phoenix of music).[4] The poems capture the feeling of absence that emerged almost the moment the singer took his last breath and that haunts descriptions of Ferri's voice from that moment through eighteenth-century commentators like Rousseau and into the twenty-first century.

Guidarelli said that he published the book "to call [Ferri] back, with such erudite voices, to more permanent life."[5] The forty-three poems in Latin and Italian read like a series of conventional encomia; they feature swans, Orpheus, Amphion, tears, gold, fame, and so forth. But within that dizzying array of familiar tropes, Biancamaria Brumana argues, the poetry traces a detailed chronicle of Ferri's life and aims to preserve the voice for future generations. An anonymous poem expresses deep frustration that you cannot record a voice, writing "voices cannot be portrayed like faces, but he has a longing to leave the world with an immortal portrait of the incomparable and angelic music of Mr. Baldassare Ferri."[6] It's as if the poet had time traveled to the twentieth century and read Jacques Rancière, who heard history as absence, as the stuff between the lines:

> And there is history because there is an absence of things in words, of the denominated in names. The status of history depends on the treatment of this twofold absence of the "thing itself" that is no longer there—that is in the past; and that never was—because it never was such as it was told.[7]

Bontempi's account of Ferri's voice, words by a castrato about a castrato, has as much phonographic potential as any record of sound from the seventeenth century. And as Martha Feldman points out, it describes the potential of the castrato voice all told. Ferri mobilized an enormous range and exquisite breath in service of perfect expression of text. He had an exceptional trill, and he could sing almost anything without moving his body.

> In addition to the clarity of his voice, the felicity of his *passaggi*, the beating of his trills, the agility with which he arrives sweetly at whatever pitch he wishes, after the extension of a very long and beautiful *passaggio* in a single measure, he had no need to take another breath. He began, without inhaling, a very long and beautiful trill, and from that passed to another *passaggio* longer and more vigorous than the first, without any movement of the forehead, the mouth or the face, immobile as a statue. The descent by trill from half-step to half-step without any insecurity, and with a voice lightly reinforced from the high octave of a‴ [a twelfth above middle C] to the same a″ [an octave below]—an operation which, if not entirely impossible, at least was of very great difficulty for any other excellent singer—[but] was nothing to Ferri, since from this he passed without taking a breath to other trills, *passaggi*, and marvels of the art.[8]

Ferri used his vocal cords to produce a two-octave chromatic scale, hitting in tune trills on every note without taking a breath. And he stood, immobile like a statue, exuding sound out of his vocal organ. He was a finely tuned machine or instrument but also a human, a reminder that instruments did not achieve their disembodied status during the modernist trend toward obsessive categorization and separation of body from machine.[9]

Bontempi described Ferri within a long discussion of modes, what in modern parlance is a history of theory. The almost three-hundred-page book quickly became a model for music history and theory and was referenced frequently throughout the eighteenth century. Like Kircher, Bontempi drew on Neoplatonic writers to produce a book that both honored ancient traditions and explained current practices. The book title is long, almost a thesis statement itself.

> The History of Music, which includes a complete knowledge of the ancient theory and practice of harmonic music, according to the words of the Greeks, who restored her to her ancient dignity, after she was first invented by Jubal before the deluge, and then, re-invented by Hermes; and as was the case with theory, it was also from ancient practice that was born later on modern practice, containing the science of counterpoint.

In this assemblage of histories, the author, himself a castrato, tells the story, embraces a universal history, and aims to rehabilitate ancient and biblical traditions. From the title itself, Bontempi linked the Judeo-Christian tradition with ancient reason. Jubal has never been a favorite of music scholars though he was "ancestor of all those who play the lyre and the pipe."[10] Music scholars tend to prefer the Greek legend of Pythagoras and the blacksmith. Jubal, a descendant of Cain, was also the brother of Tubal Cain, the "forger of all instruments." Hermes (Mercury in the Roman tradition) protected travelers, orators, and bandits. He was not just the inventor of music but was a translator and a trickster, a figure who crossed boundaries.

The volume invests in what would today be called an embodied practice of history. When historians talk about an "embodied turn," they mean capturing bodily experience and acknowledging the body as a site of history. Bontempi trained with Virgilio Mazzocchi and clearly grounded his understanding of music in the act of singing:

> But how are we to speak of our own experience, we, who among the experts in the art of singing, barely occupy last place? And what better experience to make this truth authentic can we find than the divine song of Cavalier Baldassare Ferri, our compatriot? This sublime singer has explained things with his voice that no one else can even think to show.[11]

He asserted a bodily authority; his description of the training method and of Ferri's voice is implicitly based on his own experience.

This short description became a template for writings about the castrato voice; a phonographic description that is repetition with a difference an almost lost original. As Anne Desler has demonstrated, Charles Burney used Bontempi's paragraphs as a model and called Farinelli an eighteenth-century incarnation of the seventeenth-century star.[12] Like Ferri, Farinelli could stand still as a statue, focusing audience attention on the wonders of his voice. Giovanni Battista Mancini, a castrato and famous singing teacher, wrote of Farinelli that "he could be called the Baldassare Ferri of our century without [raising] objections."[13] This seeming aside in Mancini's widely translated 1774 text *Pensieri e riflessioni pratiche sopra il canto figurato* came as a culmination of Mancini's explanation of the messa di voce. For Mancini, Farinelli "possessed such a perfect messa di voce that, in common opinion, and in my own, it was that which made him eternally famous among singers."[14]

Perhaps more potently in philosophical circles, Rousseau, who claimed that Bontempi inspired his interest in music history, riffed on this short but detailed description celebrating Ferri's voice, quoting Bontempi almost directly in his dictionary entry on the voice:

> The most extensive, the most flexible, the sweetest, the most harmonious *Voice* that has perhaps ever existed appears to have been that of the Chevalier Baldas-

> sare Ferri, a Perugian, from the last century. A unique and prodigious Singer, whom all the Sovereigns of Europe took turns in fighting over, who was heaped with goods and with honors during his lifetime, and whose talents and glory after his death all of Italy's Muses vie with one another in celebrating. All the works written to praise this famous Musician breathe with rapture, enthusiasm, and the agreement of all his contemporaries shows that a talent so perfect and so rare was even above envy. Nothing, they say, can express the brilliance of his *Voice*, nor the gracefulness of his Singing; he had, in the highest degree, all the characteristics of perfection in all the genera; he was gay, haughty, grave, tender at his will, and hearts melted at his pathos. Among the infinity of *tours de force* he made with his *Voice*, I will cite only a single one. He ascended and redescended two full Octaves in a single breath by a continual Trill marked on all the chromatic Degrees with such justness, although without Accompaniment, that if this Accompaniment was struck quickly under the note on which he was, whether Flat or Sharp, the Accord was instantly perceived with a justness that surprised all the Listeners.[15]

And for good measure, further ensuing that Ferri has a place in operatic teleology, the Spanish Jesuit Stefano Arteaga—who cited the above passage from Rousseau—gave Ferri credit for inventing the da capo aria.[16]

Bontempi and Ferri remain relatively unknown figures outside of a tiny circle of scholars, but Rousseau is quite the opposite. His influence carries with it sediments of the castrato voice. Rousseau read Bontempi and, as has already been extensively argued, was at once enamored with the castrato voice and repelled by it. His climate taxonomy was part of a centuries-long discursive process of mobilizing the castrato to racialize the south of Italy, and he used music to draw lines between human and animal. Birds whistle, humans sing.[17] By the nineteenth century, descriptions of Ferri were wrought with the absence of a whole sound world. Count Giancarlo Conestabile wrote a biography in 1846 that quoted Bontempi almost verbatim.[18] He described the system that Ferri sang in as dead: "when the performer dies his art dies with him."[19] By that time Ferri's actual bones were buried in Perugia in San Filippo, the church dedicated to him. And in that century, he had a piazza named after him that sits at the end of a street named after Bontempi. Both men are still imprinted on the cityscape.

## SECOND ENDING: WHERE THE SCHOLARS ARE READING

In 1994, around the time when castrati gained traction in music history circles, Derrida gave a lecture at Freud's house in London called "Archive Fever." He wrote of the archive: "If we want to know what this will have meant, we will only know in the times to come. Perhaps not tomorrow but in the times to come, later on or perhaps never."[20] Derrida used Freud and memory to write of the Western obsession with originals and their repetitions and with the almost erotic allure of finding lost documents, names, dates. The archive emerged as a prosthetic or theater of truth claims in which historians speak for the dead. This essay kept coming up in my mental hard drive as I finished a draft of this book in the 2020–2021. In the year of COVID-19, the end of a presidency that put truth on trial, and in years where history played out on screens, any idea of "what it will have meant" remains unknown. Doubtless, how *now* feels will likely not be accessible one hundred years from now.[21]

This is deeply relevant for the castrato. The castrato calls out the relationship between technologies of inscription and memory and issues an invitation to listen for the emotional debris of the past and the uncanny timbre of disrupted time. Or, as I have put it earlier in this book, castrati existed on and off the stage in a fluid, hyperreal space. Sometimes reading accounts feels like asking a twenty-fourth-century historian of the Trump era to distinguish between *PBS News Hour* and *Saturday Night Live*'s "Weekend Update." The voice is a site where histories are buried and recovered in ghostly form, and, as Jessica Peritz has powerfully argued, it exists in a prehistoric present where a particular sound embeds sedimented layers of affect and subjectivity.[22]

Enter Vernon Lee and her archive fever. She had a consuming passion for music, history, and aesthetics. She fell in love with archives as a teenager and grew into an obsessive archival denizen. The preface to her 1907 book *Studies of the Eighteenth Century* states that "the passion for actually seeing and touching the things of that time . . . had indeed made me hanker after archives and scores."[23] She tediously copied music manuscripts with a writing instrument: "I went on copying those few pieces for weeks, months, years, phrase by phrase; and I got to know them with the love which thinks there is nothing beyond. I filled copybooks with descriptions of utterly forgotten pieces."[24] Years later, when she mailed her files from Italy back to England, her mother played some of the music at the piano. Her "heart beat with such passion that she had to escape to the garden, barely able to stand the thought of hearing 'Pallido il sole,' an aria Farinelli had sung for the mad King Philip of Spain."[25]

Her work vehemently centered performers and the practice of history; writing was an emotional and physical experience, one animated by ghosts. Like other Anglophone travelers she found the otherness of Italy captivating, and Rome particularly felt like a space of endless decay and absence. She challenged the categories of history, fiction, and commentary. A letter to her father, written in July 1870, says that "my story (the one I am now writing) is greatly mixed up with Italian music, so I have to read up on the subject."

Looking for "Dr Burney's" history of music, she came upon a copy of Rousseau's *Dictionnaire de musique*.[26]

Lee's fascination with the castrato started early. She and the painter John Singer Sargent discovered the Accademia Filarmonica when both were just sixteen. Wearing a new pair of gloves each day, Lee obsessively copied manuscripts, and Sargent copied in watercolor the paintings of musicians. They were entranced by a portrait of Farinelli and by his ghost. Relics from the past spoke to them, whispering silent, unheard music. From that moment the figure of the castrato haunted her. Over and over again she made the case for the primacy of voice, for the voice as an instrument, and for the castrato as a present ghost in her fiction and prose.

Lee, who wrote in the years that Alessandro Moreschi was singing in the Sistine Chapel, embraced the materiality of the lost castrato voice, a tangible force object catapulting out from the singer's mouth. She chased her fervor for castrati through the material traces of portraits, instruments, manuscripts, and even recording technology. Her description of Farinelli's voice sounds very much like Bontempi, Rousseau, Mancini, Burney on Ferri, Farinelli, or some fictional amalgamation of them all. But just before the description, she wrote of the space between description and reality.

> Farinelli had a most extraordinary social career, which, magnified by time and ignorance, and fused with his immense musical fame, surrounded him with that species of mysterious effulgence which curious persons are always trying to see through, even at the risk of seeing something far smaller than they imagined. He was eminently one of those figures which leave a deep impression on the popular imagination, an impression which for a long while grows stronger in proportion as the reality is removed further away.[27]

Lee made her strongest statement on the castrato's voice as thing and instrument in her 1891 story "An Eighteenth-Century Singer: An Imaginary Portrait," which narrates the life of the young Antonio Vivarelli.[28] At every moment, Lee portrays the machinery, technology, and instrumentalization of the human that I have been arguing for in this book.

> To make a voice out of nothing at all, or at all events to make a voice into something totally different from the sort of elemental force at which it had begun, was possible to those masters and pupils who virtually knew no limits to time. The necessity of dealing largely with the now obsolete choirboys and with a class of singers preserved from mutation of voice had given the singing masters of the seventeenth and eighteenth centuries the habit of taking up their pupils

> exceedingly young, and teaching steadily on through the long period of vocal development and change. . . . This early beginning not merely enabled the master to *make* the young voice—watching it and manipulating throughout its growth and changes—instead of merely teaching certain tricks to an already made one, but enabled him to devote months to things now hurried over in as many weeks or days.[29]

Lee's 1890 story "A Wicked Voice," about a composer obsessed with a castrato, never uses the words *castrato* or *castrati* but instead uses embodied language: "O cursed human voice, violin of flesh and blood, fashioned with the subtle tools, the cunning hands of Satan."[30] She argued in her studies of eighteenth-century music that the castrato singer "had to be produced almost like a work of art; the physical powers had to be developed to the highest point, the mental powers had to control them to the utmost."[31]

Vernon Lee herself had a serious case of archive fever, plagued by what she could not hear in the sources themselves. It is precisely the impossibility of expressing in words the lost song of the castrato that inspired Lee to use an imaginary portrait. For her the castrato Gaspare Pacchierotti stood as an ultimate lack, a "faded, crumbling flower of feeling."[32] Pacchierotti was, as Jessica Peritz argues, an Orphic figure most famous not for pyrotechnics but for a deeply expressive style.[33] Lee described her search for Pacchierotti.

> In turning over the leaves of memoirs and music-books we try, we strain as it were, to obtain an echo of that superbly wasted vocal genius; nay, sometimes the vague figures of those we have never heard, and never can hear, will almost haunt us. And of all these dim figures of long-forgotten singers which arise, tremulous and hazy, from out of the faded pages of biographies and scores, evoked by some intense word of admiration or some pathetic snatch of melody, there is one more poetical than the rest—for all such ghosts of forgotten genius are poetical—that of Gasparo Pacchierotti, who flourished just a century ago.[34]

It comes as no surprise that, while wandering the outskirts of Padua, Vernon Lee "happened" upon Pacchierotti's garden. The garden for her was a theater of trees and grass, and the magic was made by

> the song of the nightingales, the dances of the fireflies, copying in the darkness below the figures which are footed by the nimble stars overhead. Into such rites as these, which the poetry of the past practices with the poetry of summer nights, one durst not penetrate, save after leaving one's vulgar flesh, one's habits, one's realities outside the gate.[35]

Crossing the threshold of Pacchierotti's secret garden, she felt the presence of a forgotten past:

> The gardener led us into the house, a battered house, covered with creepers and amphorae, and sentimental inscriptions from the works of the poets and philosophers in vogue a hundred years ago—beautiful quotations, which, in their candour, grandiloquence, and sweetness, now strike us as so strangely hollow and melancholy. He showed us into a long, narrow room, in which was a large, slender harpsichord—the harpsichord, he informed us, which had belonged to Pacchierotti, the singer. It was open and looked as if it might just have been touched, but no sound could be drawn from it. [In another room] hung the portrait of the singer, thickly covered with dust: a mass of dark blurs, from out of which appeared scarcely more than the pale, thin face—a face with deep dreamy eyes and tremulously tender lips, full of a vague, wistful, contemplative poetry, as if of aspirations after something higher, sweeter, fairer—aspirations never fulfilled but never disappointed, and forming in themselves a sort of perfection. This man must have been an intense instance of that highly-wrought sentimental idealism which arose, delicate and diaphanous, in opposition to the hard, materialistic rationalism of the eighteenth century; and the fascination which he exerted over the best of his contemporaries must have been due to his embodying all their vague ephemeral cravings in an art which was still young and vigorous—to his having been at once the beautiful soul of early romanticism and the genuine artist of yet classic music.[36]

She hoped to find some material evidence of the remarkable voice in that garden and in that house, to feel it even if she could not hear it. The garden and its vibrations keep the past as part of the materialist present.

In 2013 a collective of Italian scholars dug up Pacchierotti's skeleton. They thought his bones might reveal the secret of his voice and of the castrato's voice more generally. In 2021 a Google image search summons to the screen an oddly smiling skull juxtaposed with his deadpan in *Fortune* magazine and a flat skeleton on a turquoise background in *Nature*. In his bones they sought tangible evidence of the superhuman voice that Lee craved. He had, as it turns out, a weird mouth and an erosion of the cervical spine that looks very much like the cervical spines of well-trained opera singers today.[37] He had extra respiratory muscles and shockingly good teeth. I suspect Vernon Lee would very much have wanted to excavate those bones as another way to fill the phonographic void. But she would have found that the bones are just another kind of ghostly trace, another artifact that, in the end, fails to sing.

Vernon Lee was hungry for the sounds of eighteenth-century music and bemoaned spending more time writing about music than listening to it. She understood that live performance and recordings constitute some of that sensory present. Operas, oratorios, and songs were—and are always—encountered through the body of a performer in the present. Cecilia Bartoli, writing about her album *Farinelli*, described her body as "a vessel through which a work or a character is convincingly transmitted."[38] In the case of much early modern music, this is an assemblage of historical performance and inscription; so much music was already a reperformance and an invasion of the past into the present by a singing body. The lament of the Magdalene or any of the many Orpheus dramas were phantasmic: music dramas based on ancient stories made to rebirth a past that never happened. The archaeology is dizzying: assemblages of quotations, influences, rewritings, and false starts. Stories of the past set in the present retell past narratives, finish them, fill in the blanks.

In the winter of 2020, just days before Super Tuesday, just before COVID-19 shut down live performances across the globe, the Metropolitan Opera sent their version of Handel's *Agrippina* out to millions of viewers via a "The Met: Live in HD" simulcast. The opera premiered on December 26, 1709, in Venice, with a libretto by Cardinal Vincenzo Grimani, and it tells the story of Agrippina conspiring to put her son Nero on the throne and to put a nail in the coffin of Claudius's reign. The libretto has many contemporary allusions and may have been a satire directed at Pope Clement XI. It starred the castrato Valeriano Pellegrini as Nero, the female soprano Margherita Durastanti as Agrippina, and the alto castrato Giuliano Albertini as Narcissus.

Supposedly about an insane and despotic Roman emperor, the story was as excruciatingly relevant to the winter of Trump's impeachment as it was a satirical slam on the pope in the eighteenth century. This is an opera already caught in the fabric of memory and reperforming history; Agrippina herself never even appeared in Monteverdi's version of the story *L'incoronazione di Poppea*. Director David McVicar gave Claudius a Trump-like, ridiculously long red tie and decorated his stage with microphones, cameras, and cell phones. Kate Lindsey, as Nero, moved her voice between ethereal softness and raging coloratura, her body slithering around the stage and moving Michael Jackson–like through a series of crotch grabs. She paraded up a golden staircase that can only be likened to the golden escalator in Trump's New York hotel. She combined vocal acrobatics with a hyperphysicality that was breathtaking. Sung while rhythmically chopping what looked like pounds of cocaine, the aria "Come nube che fugge dal vento" circles through obbligato passages, pitting voice against instrument. The words say that Nero is done with erotic shenanigans, the voice clashes with the instruments in erotic dissonances that make audiences shiver. In the aria "Quando invita la donna l'amante," she sang coloratura and long notes while doing a side plank and jumping, push-up style, upstairs. In plot, body, voice, and sentiment, it was a moment of temporal disconnect. Perhaps most dramatically, when Nero ascends the throne in this production, he gets no happy ending, no Juno descending on a cloud to bless the future. Instead, he ends in a tomb, always already dead and always already a creature of memory.

## CODA: SCREENS

I forgot that Derrida's "Archive Fever" has a hint of ominous wonder at the "little portable Macintosh." Derrida explained:

> I pushed a certain key to "save" a text undamaged, in a hard and lasting way, to protect marks from being erased, so as thus to ensure salvation and indemnity, to stock, to accumulate, and, in what is at once the same thing and something else, to make the sentence thus available for printing and for reprinting, for reproduction.[39]

He warned that email would soon transform the world:

> This is not only a technique, in the ordinary and limited sense of the term: at an unprecedented rhythm, in quasi-instantaneous fashion, this instrumental possibility of production, of printing, of conservation, and of destruction of the archive must inevitably be accompanied by juridical and thus political transformations.[40]

As I finished this book, we were living this kind of transformation. My first version of this epilogue written on January 6, 2021, says "they are storming the capital." I stopped midsentence, closed my laptop, and rushed home so that my children would not have to watch alone. The little pocket computer (my phone) delivered messages, and the only way to escape the unprecedented rhythm of the moment was to turn off the screens. Just before I returned the final version to the press, the Supreme Court dropped the Dobbs decision. On my tiny screen I read the screed with a logic grounded in the centuries covered in this book that will have devastating consequences for generations.

I finished the book with a range of techniques that would have seemed like science fiction when I started this project: endless Zoom collages of friends, students, and colleagues, terrifying bits of news audible though little rubber bits stuck in my ear. Thanks to massive leaps in technology, open access and emergency scanning of books in a global pandemic that made international travel all but impossible, books—and even manuscripts I flew to Italy to look at—now pop up on my laptop and iPad. The CD-ROMs I paid a lot of money for at the Vatican library no longer talk to any reputable computer, speaking a dystopian modern technological absence. I beamed the chapters back to Rome or to Tennessee for editing and proofreading. A few years ago, I wanted to go to Perugia to look at that book of poems for Ferri that Bontempi read and whose aura echoes. I never made it. But it came as a PDF in my email just days before I turned in the book.

This book has been a sensory practice; reading in strange places, feeling the vibrations of giant churches, playing obscure tunes on piano or viola, searching for words, exploring cities, tripping on cobblestones. On one of these adventures, I wandered into Vernon Lee's garden. Running through the hills around the Villa I Tatti, I passed many decaying and boarded-up mansions. I scoped them out, which ones had spy cameras and which didn't. One caught my eye and never seemed to have anyone there. One morning I jumped the fence. The garden, it turned out, was not so decrepit; it felt magical, in part because I knew I was not supposed to be there. I later learned that the fifteenth-century villa was available for rent and that it was the garden of Violet Piaget, better known as Vernon Lee. Like her, I believe in the magic of history and crave the fantasy.

# ACKNOWLEDGMENTS

The first time I broke a promise to finish this book was to the National Endowment for the Humanities. The looming end-of-the-year grant report prompted me to write, but not submit, a snarky list of things that derailed the book. "Spent a few days in conversation with the pediatrician when the eighteen-month-old ate the head off the three-year old's batman; the head never emerged from the baby." Back at the blank screen of the actual report, I remembered the words of an undergraduate advisor, Roger Henkle, who said, "Sometimes life intercedes." And I finally wrote the real report, sheepishly explaining that between the application and the fellowship I had birthed a third kid and started teaching at a new school.

I wondered how life could do anything but intercede. This project was delayed again and again for reasons that sound like the reproductive academic woman's version of "my dog ate my homework." The global pandemic that ate my last sabbatical knocked time out of joint in ways that leave historians with archive fever. I think now that the English professor who made eighteen-year-olds write with precision and rigor but offered the occasional outload reading of "A Christmas Carol" wasn't really talking about obstructions. Scrooge cries out to the good spirit, "Your nature intercedes for me, and pities me. Assure me that I yet may change these shadows you have shown me by an altered life?" *Intercede* might have meant that even an academic book is the in-between, the emotional residue, and the tangled memories.

Acknowledgments are protosocial media. They tend to snapshot cheerful spouses and children who quietly and patiently offer dulcet tones of encouragement as the scholar toils away in a hermetically sealed studiolo. Those who know me know about my allergic reaction to such narratives and my tendency to respond by enumerating the wretched behavior of my offspring and the 810 reasons why I was never mother of the year. By design, acknowledgments feature colleagues as supportive and generous, libraries as always open, funding as a given, and teaching as always joyful. It is tempting to write a kind of anti-acknowledgments. "Thank you to the administrator who said that my work is fluffy and to everyone who gave me the opportunity to try to get decent ADA accommodations. You reminded me that rage can be production and that sublimation is not always a bad thing."

This book would no doubt have been a quicker and more boring book had it not been a part of a family. My husband and kids changed the way I feel the world, and they rerouted where I went literally and figuratively. They made research trips less about reading and writing and more about hearing the place. I gave my first paper on castrati when I was pregnant with Rebecca and Jonathan Lerdau; they will be twenty when this book goes live. The third child, the one who spiced up my NEH, Eli Lerdau, will be sixteen. If you're reading this, raising kids, serving on more committees than you knew existed, and trying to write a book, then you are correct that it sucks a lot of the time. It doesn't especially help when kids ask you at ages five, ten, fourteen when it will be done. It does help, however, when they make you laugh. Without my husband Manuel Lerdau, I'd still be writing. He read every word more than once, and he probably read some sentences at least ninety-three times. He celebrated every milestone even when I refused and was the resident natural scientist and jargon meter. We are all beyond lucky to have had my parents, Joyce and Richard Gordon, as partners in this endeavor. My mom was the coparent on more than one research trip, and together my parents made it possible for our household to have two careers and three kids. My sister, Pam Gordon, keeps it real all the time, especially in our daily pandemic phone calls.

It really does take a village to raise children, and mine were fortunate to have amazing babysitters who tolerated and loved them, drove them and me all over the place, and much more. I'm forgetting some but I owe prosecco to Rebecca Hetherington, Kathleen Flynn, Tami Morse, Kassie Hartford, Kathryn Kaiser, Lauren Hauser, Brie Meese, Elena Price, Amber Johnson, Staley Slaughter, Brian Micken, and Rachel Mink. And I'm indebted to the moms and other friends in the neighborhood who shared the craziness, especially Stephanie Bolton, Dee Cogill, Molly NcShane, and Cynthia Suchman. I'm an obsessive exerciser and runner, and the book would not have happened without so many running buddies—Sheila Blackford, Liz Wittner, Lawrie Balfour, Anna Brickhouse, Cori Field, Alicia Long, Liza Flood, Barbara Wilson—willing to brave the summer heat of Virginia and Rome. Lisa Brook is a special running buddy whom I've known since middle school, who is more encouraging than anyone I know, and whose daughters Sophia and Chloe bailed me out more than once.

To study music about which the score can only tell part of the story means that you must hear it in your head. My inner speakers work in large part because of the amazing musicians I've been able to play with over the last decade. The various iterations of Side Hustle include Liz Varon, Rich Schrager, Eric Lott, Dave Kirshner, and Vonnie Callan especially. Ten years ago I asked my colleague John D'earth to write a piece for a conference on Jefferson. He asked me to play with the band. It's been a wild improvisatory ride that helped me hear in ways I never imagined, and I'm forever grateful for his listening ears.

Teaching gets in the way of writing all the time. But it also enriches it. Hundreds of University of Virginia students have let me try out ideas and forced me to make them accessible. They remind me that history is fun and that it matters. Thanks especially to the students in my first graduate class here, who told no one, including me, that I fell asleep from exhaustion in my own seminar. I accidentally wrote a book that has Italian, Spanish, French, and Latin, and thanks to a visual disability my spelling even in English does not meet expectations. For transcription and translation help I'm grateful to Julia Victoria, Timothy Brannelly, Sergio Silva, Nataliz Perez, Emily Mellen, and David Hewett. Every early modernist should have a nephew who is a classics professor; Joe Howley rescued me multiple times.

The one thing I've always successfully negotiated for is a research assistant as an ADA accommodation. The visual disability that I was born with got harder and harder to compensate for over the two decades I worked on this project. I have no words to thank this group of amazing thought partners who not only proofread everything from emails to seven drafts of the same thing but who really changed the way I think and work. Emily Gale and Stephanie Gunst each worked with me for multiple years. Natalia Perez, Courtney Kleftis, Stephanie Doktor, and others were crucial to the project. And at the very last minute Carlehr Swanson walked into my office and said she knew how to use Finale. Ramona Martinez also came in with wildly imaginative illustrations.

My mentor and friend Deborah Wong talks about citational footprints as a practice marking a collaborative process that deliberately thinks with others and takes note of generational influences and structures. When I chose to spend my intellectual life largely in the esoteric world of gender and early modern Italy, I chose a space with brilliant, badass, and fun colleagues. Martha Feldman and Suzanne Cusick are the absolute best versions of mentors turned lifelong friends. They have read almost everything I've written for the last twenty years. Without their work, their incredible brains, and their love, this book would not be here. Many others supported this book in countless ways: Cristle Collins Judd, Cori Field, David Levin, Christopher Morris, Sara Lipton, Deborah Baker, Tess Slominsky, Emily Wilbourne, Jessica Peritz, Jane A. Bernstein, and Jessie Ann Owens. Richard Will and Joe Auner read the whole book even before it was proofread. Claudrena Harold, Deborah McGrady, Sindhu Revuluri, Oliva Bloechl, and Mark Burford have been brilliant friends, readers, advocates, and cheering sections. Julie Carrucio has been friend, work partner, and reality check. Gary Tomlinson continues to be a tremendous intellectual influence, without whom I'm certain the book would be very different.

I have four guardian angels / secret weapons. Kenny Murata has been a behind-the-scenes copy editor for a decade. We have never met, but he fixed up almost everything I wrote. Courtney Quaintance edited all my Italian and much of the rest of the book the first time I sent it in. Shawn Keener is truly magic as an editor and bookmaker. And Katherine Churchill was my last research assistant on this project and drew the short stick of helping a visually impaired person navigate a massive document and sort out images. She's a brilliant thinker and a generous reader.

My historical and cultural training have prepared me to write a book like this. I'm the product of a Talmudic tradition, and my music history PhD program emphasized close reading, rigorous argument, command of musical styles, and reading proficiency in multiple languages. The harder challenge is to identify, acknowledge, and redress current injustices. This book coincided with a massive professional shift for me toward collaborative community-driven work that allowed me to work over the years with Clarissa Foley, Rachel Caldwell and Rose Cole, and Abigail Kaiser and Brian Kaiser. Founding the Equity Center with Dayna Mathew, Susan Kools, Ben Allan, Nancy Deutsch, and Camille Burnett taught me what interdisciplinary really is. Barbara Brown Wilson has been my guide and support in all of this. And Anne Coughlin and Nomi Dave, my partners in the Sound Justice Lab, kept me ticking this last year.

Marta Tonegutti was a fantastic advocate at the University of Chicago Press and has encouraged this project from the start. Kristin Rawlings, Steven LaRue, and Stephen Twilley were an amazing team. Tim Wilkerson and Aron Teel kept my computers ticking. The book has been generously funded by the University of Virginia College of Arts and Sciences, the UVA / Università Roma Tre exchange, the Democracy Initiative, and the National Endowment for the Humanities. Previous iterations of this work appeared in *Opera Quarterly* and *New Literary History*. And almost all of this book was given first as a talk or conference paper. I'm deeply grateful to so many conveners, panel mates, and audience members. I'm tempted to include also the stuff I applied for and failed to get; but suffice it to say, behind every fellowship lies a ton of rejections. I still can't fathom my time as the Robert Leeman Visiting Professor at the Villa I Tatti and Alina Payne for making that happen. I am a fundamentally different thinker thanks to the quirky brilliant cohort there, thanks to the room of one's own, and thanks to the time spent wandering the hills. I'm especially thankful for Katharine Park, Marguerite Deslauriers, Liz Wekhurst, Chriscinda Henry, and John Hunt.

I flew home from Florence on July 5, 2017; three days before the KKK marched into Charlottesville and just over a month before a violent white supremacist rally whose opening act was a torch march past my office. I didn't finish the book that summer, and I didn't teach much of what I had planned that fall. But I became even more committed to tenure as an obligation to talk back and to be present. In 2022 I sent this book to the copy editor on the day I taught Monteverdi's *L'Orfeo*. My class met a few hours before Mike Pence was to give a talk on "how to save America from the woke left." My students asked with nervous laughter whether we were the woke liberal mob they seemed so worried about. Historians of the twenty-fourth century will struggle to square accounts of the woke liberal mob reportedly getting in Pence's way with snapshots of confused kids carrying violins trying to get to quartet rehearsal. History here in Charlottesville has never been an abstract game. While writing this book I suspect I've been doing a kind of sonic witnessing. Witnessing is by no means passive; it is active, participatory, a kind of being and acting with others. Knowledge production, it turns out, isn't just what you read; it's where and with whom you happen to be.

# NOTES

## PROLOGUE

1. The classic text is Jack [Judith] Halberstam, *In a Queer Time and Place: Transgender Bodies, Subcultural Lives* (New York: New York University Press, 2005). I am most influenced by Carla Freccero, "Queer Times," *South Atlantic Quarterly* 106, no. 3 (2007): 485–94.

2. For an account of the castrato and the Grand Tour, see Berta Joncus, "One God, So Many Farinellis: Mythologising the Star Castrato," *Journal for Eighteenth-Century Studies* 28, no. 3 (2005): 437–96. For Romantic accounts of the castrato, see James Q. Davies, "'Veluti in speculum': The Twilight of the Castrato," *Cambridge Opera Journal* 17, no. 3 (2005): 271. The most famous account of this sort is in Stendhal, *Life of Rossini*, trans. Richard N. Coe (New York: Criterion Books, 1957).

3. For a summary of recent work and an account of the sound, history, and reception of the castrato, see Martha Feldman, "Castrato Acts," in *The Oxford Handbook to Opera*, ed. Helen M. Greenwald (Oxford: Oxford University Press, 2014), 395–418. Feldman also provides a virtuosic rehearsal of the characteristic sounds of castrato voices, their rise and fall in popularity, and their reception history throughout Europe.

4. The Italian peninsula did not unify as a nation until 1861. In the seventeenth and eighteenth centuries it was still a region of separate city-states with enormous cultural, political, and linguistic differences.

5. Donna Haraway, "A Cyborg Manifesto: Science, Technology, and Socialist-Feminism in the Late Twentieth Century," in *Simians, Cyborgs, and Women: The Reinvention of Nature* (London: Routledge, 1991), 144.

6. Haraway, 144.

7. Anne McCaffrey, *The Ship Who Sang* (New York: Ballantine Books, 1969), 6.

8. McCaffrey, 11.

9. Digitized versions of Alessandro Moreschi's recordings are available on YouTube. Martha Feldman writes eloquently on the technical limitations of early recording methods and the particular challenges that frequencies of the castrato voice posed. See Martha Feldman, "Red Hot Voice," in *The Castrato: Reflections on Natures and Kinds* (Oakland: University of California Press, 2015), 79–132.

10. Jonathan Gil Harris and Natasha Korda, eds., *Staged Properties in Early Modern English Drama* (Cambridge: Cambridge University Press, 2006).

11. Sylvia Wynter, "1492: A New World View," in *Race, Discourse, and the Origin of the Americas: A New World View*, ed. Vera Lawrence Hyatt and Rex Nettleford (Washington, DC: Smithsonian Institution Press, 1995): 5–57.

12. Enrique Dussel, "The Invention of the Americas: Eclipse of 'the Other' and the Myth of Modernity," trans. Michael D. Barber (New York: Continuum, 1995), 12.

13. Carolyn Abbate, "Sound Object Lessons," *Journal of the American Musicological Society* 69, no. 3 (2016): 793–829.

14. Trevor Pinch and Karin Bijsterveld, "New Keys in the World of Sound," in *The Ox-*

*ford Handbook of Sound Studies*, ed Trevor Pinch and Karin Bijsterveld (New York: Oxford University Press, 2012), 4.

15. Rebecca Cypess, *Curious and Modern Inventions: Instrumental Music as Discovery in Galileo's Italy* (Chicago: University of Chicago Press, 2016).

16. The voice literature is expanding remarkably. See, for instance, Nina Eidsheim and Katherin Meizel, eds., *The Oxford Handbook of Voice Studies* (New York: Oxford University Press, 2019); and Martha Feldman and Judith T. Zeitlin, eds., *The Voice as Something More: Essays toward Materiality* (Chicago: University of Chicago Press, 2019).

17. For a recent discussion, see Jerry Green, "Melody and Rhythm at Plato's *Symposium* 187d2," *Classical Philology* 110, no. 2 (April 2015): 152–58.

18. Martin Heidegger, *On the Way to Language*, trans. Peter D. Hertz (New York: Harper Collins, 1971), 45.

19. See Jacques Derrida, *Of Grammatology*, trans. Gayatri Chakravorty Spivak (Baltimore: Johns Hopkins University Press, 1976), 101–40.

20. Gary Tomlinson, *The Singing of the New World: Indigenous Voice in the Era of European Contact* (Cambridge: Cambridge University Press, 2007); and "Ideologies of Aztec Song," *Journal of the American Musicological Society* 48, no. 3 (1995): 343–79.

21. Carolyn Abbate, *Unsung Voices: Opera and Musical Narrative in the Nineteenth Century* (Princeton, NJ: Princeton University Press, 1991); and Roland Barthes, "The Grain of the Voice," in *Image, Music, Text*, ed. and trans. Stephen Heath (New York: Hill and Wang, 1977), 179–89 (originally published in France in 1972).

22. Roland Barthes, "Listening," in *The Responsibility of Forms: Critical Essays on Music, Art, and Representation* (Berkeley: University of California Press, 1991), 255.

23. Adriana Cavarero, *For More Than One Voice: Toward a Philosophy of Vocal Expression*, trans. Paul A. Kottman (Stanford, CA: Stanford University Press, 2005).

24. Shaun Fitzroy Tougher, ed., *Eunuchs in Antiquity and Beyond* (London: Classical Press of Wales; Duckworth, 2002); and Kathryn M. Ringrose, *The Perfect Servant: Eunuchs and the Social Construction of Gender in Byzantium* (Chicago: University of Chicago Press, 2007).

25. Otto Meinardus, "The Upper Egyptian Practice of the Making of Eunuchs in the XVIIIth and XIXth Century," *Zeitschrift für Ethnologie* 94 (1969): 47–58.

26. Neil Moran, "Byzantine Castrati," *Plainsong & Medieval Music* 11, no. 2 (2002): 99–112.

27. Kathryn M. Ringrose, *The Perfect Servant: Eunuchs and the Social Construction of Gender in Byzantium* (Chicago: University of Chicago Press, 2007).

28. Almut Höfert, Matthew Mesley, and Serena Tolino, eds., *Celibate and Childless Men in Power: Ruling Eunuchs and Bishops in the Pre-modern World* (New York: Routledge, 2017).

29. Palmira Brummett, "Placing the Ottomans in the Mediterranean World: The Question of Notables and Households." *Osmanlı Araştırmaları* 36 (2010): 77–96.

30. Ángel Medina Álvarez, "Los atributos del Capón," *Música oral del sur: Revista internacional* 3 (1998): 9–29.

31. There is a growing literature on the castrato. See Martha Feldman, *The Castrato*; Valeria Finucci, *The Manly Masquerade: Masculinity, Paternity, and Castration in the Italian Renaissance* (Durham, NC: Duke University Press, 2003); Roger Freitas, "The Eroticism of Emasculation: Confronting the Baroque Body of the Castrato," *Journal of Musicology* 20, no. 2 (2003): 196–249; John Rosselli, "The Castrati as a Professional Group and a Social Phenomenon, 1550–1850," *Acta Musicologica* 60, no. 2 (1988): 143–79; Patrizio Barbieri, *The World of the Castrati: The History of an Extraordinary Operatic Phenomenon*, trans. Margaret

Crosland (London: Souvenir, 1996); Angus Heriot, *The Castrati in Opera* (London: Secker and Warburg, 1956); and Richard Sherr, "Guglielmo Gonzaga and the Castrati," *Renaissance Quarterly* 33, no. 1 (1980): 33–56.

32. "In che anno per la prima volta comparvero le donne sui teatri della città eterna? Non si sa: si sa invece quando lacorte pontificia vietò che vi comparissero. E fu nel 1588, regnante Sisto V, il quale, concedendo alla compagnia comica dei *Desiosi* di dare rappresentazione pubbliche, ordinò che si dessero di giorno e con uomini anche nelle parti di donna." Alessandro Ademollo, *I teatri di Roma nel secolo decimosettimo* (Bologna: L. Pasqualucci, 1888; facsimile ed., Bologna: Forni, 1969), 136.

33. Giulia de Dominicis, "The Roman Theatres in the Age of Pius VI," *Theatre History Studies* 21 (2001): 81.

34. See Frederick Hammond, *Music and Spectacle in Baroque Rome: Barberini Patronage under Urban VIII* (New Haven, CT: Yale University Press, 1994), for information about Rome. Giuseppe Gerbino has done a very thorough early history of the Sistine Chapel choir in "The Quest for the Soprano Voice: Castrati in Renaissance Italy," *Studi musicali* 33, no. 2 (2004): 303–57.

35. Emily Wilbourne, *Race, Voice and Slavery* (Oxford University Press, forthcoming). For one example of the use of enslaved labor, the d'Este Turkish slaves are discussed in the preface to Mohammad Gharipour, ed., *Gardens of Renaissance Europe and the Islamic Empires: Encounters and Confluences* (University Park: Pennsylvania State University Press, 2017).

36. Frederick Hammond, *Music and Spectacle in Baroque Rome*, 46.

37. Frederick Hammond, "More on Music in Casa Barberini," *Studi musicali* 14 (1985): 239.

38. See John Walter Hill, *Roman Monody, Cantata, and Opera from the Circles around Cardinal Montalto* (Oxford: Clarendon Press, 1977); and Keith Christiansen, "Music and Painting in Cardinal Del Monte's Household," *Metropolitan Museum Journal* 26 (1991): 213–27.

39. Hammond, *Music and Spectacle in Baroque Rome*, 240.

40. Charles d'Ancillon, *Eunuchism Display'd. Describing All the Different Sorts of Eunuchs* [. . .] (London: 1718), 30–34. Originally published as *Traité des eunuques, dans lequel on explique toutes les différentes sortes d'eunuque* [ . . . ] (Sweden: Marknadsdomstolen, 1707).

41. For a late twentieth century medical explanation of these changes, see Meyer M. Melicow and Stanford Pulrang, "Castrati Choir and Opera Singers," *Urology* 3, no. 5 (1974): 663–70; as well as Enid Rhodes Peschel and Richard E. Peschel, "Medicine and Music: The Castrati in Opera," *Opera Quarterly* 4, no. 4 (1986): 21–38. Roger Freitas, in "The Eroticism of Emasculation," 226–28, also assesses the medical literature on this subject.

42. Faccone to Ferdinando Gonzaga, Rome, February 9, 1613, cited and translated in Susan Parisi, "Acquiring Musicians and Instruments in the Early Baroque: Observations from Mantua," *Journal of Musicology* 14, no. 2 (1996): 138.

43. Parisi, 139.

44. This castrato's career is discussed in Hammond, "More on Music in Casa Barberini."

45. On Paul of Aegina, see Raffi Gurunluoglu and Aslin Gurunluoglu, "Paul of Aegina: Landmark in Surgical Progress," *World Journal of Surgery* 27, no. 1 (2003): 18–25.

46. Paul of Aegina, *Epitome of Medicine*, translated by Francis Adams as *The Seven Books of Paulus Aegineta* (London: Printed for the Sydenham Society, 1846), 2:379–80.

47. In Italian the monumental book was published as Giovanni Andrea della Croce, *Cirugia universale e perfetta di tutte le parti pertinenti all'ottimo Chirurgo* (Venice: Ziletti, 1574).

48. "Cresce ancor la carne trà le tuniche, lequali, i Greci chiamano Oscheon gli Arabi

Barichem, & i Latini Scrotum, di donde nasce una certa durezza carnosa, detta da Greci Sarcocelum, dalli Arabi Burum, & da Latini Hernia carnosa; è del tutto fuor di natura, la onde ha bisogno d'esser rimossa: nondimeno se sarà congiunta col testicolo, overo rivolta intorno, facciasi castratura, alla qual'opera si ricercano quattro cose, il castratore, il rasoio, il filo, & li cauterij." Croce, *Cirugia universale*, bk. 7, 21.

49. Thomas Gibson, "The Prostheses of Ambroise Paré," *British Journal of Plastic Surgery* 8 (1955): 3–8.

50. d'Ancillon, *Eunuchism Display'd*, 15–16.

51. Amy Brosius "Courtesan Singers as Courtiers: Power, Political Pawns, and the Arrest of Virtuosa Nina Barcarola," *Journal of the American Musicological Society* 73, no. 2 (2020): 207–67.

52. Sigmund Freud, *Essays on Sexuality*, vol. 7 of *The Standard Edition of the Complete Psychological Works of Sigmund Freud*, ed. James Strachey et al. (London: Hogarth Press, 1953–1974), 123–243.

53. Roland Barthes, *S/Z*, trans. Richard Miller (New York: Farrar, Straus and Giroux, 1974).

54. Barthes, *S/Z*; Michel Poizat, *The Angel's Cry: Beyond the Pleasure Principle in Opera* (Ithaca, NY: Cornell University Press, 1992). For examples of heavily psychoanalytic and castration-focused readings, see Joke Dame, "Unveiled Voices: Sexual Difference and the Castrato," in *Queering the Pitch: The New Gay and Lesbian Musicology*, ed. Philip Brett, Elizabeth Wood, and Gary C. Thomas, 2nd ed. (New York: Routledge, 2006), 139–54. A similar sentiment underlies the entire collection of articles published as Corinne E. Blackmer and Patricia Juliana Smith, eds., *En travesti: Women, Gender Subversion, Opera* (New York: Columbia University Press, 1995).

55. Freitas, "The Eroticism of Emasculation."

56. *The Roman History of Ammianus Marcellinus*, "The Faults of the Roman Senate and People," published in *The Surviving Books of the History of Ammianus Marcellinus*, volume 1 of *Ammianus Marcellinus*, trans. John C. Rolfe, Loeb Classical Library 300 (Cambridge, MA: Harvard University Press, 1935), http://penelope.uchicago.edu/Thayer/E/Roman/Texts/Ammian/home.html.

57. Cecilia Bartoli, *Sacrificium*, 2009, Decca 460502670325, 2009, compact disc.

58. Jan Swafford, "Nature's Rejects: The Music of the Castrato," *Slate*, November 9, 2009, http://www.slate.com/articles/arts/music_box/2009/11/natures_rejects.html.

59. Paul Henry Lang, "Performance Practice and the Voice," in *Musicology and Performance*, ed. Alfred Mann and George J. Buelow (New Haven, CT: Yale University Press, 1997), 185–98.

60. Paul Henry Lang, *George Frideric Handel* (New York: W. W. Norton, 1966), 170.

61. Heriot, *The Castrati in Opera*.

62. Heriot, 23.

63. Winton Dean, "The Castrati in Opera by Angus Heriot," *Music & Letters* 38, no. 1 (1957): 82–84.

64. Alice Hoffman, "Luxury, Sex, and Music," *New York Times*, Oct. 10, 1982, https://www.nytimes.com/1982/10/10/books/luxury-sex-and-music.html.

65. Judith A. Peraino, "The Same, but Different: Sexuality and Musicology, Then and Now," *Journal of the American Musicological Society* 66, no. 3 (2013): 825–31; Ruth A. Solie, ed., *Musicology and Difference: Gender and Sexuality in Music Scholarship* (Berkeley: University of California Press, 1993); Susan McClary, "Making Waves: Opening Keynote for the

Twentieth Anniversary of the Feminist Theory and Music Conference," *Women and Music: A Journal of Gender and Culture* 16, no. 1 (2012): 86–96; and Judith Tick, "Reflections on the Twenty-Fifth Anniversary of Women Making Music," *Women and Music: A Journal of Gender and Culture* 16, no. 1 (2012): 113–32.

66. The literature on Prince is vast. For a provocative recent article, see Saidah K. Isoke, "Dream If You Can: A Mother and Daughter's Reflections on Prince, Self-Realization, and Black Womanhood," *Journal of African American Studies* 21, no. 3 (2017): 524–27.

67. For a fascinating discussion of the historiography of castrati's sexuality and a nuanced reading, see Emily Wilbourne, "The Queer History of the Castrato," in *The Oxford Handbook of Music and Queerness*, ed. Fred Everett Maus and Sheila Whiteley (Oxford: Oxford University Press, 2018).

68. Ellen T. Harris, "Twentieth-Century Farinelli," *Musical Quarterly* 81, no. 2 (1997): 180–89.

69. Roger Freitas, *Portrait of a Castrato: Politics, Patronage, and Music in the Life of Atto Melani* (Cambridge: Cambridge University Press, 2009).

70. Martha Feldman, *The Castrato*, xxi.

71. Howard Chian, *After Eunuchs: Science, Medicine, and the Transformation of Sex in Modern China* (New York: Columbia University Press, 2018).

72. Gilles Deleuze and Felix Guattari, *A Thousand Plateaus: Capitalism and Schizophrenia*, trans. Brian Massumi (Minneapolis: University of Minnesota Press, 1987).

73. Freccero, "Queer Times."

## CHAPTER ONE

1. For theories on who might have performed which roles in the first performance, see Iain Fenlon, "Monteverdi's Mantuan *Orfeo*: Some New Documentation," *Early Music* 12, no. 2 (May 1984): 163–72; and Tim Carter, "Singing Orfeo: On the Performers of Monteverdi's First Opera," *Recercare* 11 (1999): 75–118.

2. By sound object here I reference Pierre Schaeffer, who heard in technological innovations like the radio the potential for sound to exist apart from the body that made it.

3. Claudio Monteverdi, *The Letters of Claudio Monteverdi*, trans. Denis Stevens (Cambridge: Cambridge University Press, 1980), 117.

4. Leo Schrade, *Monteverdi, Creator of Modern Music* (New York: W. W. Norton, 1950); Nino Pirrotta, *Li due Orfei da Poliziano a Monteverdi* (Turin: Einaudi, 1975); Gary Tomlinson, *Monteverdi and the End of the Renaissance* (Berkeley: University of California Press, 1987); Suzanne G. Cusick, "'There Was Not One Lady Who Failed to Shed a Tear': Arianna's Lament and the Construction of Modern Womanhood," *Early Music* 22, no. 1 (1994): 21–44; Tim Carter, *Monteverdi's Musical Theatre* (New Haven, CT: Yale University Press, 2002); Susan McClary, *Modal Subjectivities: Self-Fashioning in the Italian Madrigal* (Berkeley: University of California Press, 2004); Bonnie Gordon, *Monteverdi's Unruly Women: The Power of Song in Early Modern Italy* (New York: Cambridge University Press, 2004).

5. "Expressive Engines: Music Technologies from Automaton to Robots," February 15, 2016.

6. For a reading of primarily English Renaissance texts, see Wendy Beth, ed., *The Automaton in English Renaissance Literature* (Burlington, VT: Ashgate, 2011).

7. Jane Bennett, *Vibrant Matter: A Political Ecology of Things* (Durham, NC: Duke University Press, 2010), viii.

8. Caroline Walker Bynum, *Fragmentation and Redemption: Essays on Gender and the Human Body in Medieval Religion* (New York: Zone Books, 1992), 256.

9. Pamela H. Smith, *The Body of the Artisan: Art and Experience in the Scientific Revolution* (Chicago: University of Chicago Press, 2004), 15.

10. Elizabeth Freeman, "Time Binds, or, Erotohistoriography," *Social Text* 23, no. 3/4 (2005): 66. For a discussion of temporal drag, see Freeman, "Packing History, Count(er)ing Generations," *New Literary History* 31, no. 4 (Autumn 2000): 727–44.

11. Edmonds G. Radcliffe, *Redefining Ancient Orphism: A Study in Greek Religion*, (Cambridge: Cambridge University Press, 2013).

12. Kathleen M. Coleman, "Fatal Charades: Roman Executions Staged as Mythological Enactments," *Journal of Roman Studies* 80 (1990): 44–73.

13. William Shakespeare, *Henry the Eighth*, The Norton Shakespeare, ed. Stephen Greenblatt et al. (New York: W. W. Norton, 2016), 3.1.3–14. References are to act, scene, and line.

14. Charles Francis Adams, ed., *Memoirs of John Quincy Adams*, vol. 3 (Philadelphia: J. B. Lippincott, 1874), 441. In "The Stars of Our Flag," Schuyler Hamilton describes an Adams family heirloom that features a lyre clasped in the beak of an eagle surrounded by thirteen stars and the phrase "Nunc sidera ducit" (Now it leads the stars) from Marcus Manilius's *Astronomicon* describing Orpheus's lyre. Schuyler postulates that the stars in the image represent the astrological constellation "Lyra." Noting that the 1777 Flag Resolution was passed while John Adams was president of the board of war, Hamilton suggests that the thirteen stars of the Lyra constellation on this seal may have been the original inspiration for the stars on the American flag.

15. The Orpheus myth has influenced countless composers and songwriters since Monteverdi; some notable examples are (in chronological order) Telemann, *Orpheus* (1726); Gluck, *Orfeo ed Euridice* (1762); Haydn, *L'anima del filosofo, ossia Orfeo ed Euridice* (1791); Offenbach, *Orphée aux enfers* (1858); Harrison Birtwistle, *The Mask of Orpheus* (1973–84); Philip Glass, *Orphée* (1991); Nick Cave and the Bad Seeds, *Abattoir Blues/The Lyre of Orpheus* (2004); Judge Smith, *Orfeas* (2011); Arcade Fire, *Reflektor* (2013); and John Robertson, *Orpheus* (2015).

16. Frederick W. Sternfeld, "Orpheus, Ovid and Opera," *Journal of the Royal Musical Association* 113, no. 2 (1988): 172–202.

17. Andrew Dell'Antonio, *Listening as Spiritual Practice in Early Modern Italy* (Berkeley: University of California Press, 2011).

18. Bernice W. Kliman, "At Sea about Hamlet at Sea: A Detective Story," *Shakespeare Quarterly* 62, no. 2 (July 2011):180–204.

19. Francis Bacon, "Orpheus; or Philosophy," in *The Works of Francis Bacon, Baron of Verulam, Viscount St. Alban, and Lord High Chancellor of England*, ed. and trans. James Spedding, Robert Ellis, and Douglas Heath (London: Longmans, 1890), 6: 720–22.

20. For an article that outlines the ways the notion of the Anthropocene might change understandings of history, see Dipesh Chakrabarty, "The Climate of History: Four Theses," *Critical Inquiry* 35, no. 2 (2009): 197–222. For an approach grounded more directly in environmental science, see Simon L. Lewis and Mark A. Maslin, "Defining the Anthropocene," *Nature* 519, no. 7542 (2015): 171–80.

21. Katharine Park and Lorraine Daston, "Introduction: The Age of the New," in *The Cambridge History of Science*, vol. 3, *Early Modern Science*, ed. Katharine Park and Lorraine Daston (Cambridge: Cambridge University Press, 2003), 1–18.

22. Velluti has already attracted a lot of scholarly attention. See James Q. Davies, "'Veluti

in speculum': The Twilight of the Castrato," *Cambridge Opera Journal* 17, no. 3 (2005): 271–301. Davies also makes this argument in *Romantic Anatomies of Performance* (Berkeley: University of California Press, 2014).

23. Will Crutchfield, "Giovanni Battista Velluti e lo sviluppo della melodia romantica," *Bollettino del Centro Rossiniano di Studi* 54 (2014): 9–83.

24. John Scott and John Taylor, eds., "The Opera," *London Magazine and Review* 2 (August 1825): 516.

25. On these famous automata and automata makers, see Katherine Hirt, *When Machines Play Chopin: Musical Spirit and Automation in Nineteenth-Century German Literature* (New York: Walter de Gruyter, 2010); Daniel Cottom, "The Work of Art in the Age of Mechanical Digestion," *Representations* 66 (1999): 52–74; Jessica Riskin, "The Defecating Duck, or, The Ambiguous Origins of Artificial Life," *Critical Inquiry* 29, no. 4 (2003): 599–633; and Adelheid Voskuhl, *Androids in the Enlightenment: Mechanics, Artisans, and Cultures of the Self* (Chicago: University of Chicago Press, 2013).

26. For some representative studies, see Annette Richards, "Automatic Genius: Mozart and the Mechanical Sublime," *Music & Letters* 80, no. 3 (August 1999): 366–89.

27. Emily I. Dolan, "E. T. A. Hoffmann and the Ethereal Technologies of 'Nature Music,'" *Eighteenth Century Music* 5, no. 1 (2008): 7–26.

28. Heather Hadlock, *Mad Loves: Women and Music in Offenbach's* Les Contes d'Hoffmann (Princeton, NJ: Princeton University Press, 2000); Felicia Miller Frank, *The Mechanical Song: Women, Voice, and the Artificial in Nineteenth-Century French Narrative* (Stanford, CA: Stanford University Press, 1995); John Tresch, "The Machine Awakens: The Science and Politics of the Fantastic Automaton," *French Historical Studies* 34, no. 1 (2011): 87–123.

29. See Jonathan Sawday, *Engines of the Imagination: Renaissance Culture and the Rise of the Machine* (London: Routledge, 2007); and Allison Muri, *The Enlightenment Cyborg: A History of Communications and Control in the Human Machine, 1660–1830* (Toronto: University of Toronto Press, 2007).

30. Paolo Galluzzi, *The Italian Renaissance of Machines* (Cambridge, MA: Harvard University Press, 2020).

31. Robert Greene, *Friar Bacon and Friar Bungay* (Lincoln: University of Nebraska Press, 1963).

32. See Lorraine Daston and Katharine Park, *Wonders and the Order of Nature, 1150–750* (New York: Zone Books, 1998); and Todd Andrew Borlik, "'More than Art': Clockwork Automata, the Extemporizing Actor, and the Brazen Head in Friar Bacon and Friar Bungay," in *The Automaton in English Renaissance Literature*, ed. Wendy Beth Hyman (Burlington, VT: Ashgate, 2011), 129–44.

33. See Lorraine Daston and Katharine Park, *Wonders and the Order of Nature*; Zakiya Hanafi, *The Monster in the Machine: Magic, Medicine, and the Marvelous in the Time of the Scientific Revolution* (Durham, NC: Duke University Press, 2000); and Anthony Grafton, *Magic and Technology in Early Modern Europe* (Washington, DC: Smithsonian Institution Libraries, 2005).

34. Irma A. Richter and Thereza Wells, eds., *Leonardo da Vinci: Notebooks*, ed. (New York: Oxford University Press, 2008), 99.

35. Richter and Wells, 60.

36. The literature on della Porta is extensive. See, for example, William Eamon, *Science and the Secrets of Nature: Books of Secrets in Medieval and Early Modern Culture* (Princeton, NJ: Princeton University Press, 1994); Penelope Gouk, *Music, Science, and Natural Magic*

*in Seventeenth-Century England* (New Haven, CT: Yale University Press, 1999); Gary Tomlinson, *Music in Renaissance Magic: Toward a Historiography of Others* (Chicago: University of Chicago Press, 1994).

37. Giambattista della Porta, *Natural Magick [ . . . ] Wherein Are Set Forth All the Riches and Delights of the Natural Sciences* (London, 1658), 3.

38. Della Porta, 2.

39. Della Porta, 403.

40. Della Porta, 404.

41. Della Porta, 404.

42. Carolyn Abbate, *In Search of Opera* (Princeton, NJ: Princeton University Press, 2001), 6.

43. Abbate, 5.

44. Abbate, 5.

45. Logan K. Young, "Translation: Schoenberg's Pierrot Lunaire, 'Serenade,'" Classicalite, October 4, 2013, http://www.classicalite.com/articles/3155/20131004/translation-schoenbergs-pierrot-lunaire-serenade.htm.

46. This text is produced in full in Friedrich Kittler, *Gramophone, Film, Typewriter* (Stanford CA: Stanford University Press, 1999), 38–42.

47. Ovid, *Metamorphoses*, trans. Frank Justus Miller (Cambridge, MA: Harvard University Press, 1951), 2:123, 125.

48. "Non fuggirà, che grave suol esser più quanto più tarda scende sovra nocente capo ira celeste." Claudio Monteverdi, *La favola d'Orfeo rappresentata in musica il carnevale dell'anno MDCVII nell'Accademia degli Invaghiti di Mantova* (Mantua: Francesco Osanna, 1607). For a modern edition of the Italian and English see John Whenham, "*Orfeo*, Act V: Alessandro Striggio's Original Ending," in *Claudio Monteverdi: Orfeo* (Cambridge: Cambridge University Press, 1986), 35–42.

49. The two versions diverge at line 51. In the 1607 version, Orfeo is confronted by the dancing and screaming Bacchantes, whose Dionysian celebration ends with punishing the singer. The two most popular arguments about the reason for Monteverdi's change are that the Apollo ending had originally been the plan but that the improvised stage in the gallery did not have enough room, and that the shredded Orfeo was simply not a happy enough ending for the nobility. For various views, see, for instance, Iain Fenlon, "Monteverdi's Mantuan *Orfeo*: Some New Documentation," *Early Music* 12, no. 2 (May 1984): 163–72; Nino Pirrotta, "Theater, Sets, and Music in Monteverdi's Operas," in *Music and Culture in Italy from the Middle Ages to the Baroque*. (Cambridge: Harvard University Press, 2013), 258, and Paolo Fabbri, *Monteverdi*, trans. Tim Carter (New York: Cambridge University Press, 1994), 66–67. Barbara Russano Hanning suggests that Ottavio Rinuccini actually wrote the version with the Apollo deus ex machina; see "The Ending of *L'Orfeo*: Father, Son, and Rinuccini," *Journal of Seventeenth-Century Music* 9, no. 1 (2003), http://sscm-jscm.org/v9/no1/hanning.html. Hanning also provides a summary of the literature on the two versions.

50. Virgil, *Georgics*, 4.523–27, trans. in Shane Butler, *The Ancient Phonograph* (New York: Zone Books, 2015), 67–68.

51. Blake Wilson, *Singing to the Lyre in Renaissance Italy: Memory, Performance, and Oral Poetry* (Cambridge: Cambridge University Press, 2019), 203.

52. This was most likely the source that Striggio and his Mantuan audiences would have known best. See Lorenzo Bianconi, *Music in the Seventeenth Century*, trans. David Bryant (New York: Cambridge University Press, 1987), 218.

53. Ovid, *Metamorphoses*, trans. Miller, 2:121.

54. Elly Rachel Truitt, *Medieval Robots: Mechanism, Magic, Nature, and Art* (Philadelphia: University of Pennsylvania Press, 2015).

55. Francis Bacon, "Century II," in *Sylva Sylvarum, or, A Natural History, in Ten Centuries* [ . . . ], The Ninth and Last Edition (London, 1670), 47.

56. These grottos are discussed in detail in Kara Reilly, *Automata and Mimesis on the Stage of Theatre History* (New York: Palgrave Macmillan, 2011).

57. Du Chesne, quoted in Reilly, *Automata and Mimesis*, 59.

58. A rehearing of Orpheus can contribute to what Emily Dolan and John Tresch have called a critical organology, a corrective to the musicological marginalization of instruments and technologies. John Tresch and Emily I. Dolan, "Toward a New Organology: Instruments of Music and Science," *Osiris* 28, no. 1 (2013): 278–98.

59. "Hermes" mentions the myth of Hermes's creation of the lyre. In the work, from a collection of thirty-four ancient Greek poems (dates unknown), Hermes leaves his cradle soon after his birth and finds a tortoise, which he kills by hollowing out the shell. He then attaches reeds to the shell and strings made from sheep intestines to the shell, creating the lyre. Later, Apollo and Hermes squabble after Hermes steals Apollo's cattle. After consulting with Zeus, as the two gods are reconciling, Hermes plays a song on the lyre that greatly impresses Apollo. Hermes then presents the instrument to Apollo as a gift. Oliver R. H. Thomas, *The Homeric Hymn to Hermes* (Cambridge University Press, 2020), 97.

60. *The Etymologies of Isidore of Seville*, trans. Stephen A. Barney, W. J. Lewis, J. A. Beach, Overer Berghof, and Muriel Hall (Cambridge: Cambridge University Press, 2006), 98.

61. Wilson, *Singing to the Lyre*, 200–10.

62. Antonfrancesco Doni, translated in James Haar, "Notes on the 'Dialogo della Musica' of Antonfrancesco Doni," *Music & Letters* 47, no. 3 (1966): 220.

63. "Hora, tutte, o la maggior quantità delli sopra scritti conditioni, deve havere medesimamente uno strumentista, che suoni, o Cornetto [orig.: Cornette], o Viola da Gamba, o Violino, o flauto, o fifaro, o simili d'una parte sola. Che di quegli strumentisti, che suonano tutte le parti; dirò poi all'Altezza sua per verità e con brevità quel che io ne sento, e conosco senza pensiero di pregiudicare a nessuno, si come nel resto detto, intesi et intendo, parendomi, che ciò richiegga il comandamento dell'Altezza sua e la modestia, e sincerità virtuosa e mia." Luigi Zenobi to Francesco I de' Medici, translated by Bonnie J. Blackburn and Edward Lowinsky in "Luigi Zenobi and His Letter on the Perfect Musician," *Studi musicali* 22 (1993), 85 and 102.

64. Ercole Bottrigari, *Il desiderio* (Venice: Ricciardo Amadino, 1594). See also Michele Todini, *Dichiaratione della galleria armonica* (Rome: Francesco Tizzoni, 1676); Vincenzo Galilei, *Fronimo Dialogo / Di Vincentio Galilei Fiorentino / Nel Quale Si Contengono Le Vere* (Venice: Girolamo Scotto, 1568); Agostino Agazzari, *Del sonare sopra il basso con tutti li stromenti e dell'uso loro nel conserto* (Siena: Domenico Falcini, 1607).

65. This book was republished in facsimile; see Cosimo Bottegari, *Il libro de canto e liuto / The Song and Lute Book*, ed. Dinko Fabris and John Griffiths (Bologna: Arnaldo Forni, 2006). For a transcription and modern edition, see Carol MacClintock, *The Bottegari Lutebook* (Wellesley, MA: Wellesley College, 1965).

66. See Anthony Newcomb, "Notions of Notation around 1600," *Il saggiatore musicale* 22, no. 1 (2015): 5–31.

67. Peter Holman, "*Col nobilissimo esercitio della vivuola*: Monteverdi's String Writing," *Early Music* 21, no. 4 (1993): 577.

68. "Che principalmente siano buoni cattolici et gioveni quieti, si che si possa sperare

longa et amorevole servitù da loro. Che siano sicuri nel cantare et habbiano buona voce. Se havranno contrapunto, et sapranno sonare di leutto per cantarvi dentro saranno tanto più cari." Pietro Canal, "Della musica in Mantova: Notizie tratte principalmente dall'Archivio Gonzaga," *Memorie del Reale istituto Veneto di scienze lettere ed arti* 21 (1879): 655–744, This is reproduced in Richard Sherr, "Gugliemo Gonzaga and the Castrati," *Renaissance Quarterly* 33, no. 1 (1980): 33–56, with an English translation of some passages on page 38.

69. Andrew Dell'Antonio, *Syntax, Form, and Genre in Sonatas and Canzonas, 1621–1635* (Lucca: Libreria Musicale Italiana, 1997). See also Rebecca Cypess, "Instrumental Music and 'Conversazione' in Early Seicento Venice: Biagio Marini's 'Affetti Musicali' (1617)," *Music & Letters* 93, no. 4 (2013): 453–78; and *Curious and Modern Inventions: Instrumental Music as Discovery in Galileo's Italy* (Chicago: University of Chicago Press, 2016).

70. G. B. Doni, "Trattato della musica scenica," in *De' trattati di musica*, ed. A. F. Gori (Florence, 1763), 2:113, translated in Gloria Rose, "Agazzari and the Improvising Orchestra," *Journal of the American Musicological Society* 18, no. 3 (Autumn 1965): 390.

71. G. B. Doni, "Trattato della musica scenica," 110–11, trans. in Gloria Rose, "Agazzari and the Improvising Orchestra," *Journal of the American Musicological Society* 18, no. 3 (Autumn 1965): 390.

72. Vincenzo Galilei, *Dialogo della musica antica e della moderna* (1581), translated in Oliver Strunk, *Source Readings in Music History* (New York: W. W. Norton, 1950), 316.

73. Strunk, 313.

74. Strunk, 313.

75. Bonnie Gordon, "Orfeo's Machines," *Opera Quarterly* 24, no. 3/4 (2008): 200–22.

76. The complete text of the prologue in Italian and English can be found in a number of modern editions. I have used Claudio Monteverdi, *L'Orfeo: Favola in musica: SV 318*, ed. Claudio Gallico (New York: Eulenburg, 2004).

77. Monteverdi, p. 4, mm. 37–44.

78. For Daniel Chua, the ritornello in both of these instances signifies opera's long disenchantment and contains within it the very seeds of absolute music: "All the talk of vocal superiority, not least by Musica herself, is just rhetoric; what actually determines opera is instrumental theory. The continuo for him wields absolute power, providing the sonic backdrop for the singing that tells the story." Again my interpretation diverges from Chua's. For starters, the ritornello exists in different contexts, as part of standard theatrical practices for scene changes and in relation to the larger role of instrumental music within Monteverdi's articulated musical poetics. In theoretical terms, the ritornello points toward yet another way in which the music drama as a whole is a series of automations and animations. It's not that a human sets the instruments in motion, it's that the musical instruments set Orfeo's affective sound in motion. He is the automaton. See Daniel K. L. Chua, *Absolute Music and the Construction of Meaning* (New York: Cambridge University Press, 1999), 59.

79. "Possente spirto" has been widely analyzed in the Monteverdi literature. See, for example, Tim Carter, "*Possente Spirto*: On Taming the Power of Music," *Early Music* 21, no. 4 (1993): 517–24; Eric T. Chafe, *Monteverdi's Tonal Language* (New York: Schirmer Books, 1992); and Tomlinson, *Monteverdi and the End of the Renaissance*.

80. Thomas Forrest Kelly suggests that Caronte falls asleep to an "Apollo Sinfonia," as if the sun god actually does the work of putting the gatekeeper to sleep. He notes that the "Apollo Sinfonia" reappears when Apollo enters on a cloud. See Thomas Forrest Kelly, *First Nights: Five Musical Premieres* (New Haven, CT: Yale University Press, 2000), 47. Silke Leopold suggests that the lullaby comes from a musical representation of the lyre, that in

this moment Monteverdi evokes Orfeo playing the lyre both above and below ground. Silke Leopold, *Monteverdi: Music in Transition*, trans. Anne Smith (Oxford: Clarendon Press, 1991), 96.

81. Monteverdi, *L'Orfeo: Favola in musica*, 70, mm. 209–20.

82. As Eric Chafe has pointed out, Proserpina accomplishes her task by taking over the harmonies of "Possente spirto": "her final words and their chromatically inflected setting resemble the close of Orfeo's fifth strophe." Chafe, *Monteverdi's Tonal Language*, 151. Act 4 begins at Monteverdi, *L'Orfeo: Favola in musica*, 87.

83. For Susan McClary, this moment is about gender; she argues that Proserpina uses a rhetoric of seduction to move Pluto. Susan McClary, *Feminine Endings: Music, Gender and Sexuality* (Minneapolis: University of Minnesota Press, 1991), 44–45.

84. Monteverdi, *L'Orfeo: Favola in musica*, 90–91, mm. 58–68.

85. For Klaus Theweleit, the echo in each strophe "quietly repeats from behind what the instrument up front played loudly, thus bringing the beyond, which Orpheus seeks to enter, as a presence into the ear of the listener: as if he were already over there since the tones already are." The instrumental echoes are a protorecording technology that sends the sound directly to the listener's ear without human mediation. Klaus Theweleit, "Monteverdi's *L'Orfeo*: The Technology of Reconstruction," in *Opera through Other Eyes*, ed. David J. Levin (Stanford, CA: Stanford University Press, 1994), 156.

86. Monteverdi, *L'Orfeo: Favola in musica*, 130.

87. Ovid, *Metamorphoses*, trans. Kline, bk. 8, 183–235, http://ovid.lib.virginia.edu/trans/Metamorph8.htm#482327661.

88. Ovid, bk. 8, 183–235.

89. Monteverdi, *L'Orfeo: Favola in musica*, 92, mm. 77–82.

90. Monteverdi, 92, mm. 83–94.

91. Tim Carter, "Singing Orfeo," 84.

92. There is also an important tradition of dramatic echoes. The fifth intermedio to *La pellegrina* featured a dramatic double echo song composed and sung by Jacopo Peri as Arion. He "sang a very beautiful madrigal in the manner of an echo, at which two other echoes responded with different voices, the one after the other, and it appeared that their voices issued from caves or deep caverns." Nina Treadwell writes extensively on this in *Music and Wonder at the Medici Court: The 1589 Interludes for La pellegrina* (Bloomington: Indiana University Press, 2008).

93. For more on music about Echo, see Barbara Russano Hanning, "Powerless Spirit: Echo on the Musical Stage of the Late Renaissance," in *Word, Image, and Song*, vol. 1, *Essays on Early Modern Italy*, ed. Rebecca Cypess, Beth L. Glixon, and Nathan Link (Rochester, NY: University of Rochester Press, 2013), 193–218.

94. Lynn Enterline, *The Rhetoric of the Body from Ovid to Shakespeare* (New York: Cambridge University Press, 2000), 12–13.

95. Ovid, *Metamorphoses*, trans. Kline, bk. 3, 359–401, http://ovid.lib.virginia.edu/trans/Metamorph3.htm.

96. Ovid, bk. 3, 339–58, 359–401.

97. Ovid, *Metamorphoses*, trans. Brookes More (Boston: Cornhill, 1922), bk. 3, 337–434, http://data.perseus.org/citations/urn:cts:latinLit:phi0959.phi006.perseus-eng1:3.337–3.434.

98. Orlando di Lasso, *Libro di villanelle, moresche, et altre canzoni* (Paris, 1581).

99. Mauro Calcagno has thought deeply about the echo scene as sounding Orfeo's sta-

tus as divided subject. See Mauro Calcagno, *From Madrigal to Opera: Monteverdi's Staging of the Self* (Berkeley: University of California Press, 2012).

100. Daniel K. L. Chua, "Untimely Reflections on Operatic Echoes: How Sound Travels in Monteverdi's *L'Orfeo* and Beethoven's *Fidelio* with a Short Instrumental Interlude," *Opera Quarterly* 21, no. 4 (2005), 577.

101. Chua, 578.

102. Theweleit, "Monteverdi's *L'Orfeo*," 156.

103. Frank Lewis Dyer and Thomas Commerford Martin, *Edison, His Life and Inventions* (New York: Harper and Brothers, 1910), 208.

104. Aristotle, *On the Soul*, trans. J. A. Smith, bk. 2, pt. 8, http://classics.mit.edu/Aristotle/soul.2.ii.html.

105. Lucretius, "The Senses and Mental Pictures," in *On the Nature of Things*, trans. William Ellery Leonard, bk. 4, http://classics.mit.edu/Carus/nature_things.4.iv.html.

106. Virgil, "Eclogue I," in *Eclogues, Georgics, Aeneid*, trans. H. R. Fairclough, lines 1–5, http://www.theoi.com/Text/VirgilEclogues.html.

107. Virgil, "Eclogue IX," in *Eclogues*, lines 56–65, http://www.theoi.com/Text/VirgilEclogues.html#9.

108. Della Porta discusses speaking tubes, ear trumpets, and distortion glass in *Magia naturalis*, which was the standard book of seventeenth-century magic and thus an index to contemporary views on the subject. Published first in short form in 1558, it made an immediate splash, with at least five more Latin editions coming out over the next ten years and an expanded twenty-book version in 1589.

109. Giambattista della Porta, *Natural Magick* [ . . . ] (London, 1658), bk. 16, chap. 12, http://name.umdl.umich.edu/A55484.0001.001.

110. Della Porta, bk. 19, chap. 1.

111. Much of what historians know about Kircher came from his disciple Gaspar Schott, who was a student, spokesman, and editor. Both men existed under Barberini patronage in mid-seventeenth-century Rome. Kircher arrived in the papal city in 1634, just a year after the holy office condemned Galileo for upholding the Copernican hypothesis and right in the middle of the reign of the Barberini pope Urban VIII. In music history 1634 was just three years before the first public opera house opened in Venice.

112. "Echo ludibundae Naturae jocus, a Poetis imago vocis, juxta illud Virgilii. *Saxa sonant vocisque offensa resultat imago*. A Philosophis reflexa, repercussa, reciproca vox, ab Hebræis *Bat col.* filia vocis dicitur." Athanasius Kircher, "Phonosophia Anacamptica," Liber 1 in *Phonurgia nova* (Campidonae: Dreherr, 1673), 1.

113. "Sed eam dum persequor, fugit, dum fugio persequitur, dum blande loquor blande eludit, dum valde clamo, quasi asseclis sibi vocibus ascitis voces congeminat, cedere nescia." Kircher, 2

114. "Sit verbi gratia: *Echo* instituenda quadrisyllaba, objecta B C D E F, ita disposita sint, ut singula unam syllabam tardius reflectant. Sitque trisyllabum sequens vox CLAMORE. Sonorum autem sive vocale sit in A. Cum igitur objecta ita disposita supponantur, ut singula syllabam tardius reflectant, certum est singula objecta diversa verba significantia reflectere, ita objectum B. CLAMORE; AMORE objectum C. MORE objectum D. ORE objectum E. RE denique objectum F. reddet: *Polyphona* igitur *Echo* semper alia & alia vocabula repetet, ut si quis clamet." Kircher, 52. This passage cited and translated in Bettine Menke, "'However One Calls into the Forest . . .': Echoes of Translation," trans. Robert J. Kiss, in *Walter Benjamin and Romanticism*, ed. Beatrice Hanssen and Andrew Benjamin (New York: Continuum, 2002), 88.

115. Michael Kelly, *Reminiscences of Michael Kelly* [ . . . ] (London: Colburn, 1826), 1:151.

116. Gluck to the Duke of Braganza, October 30, 1770, trans. in Patricia Howard, *The Modern Castrato: Gaetano Guadagni and the Coming of a New Operatic Age* (Oxford University Press, 2014), 177.

117. Vernon Lee, *Althea: A Second Book of Dialogues on Aspirations and Duties* (London: Osgood, McIlvaine, 1894), 52.

118. Mladen Dolar, *A Voice and Nothing More* (Cambridge: MIT Press, 2006), begins with a ghost.

## CHAPTER TWO

1. Anne Carson, *Eros the Bittersweet* (Princeton, NJ: Princeton University Press, 1986), 141.

2. See Mario Biagioli, *Galileo, Courtier: The Practice of Science in the Culture of Absolutism* (Chicago: University of Chicago Press, 1993); and Lawrence Lipking, *What Galileo Saw: Imagining the Scientific Revolution* (Ithaca, NY: Cornell University Press, 2014).

3. Biagioli, 301–2.

4. Galileo Galilei, *Discoveries and Opinions of Galileo*, ed. and trans. Stillman Drake (Garden City, NY: Doubleday Anchor Books, 1957), 295.

5. Mark Katz, *Capturing Sound: How Technology has Changed Music* (Berkeley: University of California Press, 2004).

6. Galilei, 295.

7. The literature on radio is vast. On the shock value of radio in Weimar Germany, see, for example, Eric D. Weitz, *Weimar Germany: Promise and Tragedy* (Princeton, NJ: Princeton University Press, 2007).

8. Brian Kane, *Sound Unseen: Acousmatic Sound in Theory and Practice* (New York: Oxford University Press, 2014), 30.

9. Galilei, *Discoveries and Opinions of Galileo*, 295.

10. Galilei, 295.

11. Carolyn Abbate, "Sound Object Lessons," *Journal of the American Musicological Society* 69, no. 3 (2016): 793.

12. Plato, *Phaedrus*, trans. Robin Waterfield (New York: Oxford University Press, 2002), 45.

13. Clement of Alexandria, *Exhortation to the Greeks. The Rich Man's Salvation. To the Newly Baptized*, trans. G. W. Butterworth, Loeb Classical Library 92 (Cambridge, MA: Harvard University Press, 1919), 5.

14. Charles Darwin, *The Descent of Man, and Selection in Relation to Sex*, 2nd ed. (London: John Murray, 1875), 281.

15. "The Cicada Septendecim, or American 17 Year Locust," *Wisconsin State Register* (Portage), June 11, 1864, originally published in *Williams Monthly Miscellany* 1, no. 1 (July 1844): 26–34.

16. Rob. E. Peyton, "The Cicada Septendecim, or Seventeen Years Locust, As it appeared in Fauquier County, Va., in 1843," *Daily National Intelligencer* (Washington, DC), August 31, 1843.

17. A. R. H., "Cicada Septendence: The So Called Seventeen-Year Locust in Iowa," *Bismarck Daily Tribune* (North Dakota), June 30, 1888.

18. Christoph Cox and Daniel Warner, "Introduction: Music and the New Audio Cul-

ture," in *Audio Culture: Readings in Modern Music*, ed. Christoph Cox and Daniel Warner (New York: Continuum, 2004), xiii.

19. Claude V. Palisca, *Studies in the History of Italian Music and Music Theory* (Oxford: Clarendon Press, 1994); and Stillman Drake, "Music and Philosophy in Early Modern Science," in *Music and Science in the Age of Galileo*, ed. Victor Coelho (Dordrecht: Kluwer Academic, 1992), 3–16.

20. Antonio Goretti to Enzo Bentivoglio, Parma, November 16, 1627, in Paolo Fabbri, *Monteverdi*, trans. Tim Carter (New York: Cambridge University Press, 1994), 211.

21. Francesco Guitti to Enzo Bentivoglio, February 18, 1628, in Fabbri, *Monteverdi*, 215.

22. Antonio Goretti to Enzo Bentivoglio, Parma, February 18, 1628, in Fabbri, *Monteverdi*, 216.

23. Nino Pirrotta, "Music and Cultural Tendencies in Fifteenth-Century Italy," in *Music and Culture in Italy from the Middle Ages to the Baroque: A Collection of Essays* (Cambridge, MA: Harvard University Press, 1984), 312.

24. Claudio Monteverdi, *Madrigals, Book VIII (Madrigali guerrieri et amorosi)* (New York: Dover, 1991), xiv.

25. Monteverdi, xvii.

26. *Oxford English Dictionary*, s.v. "Sound, n. 3," https://www.oed.com/.

27. Aristotle, *On the Soul*, trans. J. A. Smith, bk. 2, pt. 8, http://classics.mit.edu/Aristotle/soul.2.ii.html.

28. Plato, *Timaeus*, trans. Donald J. Zeyl (Indianapolis, IN: Hackett, 2000), 31.

29. Plato, 43.

30. Aristotle, *On the Soul*, bk. 2, pt. 8.

31. Dante Alighieri, "Inferno," in *The Divine Comedy of Dante Alighieri*, trans. Allen Mandelbaum (New York: Bantam Classics, 1982), 113, canto 13, lines 26 and 27.

32. Galilei, *Discoveries and Opinions of Galileo*, 275.

33. Galilei, 275.

34. Carol MacClintock, "Giustiniani's 'Discorso Sopra la Musica,'" *Musica Disciplina* 15 (1961): 222.

35. Ross W. Duffin, "Just Intonation in Renaissance Theory and Practice," *Music Theory Online* 12 (2006), http://www.mtosmt.org/issues/mto.06.12.3/mto.06.12.3.duffin.html; and "Cipriano de Rore, Giovanni Battista Benedetti, and the Just Tuning Conundrum," *Journal of the Alamire Foundation* 9, no. 2 (2017): 239–66.

36. Duffin, "Just Intonation."

37. Galileo Galilei, *Dialogues Concerning Two New Sciences by Galileo Galilei*, trans. Henry Crew and Alfonso de Salvio (New York: Macmillan, 1914), 102.

38. Galilei, 103.

39. Galilei, 102.

40. Shane Butler, *The Ancient Phonograph* (New York: Zone Books, 2015), 27.

41. One of the earliest Greek works translated into Latin, this dialogue remained influential through the medieval period, the emergence of Neoplatonism in the early Renaissance, and the rise of scientific rationalism in the seventeenth century.

42. Galilei, *Discoveries and Opinions of Galileo*, 278–79.

43. Jane A. Bernstein, *Print Culture and Music in Sixteenth-Century Venice* (New York: Oxford University Press, 2001); Cristle Collins Judd, *Reading Renaissance Music Theory: Hearing with the Eyes*, Cambridge Studies in Music Theory and Analysis 14 (Cambridge Uni-

versity Press, 2000); Kate van Orden, *Music, Authorship, and the Book in the First Century of Print* (Berkeley: University of California Press, 2013).

44. Anthony Newcomb, *The Madrigal at Ferrara, 1579–1597*, 2 vols. (Princeton, NJ: Princeton University Press, 1980).

45. Bonnie Gordon, *Monteverdi's Unruly Women: The Power of Song in Early Modern Italy* (New York: Cambridge University Press, 2004).

46. Martin Kemp, ed., *Leonardo on Painting: An Anthology of Writings by Leonardo da Vinci with a Selection of Documents Relating to His Career as an Artist*, trans. Martin Kemp and Margaret Walker (New Haven, CT: Yale University Press, 1989), 35.

47. This work has been anthologized multiple times. An easily available edition is published in *Illuminations*, ed. Hannah Arendt, trans. Harry Zohn (New York: Schocken Books, 1969). In this edition the Leonardo quote appears in note 15 on page 25.

48. Kemp, 35, 37.

49. Giovanni Paolo Lomazzo, "The Second Booke of the Actions, Gestures, Situation, Decorum, Motion, Spirit, and Grace of Picture," in *A Tracte Containing the Artes of Curious Paintinge, Carvinge, and Buildinge*, trans. Richard Haydocke (Oxford: Joseph Barnes, 1598), 4; originally published as *Trattato dell'arte della pittura, scoltura, et architettura* (Milan: Paolo Gottardo Pontio, stampatore Regio. A instantia di Pietro Tini, 1585), 108.

50. Quoted in Erwin Panofsky, "Galileo as a Critic of the Arts: Aesthetic Attitude and Scientific Thought," *Isis* 47, no. 1 (1956): 5.

51. Quoted in Eileen Reeves, "Galileo, Oracle: On the History of Early Modern Science," *I Tatti Studies in the Italian Renaissance* 18, no. 1 (2015): 8–9.

52. Andrew Cunningham, *The Anatomical Renaissance: The Resurrection of the Anatomical Projects of the Ancients* (New York: Routledge, 2016); and Andrea Carlino, *Books of the Body: Anatomical Ritual and Renaissance Learning* (Chicago: University of Chicago Press, 1999).

53. "Praeterquam enim quod inter haec ipsa, & Laryngem analogia intercedat haud spernenda, etiam Laryngis actionem illustriorem reddere valent." Julius Casserius, *De vocis auditusque organis historia anatomica* (Ferrara, 1601), 116. An English translation of much of the text can be found in Julius Casserius, "The Three Books of the First Treatise on the Anatomical History of the Larynx, the Organ of Voice," trans. Malcolm H. Hast and E. B. Holtsmark, supplement, *Acta Oto-Laryngologica* 68, S261 (1969).

54. Cynthia Klestinec, *Theaters of Anatomy: Students, Teachers, and Traditions of Dissection in Renaissance Venice* (Baltimore: Johns Hopkins University Press, 2011); Antonio Gamba. "Il primo teatro anatomico stabile in Padova non fu quello di Fabrici d'Acquapendente," *Atti e memorie dell'Accademia patavina di scienze, lettere ed arti*, pt. 3, 99 (1986–1987): 157–61.

55. For some modern commentaries on these debates, see Tatsuo Sakai, "Historical Evolution of Anatomical Terminology from Ancient to Modern," *Anatomical Science International* 82, no. 2 (2007): 65–81; and Edwin L. Kaplan, George I. Salti, Manuela Roncella, Noreen Fulton, and Mark Kadowaki, "History of the Recurrent Laryngeal Nerve: From Galen to Lahey," *World Journal of Surgery* 33, no. 3 (2009): 386–93. See also Antoine Ferrein, *Mémoire de l'Académie royale des sciences* (Paris, 1754).

56. Vesalius is most famous for the publication in 1543 of the highly influential *De humani corporis fabrica*, and Colombo is most famous for "discovering" the clitoris as a sexual organ.

57. "Mox itaque longam in collo sectionem acutiori novacula duco, quae cutem & subjectos illi musculos ad asperam usque arteriam dividat: id cavens, ne sectio in latus aberret, venamque notatu dignam vulneret. Dein manibus arteriam asperam comprehendo, & illam ab incumbentibus musculis digitorum duntaxat opera detegens, ad hujus latera soporales arterias, illique attensos sexti paris nervorum cerebri nervos perquiro. Dein recurrentes nervos lateribus asperae arteriae adnatos etiam observo, quos interdum laqueis intercipio, interdum praescindo: idque primum ex altero latere, ut nervo hic intercepto praesectove dilucide observetur, qui media vox pereat, totaque ambobus nervis laesis intercidat: & si laqueos solvo, rursus redeat. Id enim cito & citra insignem sanguinis fluxum expenditur, ac pulchre auditur, quam validam efflationem animal citra vocem moliatur, recurrentibus nervis cultello divisis." Andreas Vesalius, 658. Quoted and translated in Daniel D. Lydiatt and Gregory S. Bucher, "The Historical Latin and Etymology of Selected Anatomical Terms of the Larynx," *Clinical Anatomy: The Official Journal of the American Association of Clinical Anatomists and the British Association of Clinical Anatomists* 23, no. 2 (2010): 131–44.

58. "Ubi in duos processus definunt, quibus scilicet natura veluti lingulis quibusdam uti voluit non modo ad claudendum laryngis amplitudinem & asperae arteriae meatum, ne quid, ex vomitu praesertim, quod laedat in internam illius capacitatem decidat, atque ad pulmones deferatur: sed etiam ut rimam illam moderetur variarum vocum efformandarum gratia, non secus atque in fistulis, aut tibiis lingulae quaedam imponi solent, ex duabus arundinum laminis compactae. Propterea istorum processuum id genus lingulam constituentium unio γλωττίς nuncupatur." M. R. Colombo, *De re anatomica* (Venice: Nicolai Bevilacquae, 1559), translated in Daniel D. Lydiatt and Gregory S. Bucher, "The Historical Latin and Etymology of Selected Anatomical Terms of the Larynx," *Clinical Anatomy* 23, no. 2 (2010): 142.

59. Hieronymus Fabricius ab Aquapendente, *De brutorum loquela* (Venice, 1599). There were three other editions: Venice (1601), Padua (1603; consulted), and Frankfurt am Main (1623). See James Atkinson, *Medical Bibliography: A. and B.* (London: John Churchill, 1834), 118.

60. Stefano Gensini, "Il *De brutorum loquela* di Girolamo Fabrici d'Acquapendente," *Bruniana & Campanelliana* 17, no. 1 (2011): 163–74.

61. Galen, *On Prognosis*, trans. and ed. Vivian Nutton, Corpus medicorum Graecorum, vol. 5, pt. 8, fasc. 1 (Berlin: Akademie 1979): 97.

62. Andreas Vesalius, *On the Fabric of the Human Body. Book VI, The Heart and Associated Organs. Book VII, The Brain: A Translation of* De humani corporis fabrica libri septem, trans. William Richardson and John Carman (Novato, CA: Norman, 2009), 269.

63. The posthuman turn will be the subject of the last chapter of this monograph. Important authors are Rosi Braidotti, *Nomadic Theory: The Portable Rosi Braidotti* (New York: Columbia University Press, 2011), and *The Posthuman* (Cambridge: Polity, 2013); Cary Wolfe, *Animal Rites: American Culture, the Discourse of Species, and Posthumanist Theory* (Chicago: University of Chicago Press, 2003), and *What Is Posthumanism?* (Minneapolis: University of Minnesota Press, 2010); and N. Katherine Hayles, *Unthought: The Power of the Cognitive Nonconscious* (Chicago: University of Chicago Press, 2017).

64. There is much work on animals in the early modern period. For example, see Lorraine Daston, "Intelligences: Angelic, Animal, Human," in *Thinking with Animals: New Perspectives on Anthropomorphism*, ed. Lorraine Daston and Gregg Mitman (New York: Columbia University Press, 2005), 37–58; Erica Fudge, ed., *Renaissance Beasts: Of Animals, Humans, and Other Wonderful Creatures* (Urbana: University of Illinois Press, 2004); Erica Fudge, Ruth Gilbert, and Susan Wiseman, eds., *At the Borders of the Human: Beasts, Bodies*

*and Natural Philosophy in the Early Modern Period* (New York: Macmillan; St. Martin's, 1999); and Pia F. Cuneo, ed., *Animals and Early Modern Identity* (Burlington, VT: Ashgate, 2014).

## INTERMEDIO

1. Athanasius Kircher, *Musurgia universalis*, ed. Ulf Scharlau (New York: Georg Olms, 1970), 519.

2. "Constructum non ita pridem, inquit Kircherus Lib. 6. Musurgia Par. 4 exp.1 in fine, ad melancholiam magni cuyusdam Principis depellendam ab insigni ingeniosoque histrione tale quodpiam Instrumentum. Feles vivas accepit, omnes differentis magnitudinis, & consequenter differentium acumine & gravitate vocum; quas cistae cuidam Huic negotio dedita opera fabricatae ita inclusit, ut caudae per foramina extente, certis quibus dam cana libus insererentur Hilae subidit palmulas, subtilissimus aculeis loco malleolorum instructas strusas, eo mondo, qua in seguenti Pragmatica explicabitur Feles verò juxta differentem magnitudinem tonatim ita disposuit, ut singular palmula singulis responderet felium caudis; Instrumentumque ad relaxationem Principi praeparatum oportuno loco condidit, quod deinde pulsatum. Eam harmoniam redditit, qualem felium voces reddere possunt. Nam palmula digitis Organoe di depressa aculeis suis dum caudas punyent cattorum, hi in rabiem acti, miserabili voce, nunc gravem, modò acutam intonantes, eam ex felium vocibus compositam reddiderunt harmoniam, quae et moverte homines ad risum, & vel sorices ipsos ad saltuem concitare posset. Hackenus Kircherus." Gaspar Schott, *Magiae universalis naturae et artis pars secunda: Acoustica in VII libros* (Bamberg, 1677), 372–73.

3. Rosi Braidotti, *The Posthuman* (Cambridge: Polity Press, 2013), 90.

4. Kircher, *Musurgia universalis*, 519.

5. See "Louis XI and the Pig Piano, with Special Guest Simon Cowell," in *The Horrible Histories*, series 1, episode 5, "Ruthless Rulers."

6. Michiel van Groesen, "A First Popularisation of Travel Literature: On the Methods and Intentions of the De Bry Travel Collection (1590–1634)," *Dutch Crossing* 25, no. 1 (2001): 103–31.

7. Carl Van Vechten, "The Cat in Music," *Musical Quarterly* 6, no. 4 (1920): 573–85; and Kat Eschner, "Music or Animal Abuse: A Brief History of the Cat Piano," *Smithsonian Magazine*, May 2, 2017, https://www.smithsonianmag.com/smart-news/music-or-animal-abuse-brief-history-cat-piano-180963056/.

8. Carolyn Purnell, "The Cat Piano or 'The Most Singular Music You Can Imagine,'" *Psychology Today*, June 30, 2020, https://www.psychologytoday.com/us/blog/making-sense/202006/the-cat-piano-or-the-most-singular-music-you-can-imagine.

9. Jean-Baptiste Weckerlin, *Nouveau musiciana: Extraits d'ouvrages . . . concernant la musique et les musiciens, avec illustrations et airs notés* (Paris: Garnier Frères, 1890), 348–50.

10. Claude-François Ménestrier, *Des représentations en musique anciennes et modernes* (1681; repr., Geneva: Minkoff, 1992), 180.

11. Juan Cristóbal Calvete de Estrella, *El felicísimo viaje del alto y muy poderoso príncipe don Felipe*, 2 vols. (1552; repr., Madrid: Sociedad de Bibliófilos Españoles, 1930).

12. This is detailed in Estrella, *El felicísimo viaje*, 1:29; see also Jenaro Alenda y Mira, *Relaciones de solemnidades y fiestas públicas de España* (Madrid: Rivadeneyra, 1903), 45–47.

13. "Y por buena orden y artificio sacaban todos las colas altas afuera de tal suerte, que tocando el oso el órgano tiraba de las colas a los gatos en debida proporción y medida a unos mucho, y a otros poco, y a otros medianamente a su compás, y sintiéndose lo gatos tirar por

las colas aullaban cada uno conforma como se dolía, y hacían con sus aullidos altos y bajos una música bien entonada." Juan Cristóbal Calvete de Estrella, *El felicíssimo viaje del muy alto y muy Poderoso Principe Don Phelippe [ . . . ] en quatro libros* (Antwerp, 1552), bk.2, 74r.

## CHAPTER THREE

1. See Eileen Reeves, "Hearing Things: Organ Pipes, Trumpets, and Telescopes," *The Starry Messenger, Venice 1610: From Doubt to Astonishment* (Washington, DC: Library of Congress, 2013).

2. For more on the relationship between natural philosophy and music, and particularly Galileo, see Claude Palisca, *Studies in the History of Italian Music and Music Theory* (Oxford: Clarendon Press, 1994); and Stillman Drake, "Music and Philosophy in Early Modern Science," in *Music and Science in the Age of Galileo*, ed. Victor Coelho (Dordrecht: Kluwer Academic, 1992), 3–16.

3. "Ex organo detract, vitreos varii generis orbes ad certum intervallum accomodavit, unde eventum sibi ex sententia processisse cognovit." Gian Vittorio Rossi, *Iani Nicii Erithraei Pinacotheca Altera Imaginum, Illvstrium, doctrinae vel ingenii laude, Virorum, qui, auctore superstite, diem suum obierunt* (Amsterdam, 1645, 1:279).

4. See Bonnie Gordon, "Orfeo's Machines," *Opera Quarterly* 24, no. 3/4 (2008): 200–22; Jean Paul Richter, ed., *The Notebooks of Leonardo da Vinci* (New York: Dover, 1970); and Patrizio Barbieri, "The Speaking Trumpet: Development of Della Porta's 'Ear Spectacles' (1589–1967)," *Studi musicali* 33, no. 1 (2004): 205–47.

5. The Barberini letter is translated in full in Stefano Gattei, "Galileo's Legacy: A Critical Edition and Translation of the Manuscript of Vincenzo Viviani's *Grati Animi Monumenta*," *British Journal for the History of Science* 50, no. 2 (2017): 181–228.

6. For descriptions of this event see Paolo Galluzzi, "I sepolchri di Galileo: Le spoglie 'vive' di un eroe della scienza," in *Il Pantheon di Santa Croce a Firenze* (Florence: Cassa di Risparmio di Firenze, 1993), 145–82.

7. As Katharine Park has argued, connections between the criminal and the saintly body ran deep; both had body parts imbued with difference and with quasi-magical powers. See Katharine Park, "The Criminal and the Saintly Body: Autopsy and Dissection in Renaissance Italy," *Renaissance Quarterly* 47, no. 1 (1994): 1–33.

8. Rachel Donadio, "A Museum Display of Galileo Has a Saintly Feel," *New York Times*, July 23, 2010, https://www.nytimes.com/2010/07/23/world/europe/23galileo.html.

9. This was most fully developed in Gilles Deleuze and Félix Guattari, *A Thousand Plateaus: Capitalism and Schizophrenia* (Minneapolis: University of Minnesota Press, 1987). There is now a cottage industry of writings on Deleuze and Guattari.

10. The full quotation is "Ei mi dissee che prendeva quelle due Dita, come Reliquie, perchè con essee il Galileo aveva scritto tante belle cose; ed io toccando la Fronte del Cadavere, gli replicai, che avrei vo luco poter'avere quel che già stava dentro al Cranio, alludendo al suo tanto felice Ingegno, e sapere immense." Giovanni Targioni Tozzetti, *Notizie degli aggrandimenti delle scienze fisiche: accaduti in Toscana nel corso di anni LX. del secolo XVII*, 4 vols. (Florence: Bouchard, 1780), 1:142.

11. Sigmund Freud, *Civilization and Its Discontents*, in Peter Gay, ed., *The Freud Reader* (New York: W. W. Norton, 1989), 737.

12. "Il quale conduce gl'oggetti visibili così vicini all'occhio, et così grandi et distinti gli rappresenta, che quello che è distante, v. g., nove miglia, ci apparisce come se fusse lontano

un miglio solo: cosa che per ogni negozio et impresa marittima o terrestre può esser di giovamento inestimabile." Antonio Favaro, "Galileo astrologo secondo documenti editi e inediti: Studi e ricerche," *Mente e cuore: Periodico mensile di scienze, letteratura e cose scolastiche* 8, no. 3 (1881): 99–108.

13. Reeves, "Hearing Things," 201.

14. Reeves, 201.

15. Galileo Galilei, *Sidereus nuncius, or The Sidereal Messenger*, ed. and trans. Albert van Helden (Chicago: University of Chicago Press, 1989), 26.

16. Galilei, 36.

17. Rossi, *Pinacotheca Altera Imaginum, Illvstrium*, 279.

18. Matteo Valleriani, "Galileo's Abandoned Project on Acoustic Instruments at the Medici Court," *History of Science* 50, no. 1 (2012): 1–31.

19. "In opticis specilla demonstravimus, quibus satis longae videre poteramus, nunc instrumentum construere tentabimus, quo etiam per multa miliaria audire possimus." Giambattista della Porta, *Magia naturalis*, translated by Patrizio Barbieri in Barbieri, "The Speaking Trumpet," 206.

20. For histories of the ear trumpet, see Penelope M. Gouk, "Acoustics in the Early Royal Society 1660–1680," *Notes and Records of the Royal Society of London* 36, no. 2 (1982): 155–75; Lamberto Tronchin, "The 'Phonurgia Nova' of Athanasius Kircher: The Marvellous Sound World of 17th Century," *Proceedings of Meetings on Acoustics* 4 (2008): 1–9; and Barbieri, "The Speaking Trumpet."

21. Barbieri, "The Speaking Trumpet."

22. Paolo Aproino to Galileo in Florence, Treviso, June 1, 1613, translated by Matteo Valleriani in Valleriani, "Galileo's Abandoned Project," 20.

23. Valleriani, 20.

24. Valleriani, 20.

25. Athanasius Kircher, *Phonurgia nova* (Campidonae: Dreherr, 1673; repr., New York: Broude Brothers, 1966), 250.

26. "Janitores nostri de re quadam, sive de hospitum adventu, sive de quacunque alia re me monituri, ne per varias domus ambages meum adire Musaeum illis incommodum foret, intra portam janitoriam stantes mihi loquebantur in remoto cubiculi mei recessu commoranti & tanquam si praesentes, quaecunque vellent, distincte & clare proferebant, quibus & ego mox eodem vocis tenore pro negotiorum exigentia per orificium tubi respondebam, imo nemo quicquam intra horti districtum paulo elatiori voce prolatum dicere poterat, quod intra cubiculum non audirem; quae res uti nova & inaudita Musaeum visitantibus videbatur, dum loquentes audirent, neque tamen, qui loquerentur, viderent, ita quoque, ne illos alicuius vetitae artis suspicione attonitos suspenderem clandestinam artificii structuram ostendi; quae res, dici vix potest, quam multos etiam ex Proceribus Urbis ad videndum & audiendum machinamentum attraxerit." Kircher, *Phonurgia nova*, 112. The translation is mine, but since I began the project the passages have been written about and translated by others. See Michael John Gorman, "Between the Demonic and the Miraculous: Athanasius Kircher and the Baroque Culture of Machines," in Daniel Stolzenberg, ed., *The Great Art of Knowing: The Baroque Encyclopedia of Athanasius Kircher* (Stanford, CA: Stanford University Libraries, 2001), 26. Mark A. Waddell, "Magic and Artifice in the Collection of Athanasius Kircher," *Endeavour* 34, no. 1 (2010): 30–34; Cory A. Reed, "Ludic Revelations in the Enchanted Head Episode in Don Quijote (II, 62)," *Cervantes: Bulletin of the Cervantes Society of America* 24, no. 1 (2004): 189–219; and Tronchin, "The 'Phonurgia Nova' of Athanasius Kircher."

27. Galileo Galilei, *Discoveries and Opinions of Galileo*, ed. and trans. Stillman Drake (Garden City, NY: Doubleday Anchor Books, 1957), 250–51.

28. Galilei, 251.

29. Galilei, 251.

30. Galilei, 251–52.

31. Poetry serenading Basile's voice has been much discussed. The classic nineteenth-century study is Alessandro Ademollo, *La bell'Adriana: Ed altre virtuose del suo tempo alla corte di Mantova* (Città di Castello: S. Lapi, 1888). See also Amy Brosius, "'Il Suon, Lo Sguardo, Il Canto': The Function of Portraits of Mid-Seventeenth-Century Virtuose in Rome," *Italian Studies* 63, no. 1 (2008): 17–39; and John Whenham, "The Gonzagas Visit Venice," *Early Music* 21, no. 4 (1993): 525.

32. "Forma hà di tromba, e di lunghezza in duoi / Cubiti si distende; al doppio foro / Due vetri stan; l'un ne' convessi suoi / Forma l'altro nel cavo il bel lavoro. / L'occhio al cavo s'accosta, e mostran poi, / Dove li fissi tù, gli effetti loro: / E d'appressar' ogni lontan' oggetto / Con moltiplico immenso è il lor effetto." Giulio Strozzi, *La Venetia edificata*, translated by Crystal Hall in "Galileo, Poetry, and Patronage: Giulio Strozzi's *Venetia edificata* and the Place of Galileo in Seventeenth-Century Italian Poetry," *Renaissance Quarterly* 66, no. 4 (2014): 1296–1331.

33. See Ellen Rosand, *Opera in Seventeenth-Century Venice: The Creation of a Genre* (Berkeley: University of California Press, 1991).

34. "Con gli occhi sù queste carte anco gli esteri più remoti, e ritirati, quello di c'hanno goduto gli occhi, e gli orecchi in questa Città, che in ogni sua parte eccede i confini delle meraviglie." Maiolino Bisaccioni, translated by Ellen Rosand in Rosand, *Opera in Seventeenth-Century Venice*, 95.

35. "Cominciò il giovanetto, qual era un valorosissimo cantarino da Pistoia, à cantare si delicatamente, che gli animi de gli ascoltanti quasi che usciti dalle porte de gli orecchi si sollevarono a quell'altezza per assistere al godere di tanta dolcezza." Bisaccioni, trans. Rosand, 96.

36. "Sarà nei secoli ammirabile quel trovato che per mezzo di due vetri, l'un convesso e l'altro concavo, per forza di due contrari si vedono dalla terra in cielo minutezze non prima osservate, non solo di ombre o di concavi o di stelle nella Luna, ma di minute facelle che servono di satellizio a Giove et altre nella Via Lattea. Io considerava questi giorni che la composizione del sig. Giulio Strozzi della *Finta Pazza*, le macchine ritrovate dal sig. Iacomo Torelli e la musica orditavi sopra dal sig. Francesco Sacrati eran un cielo degno d'esser contemplato da tutti, ma così lontano a gran parte dalle genti, che era un togliere il pregio a tanti, che sono concorsi a sì nobile fattura, se non si faceva comodo ad ognuno di vederla et ammirare. Maiolino Bisaccioni, *Il cannocchiale per La finta pazza drama dello Strozzi. Dilineato da M. B. C. di G* (Venice, 1641), 3–4, translated in Hall, "Galileo, Poetry, and Patronage," 1327.

37. Francis Bacon, "New Atlantis," in *The Works of Lord Bacon: Philosophical Works* (London: Henry G. Bohn, 1855).

38. Francis Bacon, *The Major Works*, ed. Brian Vickers (Oxford: Oxford University Press, 2008), 485.

39. Peter Yates used it as a prophecy for electronic music in 1967. See Peter Yates, *Twentieth Century Music: Its Evolution from the End of the Harmonic Era into the Present Era of Sound* ([New York?]: Minerva Press, 1967), 313.

40. Francis Bacon, *The New Organon* [*Novum organum*], in *The Works of Francis Bacon*, ed. and trans. James Spedding, Robert Leslie Ellis, and Douglas Denon Heath (London: Longmans, 1858), 4:47.

41. Francis Bacon used the notion of organ to put forth a new method of discovery. His 1620 *Novum organum* presents a method for revealing nature rooted in observation and experiences and stands beside texts of Galileo and Descartes in the canon of the Scientific Revolution.

42. Penelope M. Gouk, "Music in Francis Bacon's Natural Philosophy," in *Number to Sound: The Musical Way to the Scientific Revolution*, ed. Paolo Gozza (Dordrecht: Kluwer Academic, 2000), 136.

## CHAPTER FOUR

1. Charles d'Ancillon, *Eunuchism Display'd. Describing All the Different Sorts of Eunuchs* [ . . . ]. (London, 1718), 31. This passage is much discussed.

2. Vanessa Agnew, *Enlightenment Orpheus: The Power of Music in Other Worlds* (New York: Oxford University Press, 2008).

3. Rebecca Cypess, *Curious and Modern Inventions: Instrumental Music as Discovery in Galileo's Italy* (Chicago: University of Chicago Press, 2016).

4. Bartoli, in booklet accompanying *Sacrificium.*

5. Guido Adler, "The Scope, Method, and Aim of Musicology," trans. Erica Mugglestone, *Yearbook for Traditional Music* 13 (1981): 1–21.

6. Erich M. von Hornbostel and Curt Sachs, "Classification of Musical Instruments," trans. Anthony Baines and Klaus P. Wachsmann, *Galpin Society Journal* 14 (March 1961): 3–29. This chapter contributes to what Emily Dolan and John Tresch have called a critical organology, which they see as a corrective to the musicological marginalization of instruments and technologies. See John Tresch and Emily I. Dolan, "Toward a New Organology: Instruments of Music and Science," *Osiris* 28, no. 1 (2013): 278–98.

7. Galen, *Method of Medicine*, 1.6, translated by Daryn Lehoux and quoted in *What Did the Romans Know? An Inquiry into Science and Worldmaking* (Chicago: University of Chicago Press, 2012), 118.

8. Aristotle, *On the Soul*, trans. J. A. Smith, bk. 3, pt. 8, http://classics.mit.edu/Aristotle/soul.3.iii.html.

9. Aristotle, *On the Soul*, trans. J. A. Smith, bk. 2, pt. 8, http://classics.mit.edu/Aristotle/soul.2.ii.html.

10. Galen, *On the Usefulness of the Parts of the Body*, trans. Margaret Tallmadge May (Ithaca, NY: Cornell University Press, 1968), 1:69–70, 71.

11. St. Augustine, *City of God*, trans. Gerald G. Walsh et al. (New York: Doubleday, 1958), 528.

12. St. Augustine, *Expositions on the Book of Psalms*, vol. 6, *Psalm CXXVI–CL* (Oxford: John Henry Parker, 1857), 439–40.

13. See Bruce W. Holsinger, *Music, Body, and Desire in Medieval Culture: Hildegard of Bingen to Chaucer* (Stanford, CA: Stanford University Press, 2001).

14. Albert the Great, *Questions Concerning Aristotle's "On Animals,"* trans. Irven M. Resnick and Kenneth F. Kitchell Jr. (Washington, DC: Catholic University of America Press, 2008), 159. In secular song, Paul Zumthor has argued for the voice as the body's signature and stresses the similarity between the verb *cantar* (to sing) and *canso* (to make a song).

15. "Conviene etiandio a tutti quelli Istrumenti materiali, che servono a qual si voglia Arte o Scientia, con l'aiuto de i quali si può condurre in quella alcuna opera al desiderato fine." Zarlino, *Sopplimenti musicali* (Venice: Francesco de Franceschi, 1588), 288.

16. "Il Martello, che adopera il Fabro nel fare i chiodi, & la Sega, che adopera il Legnaiuolo a segare & fender l'Asse, sono detti Istrumenti. Il Denaro anco, col quale, comperiamo le cose necessarie al vivere humano, è detto Istrumento. Et non pur le cose materiali, c'hanno la forma loro permanente; ma quelle che non hanno cotal forma, com'è la Logica: diciamo Istrumento." Zarlino, 288.

17. The letters were published in 1613 as "Istoria e dimostrazioni intorno alla macchie solari e loro accidenti" (A history and some demonstrations with respect to the sunspots) in Galileo Galilei, *Opere di Galileo Galilei*, ed. Antonio Favaro (Florence: Le Monnier, 1883), 5:96–97. See also Galileo Galilei, *Discoveries and Opinions of Galileo*, ed. and trans. Stillman Drake (Garden City, NY: Doubleday Anchor Books, 1957), 91–92n24; and Galileo Galilei and Christoph Scheiner, *On Sunspots* (Chicago: University of Chicago Press, 2010).

18. "Non per questo voglio disperarmi ed abbandonar l'impresa, anzi voglio sperar che queste novità mi abbino mirabilmente a servire per accordar qualche canna di questo grand'organo discordato della nostra filosofia; nel qual mi par vedere molti organisti affaticarsi in vano per ridurlo al perfetto temperamento, e questo perchè vanno lasciando e mantenendo discordate tre o quattro delle canne principali, alle quali è impossibile cosa che l'altre rispondino con perfetta armonia." Translation from Galileo Galilei, *On Sunspots*, trans. and introduced by Eileen Reeves and Albert Van Helden (Chicago: Chicago University Press, 2010), 110.

19. Adriano Banchieri, *Conclusions for Playing the Organ (1609)*, trans. Lee R. Garrett (Colorado Springs: Colorado College Music Press, 1982), 2.

20. Rebecca Cypess uses Marino as well. See Cypress, *Curious and Modern Inventions*, chap. 1.

21. "Ma io non so per certo vedere se senso alcuno vi sia, il cui ufficio di piú machine e piú maravigliosi arnesi abbia la Natura proveduto di quel che nella bocca ha fatto, la qual propriamente all'uso della musica fu deputata da lei, là dove tutto l'ingegno suo, tutte le sue forze impiegò. Quivi tanti sono gli stromenti [sic], con tanta cura e sottilità lavorati e tanto di lontano condotti, che quante membra sono in tutto l'universo corpo, par che solo per servire alla musica fatte sieno, talché, i piú chiari intelletti filosofando e le piú dotte mani scrivendo, a sí fatta considerazione si stancano." Giovanni Battista Marino, *Dicerie sacre e la Strage de gl'innocenti*, ed. Giovanni Pozzi (Turin: G. Einaudi, 1960), 255–56.

22. "Quest'organo medesimamente nella bocca dell'uomo si ritrova: la voce ottiene il luogo del suono; li polmoni sostengono la vece de' mantici, i quali il petto comprime per render l'aria che ricetta; l'arteria è come il cannone, per cui discorre lo spirito; con l'ordine delle canne disuguali si conforma la varia disposizione de' denti a' quale s'appartiene frangere e figurar la voce e dividere gli articoli del canto. Volete poi l'artefice o il sonatore? Ecco l'intelletto, il qual servendosi della lingua in cambio della mano, corregge e il fiato incomposto e dà norma e forma alla voce che vien senza regola e senza legge." Marino, 261–62.

23. "Per la qual cosa dico, che l'Organo proposto s'acquistò questo nome universale & commune d'Organo proprio & particolare, per una certa eccellenza dalle parti naturali, che formano la Voce, che si chiamano Istrumenti naturali: percioche fu fabricato all guisa del Corpo humano, corrispondendo le Canne alla Gola, i Mantici al polmone, i Tasti a i Denti, & colui che sona alla Lingua, & cosi l'altre parti di esso a quella che sono nell'Huomo. Ma veramente l'Organo nostro in quanto ad una parte della forma materiale, non è molto antico, anzi moderno: percioche sono aggiunti ne i Moderni i Mantici, i quali dalla Cassa che conteneva l'Acqua detta hora Sommiero somministrano il Vento, che passa nelle Canne: come nel sudetto luogo dipinge Vitruvio; dal che s'acquistò il nome di Hidraulica; il perche si può

vedere, che'l nostro Organo non è Istrumento moderno, se non in quanto all'alteratione della sua prima forma: percioche il Vento, che hora si fa con i Mantici, è posto in luogo di quello, che si facea col mezo dell'acqua." Zarlino, *Sopplimenti musicali*, 288–89.

24. Giambattista della Porta, *Natural Magick [ . . . ] Wherein Are Set Forth All the Riches and Delights of the Natural Sciences* (London, 1658), 386.

25. della Porta, 386.

26. "Il tremolo nella musica non è necessario; ma facendolo oltra che dimostra sincerità, et ardire; abbellisce le cantilene . . . dico ancora, che il tremolo, cioe la voce tremante è la vera porta d'intrar dentro a passaggi, & di impataonirse [*sic*] delle gorgie: perche con piu facilità se ne va la Nave quando che prima e mossa; che quando nel principio la si vuol muovere: & il saltatore meglio salta, se prima che salta si promove al salto. Questo tremolo deve essere succinto, & vago; perché l'ingordo, & forzato tedia, & fastidisce: Et è di natura tale che usandolo, sempre usar si deve; acciochè l'uso si converti in habito; perchè quel continuo muover di voce, aiuta, & volontieri spinge la mossa delle gorgie, & facilita mirabilmente i principij de passaggi." Lodovico Zacconi, *Prattica di musica* (Bologna: Forni, 1592), bk. 1, 62, p. 55, 60. Segments of Zacconi are translated in Lodovico Zacconi, "Prattica di Musica (1596): Part 1, Book 1, Chapter 66," trans. Sion M. Honea, Brisch Center for Historical Performance, Historical Translation Series, University of Central Oklahoma, https://www.uco.edu/cfad/files/music/zacconi-prattica.pdf. This passage is also translated in Edward Vaught Foreman, "A Comparison of Selected Italian Vocal Tutors of the Period circa 1550 to 1800" (DMA thesis, University of Illinois, 1969), 113–14.

27. Claudio Monteverdi, *The Letters of Monteverdi*, trans. Dennis Stevens (Cambridge: Cambridge University Press), 381.

28. "Nunc submissa occulta voce clam in ludicris oraculis fictisque consultationibus peragat eo artificio, ut nemo adstantium de secreto, reciproca colloquentium mussitatione instituto percipere quicquam valeat, quod & advenis in hunc usque diem exhibetur non sine daemonis alicuius latentis suspicione, eorum qui machinam non capiunt; nam & statua os ad normam loquentis aperit & claudit, oculos movet." Monteverdi, 113.

29. Paula Findlen, *Possessing Nature: Museums, Collecting, and Scientific Culture in Early Modern Italy* (Berkeley: University of California, 1994).

30. On the various ceremonial visits see "Origine del Collegio Romano e suoi progressi," MS 42, Archivio della Pontificia Università Gregoriana, Rome. This manuscript forms the basis of the descriptions of ceremonial receptions given in the Collegio Romano provided in R. Garcia Villoslada, *Storia del Collegio Romano dal suo inizio alla soppressione della compagnia di Gesù* (Rome: Typis Pontificiae Universitatis Gregorianae, 1954), 263–96. On theatrical productions in the Collegio Romano during this time, see Irene Mamczarz, "La trattatistica dei Gesuiti e la pratica teatrale al Collegio Romano: Maciej Sarbiewski, Jean Dubreuil e Andrea Pozzo," in *I Gesuiti e i primordi del teatro barocco in Europa*, ed. Maria Chiabò and Federico Doglio (Rome: Centro studi sul teatro medioevale e Rinascimentale, 1995), 349–87. See also Jean-Yves Boriaud, "La Poésie et le théâtre latins au Collegio Romano d'après les manuscrits du Fondo Gesuitico de la Bibliothèque Nationale Vittorio Emanuele II," *Mélanges de l'École française de Roma, Italie et Méditerranée* 102, no. 1 (1990): 77–96.

31. Translated in Michael John Gorman and Nick Wilding, "Technica Curiosa: The Mechanical Marvels of Kaspar Schott (1608–1666)," in *La technica curiosa di Kaspar Schott*, ed. Maurizio Sonnino (Rome: Edizione dell'Elefante, 2000), 261.

32. "Avidé-que scire desiderant, & Machinarum constructarum rationes, & Machinalium motionum causas. Horum desiderio ut satisfacerem, omnium dicti Musei Machinarum

fabricam & quasi anatomiam edocere, aut alicubi jam ab ipso Auctore edoctam enarrare, brevi opusculo aggressus sum." Gaspar Schott, *Mechanica hydraulico-pneumatica* (Würzburg: Pigrin, 1657), 4.

33. "Nec oculos tantum modo pascere satagimus; auribus etiam sua paramus delinimenta, variaque Organa atque Instrumenta automata, & autophona, solo aquarum lapsu, aërisque allapsu, in motum ac sonum concitamus, non minori facilitate, quam arte." Schott, *Mechanica hydraulico-pneumatica*, 5.

34. On the museum, see Stewart Pollens, "Michele Todini's Golden Harpsichord: Changing Perspectives," *Music in Art: International Journal for Music Iconography* 32, no. 1/2 (Spring–Fall 2007): 142; and Patrizio Barbieri, "Michele Todini's *galleria armonica*: Its Hitherto Unknown History," *Early Music* 30, no. 4 (November 2002): 565–82.

35. Michele Todini, *Dichiaratione della galleria armonica* [ . . . ] (Roma: F. Tizzoni, 1676).

36. "Che il suono d'una canna d'Organo, o d'altra cosa concava, passando per meati ritorti, composti di cose morbide, come di pelli d'animali, o simili, fossero bastanti a fare quello, che con sì grand'arte, anzi miracolosa, si sa appena dall'organo naturale, guidato dalla volontà, quale fa che si muovino tutte le parti, per tal effetto necessarie." Todini, *Dichiaratione della galleria armonica*, 21.

37. Denis Diderot, *Rameau's Nephew and Other Works*, trans. Jacques Barzun and Ralph H. Bowen (Indianapolis, IN: Hackett, 2001), 101; "Entretien entre d'Alembert et Diderot" (1769), in *Oeuvres complètes de Diderot, revues sur les éditions originales*, ed. J. Assézat (Paris: Garnier Frères, 1875), 2:105–21. Carolyn Abbate writes on this passage in "Outside Ravel's Tomb," *Journal of the American Musicological Society* 52, no. 3 (1999): 465–530.

38. Roger Moseley, "Digital Analogies: The Keyboard as Field of Musical Play," *Journal of the American Musicological Society* 68, no. 1 (2015): 151–228. Also see Emily I. Dolan, "Toward a Musicology of Interfaces," *Keyboard Perspectives* 5 (2012): 1–12.

39. The spinettone was invented by Bartolomeo Cristofori and was very large. It could make multiple strings sound at once.

40. *Tiorbino* was assumed to be a small theorbo, but Francesco Nocerino and Grant O'Brien argue that it is a special kind of keyboard instrument. Francesco Nocerino, "Il tiorbino fra Napoli e Roma: Notizie e documenti su uno strumento di produzione cembalaria," *Recercare* 12 (2000): 95–109; and Grant O'Brien, "The Tiorbino: An Unrecognised Instrument Type Built by Harpsichord Makers with Possible Evidence for a Surviving Instrument," *Galpin Society Journal* 58 (2005): 184–208.

41. "Sono posti tre strumenti da penna; una Spinettina all'ottava alta, un Spinettone, & un Tiorbino, nella facciata dirimpetto a quello, che suona, situati con tale disinvoltura, che pare impossibile habbino communicatione con le machine intrinseche, non conoscendosi, ne come siano le operationi, ne come possono essere cosi stabili, facendo il suo effetto, quasi invisibilmente. Stanno detti strumenti separati l'uno dall'altro, più d'un palmo, sopra suoi tavolini, col suo piede per ciascheduno, nè toccano il muro, nè altra cosa, che apparisca stabile. Vi è un'Organo, come si disse, che pur suona con la detta Tastatura, quale hà molti registri, & uno in particolare è raro, per la sua dolcezza, fatto di legno d'India." Todini, *Dichiarazione della galleria armonica*, 10–11. This paragraph is also cited and translated in O'Brien, "The Tiorbino."

42. "Bisogna che facci prima buona riflessione, che la Matematica mecanica prattica è molto più difficile della mecanica specolativa, perche in questa basta la lettura de' libri di tali materie, e sapere nel discorrere, ò scrivere quello, che probabilmente può ridurli all'atto, ma in quella contrastano sempre il peso, il numero, e la misura, quali non si possono ingannare in modo alcuno." Todini, *Dichiarazione della galleria armonica*, 27–28.

43. "Nella seconda stanza si vede rappresentata la favola di Polifemo con molte statue messe à oro, e trà le altre Galatea, che mostra palleggiar per il mare portata da due Delfini imbrigliati da un Cupido, sedendo lei in una conchiglia cortegiata [*sic*] da Ninfe marine, e servita da Tritoni grandi al naturale, che li portano un Cimbalo; la cassa del quale è ricca d'intagli, rappresentante, in basso rilievo pur messo à oro, il Trionfo di detta Galatea, con Mostri marini, che li porgono per tributo diversi frutti di mare. Polifemo siede alle falde d'un monte, nel quale hà la sua abitazione, come dice la favola, in atto di sonare una Sordellina, o Musetta per compiacere a Galatea; e dentro al detto monte stanno le Machine per far sonare la detta Sordellina, quale si suona con una tastatura posta sotto à quella del già nominato Cimbalo. Le statue sono fatte da valent'huomini, come anche tutti gl'altri materiali, quali sono secondo richiede l'opportunità di rappresentare, o mare, o monte, o aria. Detta Machina occupa da terra fino al soffitto; le sue difficoltà per esser state molte, si sono descritte altrove verso il fine, per non impedire il racconto succinto di tutte l'operationi [*sic*], come dissi." Michele Todini, *Dichiaratione della galleria armonica*, ed. Patrizio Barbieri (Lucca: Libreria musicale italiana, 1988), 5–6. The translation of this passage is adapted from Wendy Powers, "The Golden Harpsichord of Michele Todini (1616–1690)," Heilbrunn Timeline of Art History, New York, The Metropolitan Museum of Art, October 2003, https://www.metmuseum.org/toah/hd/todi/hd_todi.htm.

44. On the demise of Todini, see Barbieri, "Michele Todini's *galleria armonica*; and Herbert Heyde, "Todini's Golden Harpsichord Revisited," *Journal of the American Musical Instrument Society* 39 (2013): 5–61.

45. Athanasius Kircher, *Phonurgia nova* (New York: Broude Brothers, 1966), 166–70.

46. "Non solum aures auditorum mirum in modum solicitat, sed et marculorum subsultu oculos Vidius attonitos reddit; magicum incantamentum jure diceres." Kircher, 169

47. "Quomodò verò hoc harmonicum opus constructum." Kircher, 169.

48. Charles Burney, *The Present State of Music in France and Italy* (London, 1773), 392–93.

49. Charles Burney, *A General History of Music: From the Earliest Ages to the Present Period*, critical and historical notes by Frank Mercer (New York: Dover, 1957), 2:790.

50. For more detail on this aria, see Martha Feldman, *The Castrato: Reflections on Nature and Kinds* (Oakland: University of California Press, 2015), 114.

51. Feldman, 114.

## CHAPTER FIVE

1. Kircher, 346.

2. Anton Francesco Doni, *Le ville del Doni* (Bologna, 1566). For a classic study on gardens, and the one from which I first learned of this in Doni's corpus, see Elisabeth B. MacDougall, *Fountains, Statues, and Flowers: Studies in Italian Gardens of the Sixteenth and Seventeenth Centuries* (Washington, DC: Dumbarton Oaks Research Library and Collection, 1994).

3. Daniello Bartoli, "Le fontane di Roma, di Tivoli, di Frascati," in *Prose scelte* (Turin: Marietti, 1836), 2:263–65.

4. Bartolomeo Taegio, *La villa* (Milan: Francesco Moscheni, 1559), as cited in Claudia Lazzaro and Ralph Lieberman, *The Italian Renaissance Garden: From the Conventions of Planting, Design, and Ornament to the Grand Gardens of Sixteenth-Century Central Italy* (New Haven, CT: Yale University Press, 1990), 9. On the third nature, see Thomas E. Beck, "Gardens as a 'Third Nature': The Ancient Roots of a Renaissance Idea," *Studies in the History of Gardens & Designed Landscapes* 22, no. 4 (2002): 327–34.

5. "Si vedono li Organi stupendi et maravigliosi fatti et disegnati da un certo ms. Claudio Venardo di nation franzese, il quale organo si ritrova dentro di una gran nicchia . . . l'organo, il quale, quando che da alcuni si dà ordine per sonar, primeramente si sentono doi trombe che suonano alquanto et dopo sequita la consonantia, al solito della musica, di tal sorte ben ordinato et con misura come quel di S. Salvatore del Lauro in la città Roma non puol sonare più gratiosamente che questo con misura; et che si habbia da credere quanto gli dico, quando che la S.tà di N. S. Papa Gregorio XIII si ritrovò in la città di Tivoli l'anno 1573 si dede l'ordine che dovessi sonare, sonando sua S.tà restò tanto satisfatta et maravigliata che non solo lo volse sentire una volta ma doi et tre, et volse in ogni modo parlare al detto ms. Claudio inventore di esso, vista et intesa tal magnificenza rara con la presenza di tanti cardinali et principi signori, sono stati infeniti signori che non possevano credere questo organo sonasse de sè stesso temperatamente li registri con l'acqua, ma più tosto credevan che vi fussi alcuno dentro, perciò ne restorno certificati; dopo sonato l'organo si dà ordine nel medemo tempo dare l'acqua a 15 lampolli dalli quali scatoriscano l'acqua con tanta gran forza et violenza che salle su l'acqua alta una piccha di tal sorte che si fa chiamare il Diluvio." Giovanni Maria Zappi, *Annali e memorie di Tivoli*, ed. Vincenzo Pacifici (Tivoli: n.p., 1920), 61. Patrizio Barbieri says that the papal visit probably happened in 1572, not 1573. See Patrizio Barbieri, "Organi e automi musicali idraulici di Villa d'Este a Tivoli," *L'organo* 24 (1986): 3–61.

6. The literature on this garden is massive. David R. Coffin, *Pirro Ligorio: The Renaissance Artist, Architect, and Antiquarian* (University Park: Pennsylvania State University Press, 2004); David R. Coffin, "John Evelyn at Tivoli," *Journal of the Warburg and Courtauld Institutes* 19, no. 1/2 (1956): 157–58; John Dixon, *Garden and Grove: The Italian Renaissance Garden in the English Imagination, 1600–1750* (Philadelphia: University of Pennsylvania Press, 2016); Simone M. Kaiser and Matteo Valleriani, "The Organ of the Villa d'Este in Tivoli and the Standards of Pneumatic Engineering in the Renaissance," in *Gardens, Knowledge and the Sciences in the Early Modern Period*, ed. Hubertus Fischer, Volker R. Remmert, and Joachim Wolschke-Bulmahn (Cham, Switzerland: Springer International, 2016), 77–102; and Patrizio Barbieri, "Giambattista Della Porta's 'Singing' Hydraulis and Other Expressive Devices for the Organ, c. 1560–1860," *Journal of the American Musical Instrument Society* 32 (2006): 145–67.

7. Claudian, *Claudian*, trans. Maurice Platnauer, Loeb Classic Library135, 136 (Cambridge, MA: Harvard University Press, 1922), 1:361.

8. On sound and baroque gardens, see Anatole Tchikine, "Giochi d'acqua: Water Effects in Renaissance and Baroque Italy," *Studies in the History of Gardens & Designed Landscapes* 30, no. 1(2010): 57–76.

9. "Il y a encore une autre qualité de la voix qui la rend plaine, & solide & qui augmente son harmonie, ce que l'on peut expliquer par la comparaison d'un canal qui est toujours plain d'eau, quand elle coule, ou par celle d'un corps & d'un visage charnu & en bon point; au lieu que les voix qui sont privées de cette qualité, sont semblables à un filet d'eau qui coule par un gros canal, & un visage maigre, & décharné." Mersenne, *Harmonie universelle* (Paris: S. Cramoisy, 1636–1637), ed. François Lesure (Paris: Centre national de la recherche scientifique, 1965), 2:354.

10. Athanasius Kircher, *Musurgia universalis*, ed. Ulf Scharlau (New York: Georg Olms, 1970), 346.

11. Kircher, 346.

12. On the cost of the garden and other details, see Ronald W. Lightbown, "Nicolas

Audebert and the Villa d'Este," *Journal of the Warburg and Courtauld Institutes* 27 (1964): 164–90; Kaiser and Valleriani, "Organ of the Villa d'Este," 77–102.; Patrizio Barbieri, "The New Water Organ of the Villa d'Este, Tivoli," *Organ Yearbook* 33 (2004): 33; and John Dixon Hunt, "Ovid in the Garden," *AA Files*, no. 3 (1983): 3–11.

13. Zappi, *Annali e memorie di Tivoli*, 78.

14. The text is transcribed in Lightbown, "Nicolas Audebert." Parts of Audebert's description are also transcribed and translated in Kaiser and Valleriani, "Organ of the Villa d'Este," 93–97.

15. See Katelijne Schiltz, "Gioseffo Zarlino and the *Miserere* Tradition: A Ferrarese Connection?," *Early Music History* 27 (2008): 181–216.

16. "Ma la simiglianza, che da questa descrittione si scorge tra l'Hidraulica & l'Organo nostro, mosse l'Illustr. S. Suor Leonora d'Este, l'Anno 1571. nel mese di Novembre, a farmi richiedere da Francesco Viola gia mio singolare amico, Se cotale organo era antico o pur moderno, & dov'ei prendesse il suo nome; alla quale havendo primo risposto, mosso da questa richiesta, deliberai di Scrivere i presenti Sopplementi." Gioseffo Zarlino, *Sopplimenti musicali* (Venice: Francesco de Franceschi, 1588), 288.

17. Michel de Montaigne, *Montaigne's Travel Journal*, trans. Donald M. Frame (San Francisco: North Point Press, 1983), 99.

18. *Annali e memorie di Tivoli.*

19. Montaigne, *Montaigne's Travel Journal*, 99.

20. "Chasque oyseau auec une telle melodie qu'il semble ny auoir difference entre cest' artifice, et le naturel: Ce pendant venant tout doulcement a apparoir *une Cheueche* entre le pied d'une des statues & le museau du sanglier; en un instant tous les oyseaux se taisent, et cessent de chanter, tant que ceste cheueche demeure: laquelle puis apres s'estant retiree et disparue ilz recommencent a chanter comme deuant, non toutesfoys tout a coup & tous ensemble, maus un seul commence le tintin d'un Pinson, puis un aultre murmure a demy un petit gasoullement interrompu & fort bas, comme craignant encores ce qu'ilz ont veu; un tiers vient aussy a siffler, & lors ilz se reueillent tous, et commencent a degoyser comme au parauant, continuant tous jours jusques a ce que la cheueche retourne." Transcribed in Lightbown, "Nicolas Audebert," 182.

21. "Les moyens de ce faire sont telz. II ỳ a une quantité de petits siffiets dont les oyseleurs ont accoustumé d'user pour imiter le chant de diuers oyseaux: à iceux respond un canal qui se remplit de vent par le moyen de l'eau, qui faict que tous les siffietz rendent chascun leur chant: mais quand la cheueche apparoist elle se trouue en un lieu ou elle bousche tout a coup le conduict qui enuoye le vent: puis s'en retournant & retirant peu a peu elle laisse premierement desbouché le tuyau qui seul va au Pinson, & se retirant dauantage laisse aussy uenir le vent a un aultre siffiet qui gazoulle fort bas n'ayant du vent a suffire, En fin il se faict le semblable d'un aultre qui siffie, puis s'estant la cheueche du tout retiree & les conduictz demeurez entierement ouuerts, tous les oyseaux recommencent leur premier gazoullement, et a se desgoyser fort hault comme chantans a l'enuie l'un de l'aultre mais confusement ainsy que lon oit un grand nombre d'oyseaux en une volliere." Lightbown, 182.

22. "Ces orgues sans ayde de personne sonnent une chancon de musique auec toutes ses partyes, non moins bien & melodieusement auec ses mesures et fredons, que pourroit faire le plus excellent joueur: Mais pour rendre cela plus admirable on desnye a la plus grand part de ceux qui y vont de leur faire veoir l'artifice, toutesfoys J'euz bien ceste faueur que tout me fut communiqué & allay aux lieux plus secretz ou Je remarquay curieusement tous les mouuements & moyens de ceste ingenieuse & admirable inuention, que je descriray icy

selon ce que Jay veu. Lightbown, 185. This passage is also translated in Kaiser and Valleriani, "Organ of the Ville d'Este," 93.

23. "Pendant que ceste musique dure, il ŷ a un *Rossignol* qui ne cesse de chanter par le moyen d'un tuyau faict en subleau lequel par un bout recoit du vent de ceste cauerne, & par l'aultre trempe en un petit vase plein d'eau. Ces orgues ne sonnent qu'une seule chançon de musique, qui est a cinq partyes, fort longue, et bien musicale." Quoted in Lightbown, "Nicolas Audebert," 186. Also translated and transcribed Kaiser and Valleriani, "Organ of the Ville d'Este," 95.

24. Sources about this event are in Alessandro Ademollo, *I teatri di Roma nel secolo decimosettimo* (Bologna: L. Pasqualucci, 1888), 106–8; Maurizio Fagiolo dell'Arco and Silvia Carandini, *La festa barocca* (Rome: De Luca, 1997), 460–63; Jean Lionnet, "Les activités musicales de Flavio Chigi cardinal neveu d'Alexandre VII," *Studi Musicali* 9 (1980): 289–90, 300; Carlo Fontana, Pietro Santi Bartoli, and Teresa Del Po, *Risposta del Signor Carlo Fontana alla lettera dell'Illustrissimo Signor Ottavio Castiglioni* (Rome: Angelo Bernabò, 1668). Colleen Reardon's work is invaluable for thinking about the Chigi dynasty and seventeenth-century musicking. See especially Colleen Reardon, *A Sociable Moment: Opera and Festive Culture in Baroque Siena* (New York: Oxford University Press, 2016).

25. Robert Fahrner and William Kleb, "The Theatrical Activity of Gianlorenzo Bernini," *Educational Theatre Journal* 25, no. 1 (1973): 5–14; Genevieve Warwick, *Bernini: Art as Theatre* (New Haven, CT: Yale University Press, 2012). See also Reardon, *A Sociable Moment*; Frank D'Accone, "Cardinal Chigi and Music Redux," in *Music Observed: Studies in Memory of William C. Holmes*, ed. Colleen Reardon and Susan Helen Parisi (Warren, MI: Harmonie Park Press, 2004), 65–100.

26. "Roma 18 agosto 1668. Seguì martedì mattina lo scritto Banchetto fatto da' Chigi a' ss.ri, e ss.re Rospigliosi nel Giardino di S.ta M. Maggiore, ov'anche intervennero le Principesse di Rossano, e Farnese. Le sontuosità preparate furono, che entrati gli invitati per la Porta del Giardino trovarono al di dentro con vaghe prospettive un altra Porta, ove battendo comparve un Musico in habito rusticale, il quale con melodia dimandò, che cosa volevano, e risposto da convitati di voler vedere il Giardino, non solo gli fù concesso, mà anco apparechiata [*sic*] una tavola con ricotte, et altri cibi rusticali, il che penetrato da quattro Ninfe Boscareccie Padrone del luogo, frettolose v'accorsero, e dopo molti rimproveri contro il rustico Guardiano, riversciarono la mensa per terra, dopo di che ne comparve un altra regalmente imbandita, ove appena assisi i Convitati venne una pioggia d' acque odorose, et una tempestosa grandine di confetture, quale cessata si diede principio al Banchetto con suoni, e Musiche dopo di che successe un eco canoro di diversi Augelli. Il tutto riuscì in perfettione per esser stato disposto dal Cavalier Bernino." Quoted in Maurizio Fagiolo dell'Arco and Silvia Carandini, *La festa barocca* (Rome: De Lucca, 1997), 460–61. Dell'Arco sites this as a reliable account discovered by Marc Worsdale.

27. Vernon Lee, "Old Italian Gardens," in *In Praise of Old Gardens* (Portland, ME: Thomas B. Mosher, 1912), 29.

28. Lee, 57.

## CHAPTER SIX

1. Luigi Manzini, *Applausi festivi fatti a Roma per l'elezzione di Ferdinando III al Regno de' Romani dal Ser.mo Princ. Maurizio Card. Di Savoia, descritti al Ser.mo Francesco d'Este Duca di Modena da D. Luigi Manzini* (Rome: Facciotti, 1637).

2. "Per svegliare le penne più erudite a celebrare le glorie di questa casa, appunto come il Tamburo, il più vile tra i Soldati, sveglia non solo i generosi Guerrieri, ma tutti i capi della milizia." Giuseppe Elmi, *Relazione de' fuochi artificiati e feste fatte in Roma per la Coronazione del novello Cesare Leopoldo Primo dall'Eminentissimo, e Reverendissimo Sig. Cardinal Colonna Protettore del S. Romano Imperio* [ . . . ] (Rome: Francesco Cavalli, 1658), 12.

3. "Si sentirono in ogni quartiere i tamburi di radunanza dei cittadini, i quali si raggrupparono in squadrone in diverse piazze e in modo particolare attorno ai cancelli della detta Piazza del Senato, dal principio della quale, fino alla Chiesa venne eretto un ponte di legno coperto da 4000 braccia di panno nero, giallo e bianco (dei quali è l'aurea Cesarea)." Elmi, 12.

4. There is a large literature on baroque festive culture and on early modern stagecraft. See, for instance, Kristiaan Aercke, *Gods of Play: Baroque Festive Performances as Rhetorical Discourse* (Albany: State University of New York Press, 1994); Jessica Wolfe, *Humanism, Machinery, and Renaissance Literature* (Cambridge: Cambridge University Press, 2004); and James M. Saslow, *The Medici Wedding of 1589: Florentine Festival as Theatrum Mundi* (New Haven, CT: Yale University Press, 1996).

5. Gilles Deleuze, *The Fold: Leibniz and the Baroque*, trans. Tom Conley (Minneapolis: University of Minnesota Press, 1993).

6. Martha Feldman, *Opera and Sovereignty: Transforming Myths in Eighteenth-Century Italy* (Chicago: University of Chicago Press, 2007).

7. Feldman, 24.

8. Olivia Bloechl, *Opera and the Political Imaginary in Old Regime France* (Chicago: University of Chicago Press, 2017).

9. Simon Werett, *Fireworks: Pyrotechnic Arts and Sciences in European History* (Chicago: University of Chicago Press, 2010); and "Watching the Fireworks: Early Modern Observation of Natural and Artificial Spectacles," *Science in Context* 24, no. 2 (2011): 167.

10. Jonathan Sawday, *Engines of the Imagination: Renaissance Culture and the Rise of the Machine* (London: Routledge, 2007).

11. William Gibson, *Neuromancer* (New York: Berkley Publishing Group, 1989), 128.

12. Feldman, *Opera and Sovereignty*, esp. chap. 4, 141–81.

13. Feldman, 154.

14. Valeria De Lucca, *The Politics of Princely Entertainment: Music and Spectacle in the Lives of Lorenzo Onofrio and Maria Mancini Colonna* (New York: Oxford University Press, 2020).

15. Genevieve Warwick, "Ritual Form and Urban Space in Early Modern Rome," in *Late Medieval and Early Modern Ritual: Studies in Italian Urban Culture*, ed. Samuel Cohn Jr., Marcello Fantoni, Franco Franceschi, and Fabrizio Ricciardelli (Turnhout: Brepols, 2013), 297–328.

16. On fountains in particular, see Rose Marie San Juan, *Rome: A City Out of Print* (St. Paul: University of Minnesota Press, 2001); and Katherine Wentworth Rinne, *The Waters of Rome: Aqueducts, Fountains, and the Birth of the Baroque City* (New Haven, CT: Yale University Press, 2010).

17. Feldman, *Opera and Sovereignty*, 156.

18. Valeria De Lucca, and Christine Jeanneret, eds., *The Grand Theater of the World: Music, Space, and the Performance of Identity in Early Modern Rome* (London: Routledge, 2019).

19. Grazioso Uberti, *Contrasto musico: Opera dillettevole* (Rome: Lodovico Grignani, 1630).

20. Frederick Hammond, *Music and Spectacle in Baroque Rome: Barberini Patronage*

*under Urban VIII* (New Haven, CT: Yale University Press, 1994); Margaret Murata, *Operas for the Papal Court 1631–1668* (Ann Arbor, MI: UMI Research Press, 1981); Maurizio Fagiolo dell'Arco and Silvia Carandini, *La festa barocca* (Rome: De Luca, 1997).

21. The most detailed account of these bulls in the musicological literature is Giuseppe Gerbino, "The Quest for the Soprano Voice: Castrati in Renaissance Italy," *Studi musicali* 33, no. 2 (2004): 303–57.

22. Giulia de Dominicis, "I teatri di Roma nell'età di Pio VI," *Archivio della Società romana della storia patria* 46 (1923), 74–76.

23. Charles Burroughs, "Opacity and Transparence: Networks and Enclaves in the Rome of Sixtus V," *RES: Anthropology and Aesthetics* 41, no. 1 (2002): 56–71; Jeffrey Collins and Anthony Grafton, "Obelisks and Empires of the Mind," *American Scholar* 71, no. 1 (2002): 123–27; and Rinne, *The Waters of Rome*.

24. Paolo Prodi, *The Papal Prince: One Body and Two Souls; The Papal Monarchy in Early Modern Europe*, trans. Susan Haskins (Cambridge: Cambridge University Press, 1987).

25. Laurie Nussdorfer, "Masculine Hierarchies in Roman Ecclesiastical Households," *European Review of History: Revue européenne d'histoire* 22, no. 4 (2015): 620–42.

26. On the gendered consequences of this turbulence in early modern Rome, see Laurie Nussdorfer, "Men at Home in Baroque Rome," *I Tatti Studies in the Italian Renaissance* 17, no. 1 (2014): 103–29; Elizabeth S. Cohen, "Open City: An Introduction to Gender in Early Modern Rome," *I Tatti Studies in the Italian Renaissance* 17, no. 1 (2014): 35–54; John M. Hunt, "Carriages, Violence and Masculinity in Early Modern Rome," *I Tatti Studies in the Italian Renaissance* 17, no. 1 (Spring 2014), 175–96.

27. Joseph Connors, "Alliance and Enmity in Roman Baroque Urbanism," *Römisches Jahrbuch der Bibliotheca Hertziana* 25 (1989): 207–94; Laurie Nussdorfer, "The Politics of Space in Early Modern Rome," *Memoirs of the American Academy in Rome* 42 (1997): 161–86.

28. John M. Hunt, *The Vacant See in Early Modern Rome: A Social History of the Papal Interregnum* (Boston: Brill, 2016).

29. Peter Burke, "Varieties of Performance in Seventeenth-Century Italy," in *Performativity and Performance in Baroque Rome*, ed. Peter Gillgren and Mårten Snickare (London: Routledge, 2017), 29–38.

30. Giacinto Gigli, *Diario di Roma*, ed. Manlio Barberito (Rome: Colombo, 1994), 2:584.

31. For more on these festivals, see Fernando Checa Cremades, and Laura Fernández-González, *Festival Culture in the World of the Spanish Habsburgs* (New York: Routledge, 2016).

32. Gigli, *Diario di Roma*, 2:585

33. Susan Parisi, "Acquiring Musicians and Instruments in the Early Baroque: Observations from Mantua," *Journal of Musicology* 14, no. 2 (1996): 117–50.

34. Faccone to Ferdinando Gonzaga, Rome, March 27, 1615, as cited in Parisi, "Acquiring Musicians and Instruments," 138.

35. As cited in Stuart Reiner, "Preparations at Parma 1618, 1627–28," *Music Review* 25 (1964): 276.

36. On the festival, see Hammond, *Music and Spectacle in Baroque Rome*, 214–24; Frederick Hammond, "The Artistic Patronage of the Barberini and the Galileo Affair," in *Music and Science in the Age of Galileo*, ed. Victor Coelho (Dordrecht: Springer, 1992), 67–89; Warwick, "Ritual Form and Urban Space."

37. Vitale Mascardi, *Festa fatta in Roma alli 25 di Febraio MDCXXXI e data in luce da*

*Vitale Mascardi* (Rome: Vitale Mascardi, 1635). On festival books, see Laurie Nussdorfer, "Print and Pageantry in Baroque Rome," *Sixteenth Century Journal* 29, no. 2 (1998): 439–64.

38. "Condotto sopra quattro ruote messe à oro." Mascardi, *Festa fatta in Roma*, 4.

39. "La Fama in un vago Carro, il quale da una grand'aquila condotto sopra quattro ruote messe à oro, appresentossi nel mezzo della sala, dove erano adunate le Dame, e diversi altri Cavalieri. Si scompartiva il Corpo del Carro in molti scanellamenti adornati con fogliami, e fregi d'oro, che in campo verde maggiormente spiccavano. Ma dal corpo del medesimo Carro s'alzava sopra due Arpie d'argento il seggio della Fama, il quale pure da un grand'Arpia d'argento per la parte di dietro veniva sostenuto. Salivasi al detto seggio per due gradi d'argento tutti lavorati di varii arabeschi & intagli, e sù l'estremo del piano, ove l'aquila haveva i legami per tirarlo, due leggiadri vasi d'argento adornavano il pavimento del Carro. La Fama, che maestosa sedeva sù la sommità di esso comparve poi superbamente vestita, e la sua veste, che di varij colori era tutta con oro tessuta veniva ancora da una moltitudine d'occhi, di bocche, di orecchie tempestata. Portava una tromba d'oro in mano, & alle spalle spiegava due ali anch'esse ripiene d'occhi, d'orecchie, e di bocche. Fermossi il Carro quando fù di bisogno, e mentre si stava aspettando d'intendere quello che la Fama fusse per apportare, ella accompagnata con un'armonioso concerto d'instrumenti in queste note con suavissimo canto spiegò la cagione della sua venuta." Mascardi, 4–5.

40. Virgil, *The Aeneid*, trans. Robert Fitzgerald (New York: Vintage Books, 1983), 101.

41. Ovid, *Metamorphoses*, trans. Frank Justus Miller (Cambridge, MA: Harvard University Press, 1951), 2:183.

42. Hammond, *Music and Spectacle in Baroque Rome*, 215.

43. Mascardi, *Festa fatta in Roma*, 6.

44. For a visual representation of the joust, see Andrea Sacchi's painting *Giostra del Saracino a Piazza Navona nel Carnevale del 25 febbraio 1634* (The Saracen joust in Piazza Navona for carnival on 25 February 1634) in the Museo di Roma.

45. "Al cominciare d'un suavissimo suono di strumenti cessò ad un tratto ogni susurro nel Theatro, il quale ben presto riempissi di angeliche voci. Fù il primo a cantare il Dio Bacco, seguitando poi il Choro delle Ninfe, e de i Pastori; e dal Riso finalmente con gratia soprahumana terminossi la musica, la quale però venne tramezzata da un gentilissimo balletto di Pastori, che secondato da ben concertati strumenti, mentre diletta la vista, e lusinga l'udito, insensibilmente a' riguardanti rapisce il cuore." Mascardi, 123.

46. On architectural theory in the Renaissance, see Alina A. Payne, *The Architectural Treatise in the Italian Renaissance: Architectural Invention, Ornament, and Literary Culture* (Cambridge: Cambridge University Press, 1999); and Tod A. Marder, "Vitruvius and the Architectural Treatise in Early Modern Europe," in *Companions to the History of Architecture*, vol. 1, *Renaissance and Baroque Architecture*, ed. Alina Payne (Hoboken, NJ: Wiley-Blackwell, 2017), 1–31.

47. Luigi Allegri, "Introduzione allo studio degli apparati a Parma nel Settecento," in *La Parma in festa: Spettacolarità e teatro nel Ducato di Parma nel Settecento*, ed. Luigi Allegri and Renato di Benedetto (Modena: Mucchi, 1987), 13–34. Thanks to Martha Feldman for this reference.

48. Jake Morrissey, *The Genius in the Design: Bernini, Borromini, and the Rivalry That Transformed Rome* (London: Duckworth, 2005); Dorothy Metzger Habel, *"When All of Rome Was Under Construction": The Building Process in Baroque Rome* (University Park: Pennsylvania State University Press, 2013); Ingrid Rowland, "'Th' United Sense of th' Uni-

verse': Athanasius Kircher in Piazza Navona," *Memoirs of the American Academy in Rome* 46 (2001): 153–81.

49. Maurizio Fagiolo dell'Arco and Silvia Carandini, *L'effimero barocco: Strutture della festa nella Roma del '600* (Rome: Bulzoni, 1978), 2:179. For a list of descriptions, see Fagiolo dell'Arco, *La festa barocca*, 349–52.

50. Athanasius Kircher, *Obeliscus Pamphilius, hoc est, Interpretatio nova & hucusque intentata obelisci hieroglyphici quem non ita pridem ex veteri hippodromo Antonini Caracallae caesaris* [ . . . ] (Rome: Grignani, 1650).

51. "In Piazza Navona furno fatti ornamenti, come già si faceva prima, et anco maggiori. Le due Fontane, che sono nella Piazza furno rinchiuse dentro un Arco di quattro facciate con colonne altissime, et sopra l'Archi vi erano torri, et Cuppole che pareva ogni cosa di pietre et marmi colorate. Dentro vi erano palchi nelli quali al tempo della processione erano Chori di Musici. Nel mezzo della Piazza, dove hora è la Guglia (l'ornamento della quale ancora non è finito) fu fatto un gran serraglio di legname riquadrato coperto con tele dipinte a maraviglia, et nelle quattro cantonate furono fatte quattro torri con palchi dentro per i musici, et a filo della Guglia per il mezzo della piazza di quà, et di là erano fatte doi altre Guglie dipinte di fuochi artifitiali, et altre machine tutte piene di fuochi. Tutto il Teatro della Piazza era cinto da Archi di legname dipinto tutti pieni di lumi accesi, et tutte le torri, et tutti li altri ornamenti erano ripieni di lampadi accese. Incontro alla Guglia dove è la Chiesa di S. Agnese fu fatto un'Altare molto bello con colonne et cornicione di sopra dipinto, et indorato, sopra il qual Altare doveva posarsi il SS.mo Sacramento." Gigli, *Diario di Roma*, 2:585–86.

52. See San Juan, *Rome*, 187–218.

53. San Juan, 187–218.

54. Gigli, *Diario di Roma*, 2:533–34. August 1648. The verse is also quoted in an avviso by Francesco Mantovani to the Duke of Modena on July 18, 1648. See Stanislao Fraschetti, *Il Bernini, la sua vita, la sua opera, il suo tempo* (Milan: Hoepli, 1900), 182.

55. Quoted and translated in San Juan, *Rome*, 198. This episode has been much discussed and is detailed in Cesare d'Onofrio, *Le fontane di Roma* (Rome: Romana Società, 1986), 210. See also John M. Hunt, "Ritual Time and Popular Expectations of Papal Rule in Early Modern Rome," *Explorations in Renaissance Culture* 45, no. 1 (2019): 29–49; Laurie Nussdorfer, *Civic Politics in the Rome of Urban VIII* (Princeton, NJ: Princeton University Press, 2019); Genevieve Warwick, *Bernini: Art as Theater* (New Haven, CT: Yale University Press, 2012).

56. The literature on this fountain is as vast as the literature on the piazza. A few helpful examples are Rose Marie San Juan, "The Transformation of the Río de la Plata and Bernini's Fountain of the Four Rivers in Rome," *Representations* 118, no. 1 (2012): 72–102; D'Onofrio, *Le fontane di Roma*; and Nussdorfer, "Print and Pageantry." On the festivities for the fountain, see Rose Marie San Juan, "Enclosures and Exchanges in Piazza Navona," in *Rome*, 187–217.

57. Francesco Ascione, *Le lodi e grandezze della Aguglia e Fontana di piazza Navona* (Rome: Francesco Cavalli, 1651); and Emilio Meli, *La Fontana Panfilia comedia del Sig. D. Emilio Meli* (Rome: Francesco Moneta, 1652).

58. Maria Portia Vignoli, *L'Obelisco di Piazza Navona* (Rome: Francesco Moneta, 1651).

59. "Ma chi poi fu di sì bell'opera il fabro? / Fu l'esperto Bernini, / che come un nuovo Archimede / con il suo valore ogni valore eccede." Vignoli, 4.

60. "In modo tale, che messosi fra Cittadini un gran bisbiglio, l'uno restava non men' attonito dell'altro, mentre niun di loro la cagione del prodigioso portento investigar poteva, e scorgendosi cossì sospesi l'animi dell'ammirato popolo, ecco che all'improvviso comparir si

vidde un'altra donna la cui veste brillava di duoi vivaci colori, rosso, e azzurro." Antonio Bernal di Gioya, *Copiosissimo discorso della fontana, e guglia eretta in Piazza Navona per ordine della Santità di Nostro Signore Innocentio X* (Rome, 1651), 1.

61. "À che si bada, sù volga ogn'un le piante verso quel lugo che un tempo era da voi nomato d'Alessandro il Circo, che ivi vedrete scoverta la più superba meraviglia di quante vantarsi possa a nostri tempi il mondo tutto, la quale non poca maestosa la rende l'Innocente Colomba della Pamphilia Casa, che la gran macchina tutta, benché magnifica, è superba premendola col suo legiadro piede la custodisce, e difende; A tale voci si sveglìorno l'animi della sbigottita gente; e ravisandosi esser quella la Fama, & questa la Curiosità che li spronava, avida di vedere meraviglia tale con precipitosa carriera si vidde incaminar tutta verso là, dove ogni persona, che vi giongeva scorgendosi spettatore di sì maestoso edificio, in tal guisa ammirata restava, ch'a'pena scorger si poteva qual differenza fusse tra loro, e quei raviviti marmi, che con occhi fissi, così intenti miravano." Bernal di Gioya, 1–2.

62. Avvisi di Roma, February 18, 1634. "La sera domenica et anco mercordi fu di nuovo recitata nel Palazzo dell'Ecc.mo Prefetto di Roma la rappresentatione di Sant'Alessio in musica a compiacenza de' Signori et altra nobilta che per prima non l'havevano potuto vedere, riuscendo cosa bellissima cosi per l'interlocutori che ciascuno fece eccellentemente la parte sua, come per la vaghezza degli habiti, e diversita dell'apparenza delle scene, et intermedij." As cited in Murata, *Operas for the Papal Court*, 225.

63. "A giudicio mio, non era di quella riputazione che si conviene a questa carica che vesto." Fulvio Testi, *Lettere*, ed. Maria Luisa Doglio, vol. 2 (Bari: Giuseppe Lateranza e Figli, 1967), 74.

64. "Non è dubio che quando si rappresentò al Signor Principe di Pollonia, le macchine sarebbero potute andar meglio, ma ci fu pochissimo tempo. Et io dubitavo di peggio, perchè a pena si erano provate mai. Oltre che rispetto alla moltitudine e grandezza delle macchine, il palco dove si rigira ogni cosa, è così angusto che è da maravigliarsi che non seguisse sempre mille disordini." Giulio Rospigliosi to his brother Camillo, Rome, January 11, 1634, fol. 61v, MS Vat. lat. 13362, Biblioteca Apostolica Vaticana, Rome. Transcribed in Murata, *Operas for the Papal Court*, 224. I have made minor revisions to Murata's transcription.

65. "ROMA sopra un Trofeo di spoglie, circondata da diversi Schiavi dopo haver sentito le lodi del Sereniss. [imo] Principe ALESSANDRO CARLO di Polonia, & il giubilo commune per la venuta di S.[ua] Altezza, risolve di rappresentarle i casi di Santo Alessio, quale tra i suoi Cittadini fù non meno conspicuo nella gloria della santità, di quello che fussero molti nel valore dell'armi. E per accennare, come ella stima più d'ogn'altro dominio l'effer Regina de' cuori, ordina, che i medisimi Schiavi rimanghino liberi dalle catene." Stefano Landi, *Il S. Alessio: dramma musicale Alessio* (Rome: Paolo Masotti, 1634), 8.

66. "La cui penna fece senza colori un ritratto dell'opera." Landi, [3].

67. "Che anche i gesti, e le movenze parevano armoniosi, e consonanti, come le voci." Landi, [4].

68. "La prima introdutione di Roma nuova, il volo dell'Angelo tra le nuvole, l'apparimento della Religione in aria, opere furono d'ingegno e di macchina ma gareggianti con la natura. La scena artifitiosissima; le apparenze del Cielo e dell'Inferno meravigliose; le mutazioni dei lati e della prospettiva sempre più belle, ma l'ultima della sfuggita e del cupo illuminato di quel portico con l'apparenza lontanissima del giardino, incomparabile." Landi, [4].

69. On Roman theater in particular see Saverio Franchi, "Osservazioni sulla scenografia dei melodrammi romani nella prima metà del Seicento," in *Musica e immagine: Tra iconogra-*

*fia e mondo dell 'opera: Studi in onore di Massimo Bogianckino*, ed. Biancamaria Brumana and Galliano Ciliberti (Florence: Olschki, 1993), 151–75.

70. Carol MacClintock, *Readings in the History of Music in Performance* (Bloomington: University of Indiana Press: 1979), 190.

71. "Noi per l'arte del corago intenderemo qui quella facoltà mediante la quale l'uomo sa prescrivere tutti quei mezzi e modi che sono necessarii acciò che una azione drammatica già composta dal poeta sia portata in scena con la perfezione che si richiede per insinuare con ammirazione e diletto quella utilità e frutto anche morale che la poesia richiederà." Paolo Fabbri and Angelo Pompilio, eds., *Il Corago: Overo alcune osservazioni per metter bene in scena le composizioni drammatiche* (Firenze: Olschki, 1983), 21. Translated in Roger Savage and Matteo Sansone, "'Il Corago' and the Staging of Early Opera: Four Chapters from an Anonymous Treatise circa 1630," *Early Music* 17, no. 4 (1989): 499.

72. "Per dovere eglino tenere alzati gli occhi più della naturale e commoda maniera di guardare, raddoppiandosi la poca sodisfazione della veduta per apparergli l'oggetto più alto di quello che il perfetto modo di posseder con l'occhio una cosa richiede. Il quale incommodo non si pole evitare nel palco moderno se non fusse d'una grandezza smisurata sopra ogni uso nostrale. Imperoche' se l'attore si vol trattenere lontano dallo spettatore a canto alla prospettiva, subbito egli comparisce molto maggiore non solo delle porte de' palazzi vicini ma dei tetti e camini ancora delle case nella prospettiva dipinte, la quale cosa appare molto sconcia all'occhio de' rimiranti e si scopre troppo chiaramente la finzione de l'artificio." Fabbri and Pompilio, 27.

73. "Non si può negare che in quanto alla veduta non ricrei molto più la nostra scena dell'antica, ancorche' ciò con qualche fallacia delli ochi apparisca, come le vedute anche finte per mezzo di artificii cristalli molto più delettano che i semplici riguardi dei veri oggetti." Fabbri and Pompilio, 30.

74. "Di quest'artificij a me pare il migliore quello delle Tromba o d'altro instrumento, poiché quello di far la rissa e la finzione che si rompa lo scalone porta seco molti pericoli, come di far nascere qualche gran tumulto il quale non si acquieti poi così di leggieri; ma per lo contrario sentito solo il tocco dell'instromento, come si disse di sopra e finito quello, le genti si rivolgano subito verso la scena, come erano prima acquetandosi, e con maraviglia, e con gusto ammirando il nuovo apparato che si rappresenta a gli occhi loro." Nicola Sabbatini, *Pratica di fabricar scene e machine ne' teatri* (Ravenna: Pietro de' Paoli and Gio. Battista Giovannelli, 1638), 61–62.

75. "Anche D. Paolo Sforza fu quasi constretto di duellare con un francese che voleva maltrattare Marc'Antonio il castrato." Transcribed in Murata, *Operas for the Papal Court*, 305.

76. "Il medesimo cardinale di suo mano bastonò alcuni servitori che fuori della porta facevano troppo strepito nell'aspettare i padroni. Minacciò altri che dicevano alli recitanti che parlassero più forte; et continuamente si è inserito in cose poco proporzionate al grado che sostiene." Transcribed in Murata, 305.

77. Irving Lavin demonstrated this forty years ago, and it has been much written about since. Irving Lavin, "Bernini and the Theater," in *Bernini and the Unity of the Visual Arts* (New York: Oxford University Press, 1980), 1:146–57.

78. The literature on Bernini and the theater is vast. See, for example, D. A. Beecher, "Gianlorenzo Bernini's *The Impresario*: The Artist as the Supreme Trickster," *University of Toronto Quarterly* 53, no. 3 (1984): 236–47; Jackson I. Cope, "Bernini and Roman *Commedie Ridicolose*," *PMLA* 102, no. 2 (1987): 177–86; Irving Lavin, *Bernini and the Unity of the Visual Arts*, 2 vols. (New York: Oxford University Press, 1980); Lavin, "Bernini and the Art of

Social Satire," *History of European Ideas* 4, no. 4 (1983): 365–420; and Genevieve Warwick, *Bernini: Art as Theater* (New Haven, CT: Yale University Press, 2012).

79. Filippo Baldinucci, *The Life of Bernini* (University Park: Pennsylvania State University Press, 2007), 39.

80. "Nel calar delle tende comparve una scena mirabile con una prospetiva che mostrava fabriche lontanissime, e principalmente la Chiesa di S. Pietro, Castel Sant'Angelo, et molti altri edifizi molto ben noti à chi habita in Roma. Più da vicino si vedeva il Tevere, il quale con modi finti, et con rara invenzione andava crescendo, volendo il Cavaliere dimostrare quegli effetti che l'anno passato pur troppo s'eran veduti quando il Tevere stette per indondar la Città. Più propinqua al Palco dove se recitava era Acqua vera sostenuta da certi ripari ch'erano stati distribuiti appostatamente per tutto il giro della scena; et si vedevano huomini reali i quali traghettavono altre persone da una parte all'altra, come se il fiume avendo occupati i luoghi più bassi della Città, havesse impedito il commercio, come appunto successe l'anno antecedente. Mentre ogn'uno stava attonito per questo spettacolo, andavano diversi Ministri rivedendo l'argine, accomandando travi e ripari, affinché il fiume non sommergesse la Città. Ma all'improviso cascò l'argine, e l'acqua sormontando sopra il palco, venne à correr furiosamente verso l'Auditorio, e quei ch'erano più vicini dubitando veramente che li rovinasse, si alzarono in piedi per fuggirsene; ma quando l'Acqua stava per caderli addosso si alzò all'improviso un riparo nel finire del Palco et si disperse la medesima Acqua senza far danno à persona alcuna." Massimiliano Montecucoli, letter to the Duke of Mantua, Febuary 17, 1638. Quoted in Fraschetti, *Il Bernini*, 264–65.

81. "Nel mezzo della Comedia fece vedere un altra scena maravigliosa con occasione che si sentiva una musica: attaccato dunque al Palco dove si recitava fece comparire una strada piena di carrozze, di cavalli, e di gente vera che si fermarono a sentire la musica. Più lontano si vedevano fabbriche nobilissime, e Piazze grandi che venivano frequentate da carrozze finte che correvano e da gente simulata. Ci era ancora il cielo con la luna e le stelle che talvolta si movevano, e questa scena fornì con la comparsa d'un somar che ragliò opportunamente." Montecucoli to the Duke of Mantua, February 17, 1638.

82. For many details, see Frederick Hammond, "Bernini and the 'Fiera di Farfa,'" in *Gianlorenzo Bernini: New Aspects of His Art and Thought. A Commemorative Volume*, ed. Irving Lavin (University Park: Pennsylvania State University Press, 1985), 115–78; Susan Lewis Hammond, "'Chi soffre speri' and the Influence of the *Commedia dell'arte* on the Development of Roman Opera" (Master's thesis, University of Arizona, 1995).

83. Girolamo Teti's description of this is translated in Murata, *Operas for the Papal Court*, 206n64.

84. Gian Lorenzo Bernini, *Fontana di Trevi: Commedia inedita*, ed. Cesare D'Onofrio (Roma: Staderini, 1963). For an English translation, see Bernini, *The Impresario (Untitled)*, trans. Donald Beecher and Massimo Ciavolella (Ottawa, ON: Dovehouse Editions Canada, 1994). My translations are slightly different.

85. "O! belle nuvole vedo andar per aria. Infatt dov'è naturalezza è artifitio." Bernini, *Fontana di Trevi*, 60.

86. "L'inzegn, el desegn, è l'Arte Mazica per mezz dei quali s'arriva à ingannar la vista in modo da fere stupier, e di fere spiccar una nuvola dall'orizzonte e venir inanz sempre spiccada con un moto naturel, e a man, a man che la s'avvizina alla vista dilatandose apparir più grand. Mostrer che 'l vento l'azisi e la trasporti via in zà, e in là e poi, se ne vada in sù, e non calarla zù comuod fan i contrapis." Bernini, *Fontana di Trevi*, 74.

87. Hunt, *Vacant See in Early Modern Rome*, 201.

88. The literature on Bernini is vast and much has been written about this fountain; see especially Matthew Knox Averett, "'Redditus orbis erat': The Political Rhetoric of Bernini's Fountains in Piazza Barberini," *Sixteenth Century Journal* 45, no. 1 (2014): 3–24; William Collier, "New Light on Bernini's Neptune and Triton," *Journal of the Warburg and Courtauld Institutes* 31 (1968): 438–40; Connors, "Alliance and Enmity"; and J. H. Larson, "The Conservation of a Marble Group of Neptune and Triton by Gian Lorenzo Bernini," supplement, *Studies in Conservation* 31, no. S1 (1986): 22–26. An important book on Bernini in baroque Rome that covers the fountain is Warwick, *Bernini*.

89. Ovid, *Metamorphoses*, trans. Miller, 1:324–48.

90. On Bernini and Ovid see Paul Barolsky, "Bernini and Ovid," *Source: Notes in the History of Art* 16, no. 1 (1996): 29–31; and Ann Thomas Wilkins, "Bernini and Ovid: Expanding the Concept of Metamorphosis," *International Journal of the Classical Tradition* 6, no. 3 (1999): 383–408.

91. These are described in detail in *Vera Relatione del viaggio fatto dalla Maestà della regina di Svetia per tutto lo Stato Ecclesiastico, del suo ricevimento et ingresso nell'alma città di Roma il dì 20 di dicembre 1655* (Rome: Francesco Cavalli, 1655); and *Lettera di raguaglio scritta da N.N. ad un suo amico toccante la cavalcata solenne fatta per il ricevimento della Maestà della Regina di Svetia in Roma* (Rome: Coligni, 1655),

92. Gualdo Priorato, *Historia della Sacra Real Maestà di Christina Alessandra Regina di Svetia* (Rome: Stamperia della Rev. Camera Apostolica, 1656); *The History of the Sacred and Royal Majesty of Christina Alessandra Queen of Swedland* [ . . . ], trans. John Burbury (London, 1658).

93. "Indi frà suono incessante di Trombe, Tamburi e Campane con festivissime salve passò la Maestà Sua per doppia spalliera d'Infantaria. . . . Sù la Porta della Città trà lo sfavillar delle fiaccole, e torce, il Signor Fabio Compagnoni Commissario a ripartir la Militia si presentò à dedicar al Real servigio di Sua Maestà in qualità de Paggi una schiera di Giovanetti per sangue, per indole de più conspicui vestiti ad una divisa con calzone nero di velluto piano bottonato d'argento, con giuppone di lama d'argento, calzette bianche abbigliati con ogni attillatura, e concerto. . . . Tolta in mezzo da Paggi con torce accese la Regia Lettiga, passò per strade infiorate di verdure, ornate, & illustrate alle finestre, e che bollivano per le ondate di numerosissimo popolo." *Lettera di raguaglio scritta*, n.p.

94. There is a large literature on this series of spectacles. See, for example, Frederick Hammond, "The Creation of a Roman Festival: Barberini Celebrations for Christina of Sweden," in *Life and the Arts in the Baroque Palaces of Rome: Ambiente Barocco*, ed. Stephanie Walker and Frederick Hammond (New Haven, CT: Yale University Press, 1999), 53–69; and Clementi, *Il carnevale romano nelle cronache contemporanee* (Rome, 1899).

95. Hammond, "Creation of a Roman Festival," 61.

96. Hammond, 61.

97. "In mezzo di loro si ergeva un magnifico portone all'incontro del posto della Regina abbellito di varie figure, che intorno all'Arme di Sua Maestà invaghivano una prospettiva di nobili imprese. Nella sommità di detto portone in quattro gran fenestre, con finte gelosie, si stendeva un Choro per i musici, che con varietà d'instrumenti, fecero melodie isquisite." Priorato, *Historia*, 247–48.

98. "Il coro de' Musici situato, come disse, sopra l'arco eretto incontro à sua Maestà, come di quando in quando faceva armoniose sinfonie, così cedette al suono delle trombe. Da queste dunque furono svegliati tutti alla battaglia." Priorato, *Historia*, 25.

99. Deleuze, *The Fold*, 125.

## CHAPTER SEVEN

1. Lewis C. Seifert, "Masculinity and Satires of 'Sodomites' in France, 1660–1715," *Journal of Homosexuality* 41, no. 3/4 (2002): 37–52; Rebecca E. Zorach, "The Matter of Italy: Sodomy and the Scandal of Style in Sixteenth-Century France," *Journal of Medieval and Early Modern Studies* 28, no. 3 (1998): 581.

2. Mary Louise Pratt, "Arts of the Contact Zone," *Profession* (1991): 33–40.

3. Martha Feldman, "Denaturing the Castrato," *Opera Quarterly* 24, no. 3/4 (2008), 178–99.

4. On Pliny and monstrous races, see Mary Beagon, "Situating Nature's Wonders in Pliny's *Natural History*," in "*Vita vigilia est*: Essays in Honour of Barbara Levick," supplement, *Bulletin of the Institute of Classical Studies*, no. 100 (2007): 19–40.

5. Serena Guarracino, "Voices from the South: Music, Castration, and the Displacement of the Eye," in *Anglo-Southern Relations: From Deculturation to Transculturation*, ed. Luigi Cazzato (Lecce: Negroamaro: 2012), 40. See also Giovanni Sole, *Castrati e cicisbei: Ideologia e moda nel Settecento italiano* (Soveria Mannelli, Italy: Rubbettino, 2008).

6. Walter Mignolo, *The Darker Side of the Renaissance: Literacy, Territoriality, and Colonization* (Ann Arbor: University of Michigan Press, 2003).

7. The Italian peninsula did not unify as a nation until 1861. In the seventeenth and eighteenth centuries it was still a region of separate city-states with enormous cultural, political, and linguistic differences.

8. Quoted in Nelson Moe, *The View from Vesuvius: Italian Culture and the Southern Question* (Berkeley: University of California Press, 2002), 37.

9. Walter Benjamin, *Selected Writings*, vol. 1, *1913–1926*, ed. Marcus Bullock and Michael W. Jennings (Cambridge, MA: Harvard University Press, 1996), 419.

10. Jean-Jacques Rousseau, *Complete Dictionary of Music*, trans. William Waring (London, 1779), 30.

11. Charles de Brosses, *Lettres familières écrites d'Italie en 1739 et 1740* (Paris: È. Perrin, 1885).

12. Felicity Nussbaum, *The Limits of the Human: Fictions of Anomaly, Race and Gender in the Long Eighteenth Century* (New York: Cambridge University Press, 2003).

13. The literature on Frantz Fanon is vast. Two readings that have been helpful to me are Rey Chow, "Female Sexual Agency, Miscegenation, and the Formation of Community in Frantz Fanon," in *Ethics after Idealism: Theory, Culture, Ethnicity, Reading* (New York: Routledge, 1999); and Anthony C. Alessandrini, ed., *Frantz Fanon: Critical Perspectives* (New York: Routledge, 1999).

14. Michel Foucault, *The History of Sexuality: An Introduction*, trans. Robert Hurley (New York: Vintage, 1990).

15. Roger Freitas, "Singing and Playing: The Italian Cantata and the Rage for Wit," *Music & Letters* 82, no. 4 (2001): 509–42; Roger Freitas, "The Eroticism of Emasculation: Confronting the Baroque Body of the Castrato," *Journal of Musicology* 20, no. 2 (2003): 196–249; Valeria De Lucca, *The Politics of Princely Entertainment: Music and Spectacle in the Lives of Lorenzo Onofrio and Maria Mancini Colonna* (New York: Oxford University Press, 2020); Margaret Murata, "Why the First Opera Given in Paris Wasn't Roman," *Cambridge Opera Journal* 7, no. 2 (1995): 87–105; and Amy Brosius, "Courtesan Singers as Courtiers: Power, Political Pawns, and the Arrest of Virtuosa Nina Barcarola," *Journal of the American Musicological Society* 73, no. 2 (2020): 207–67.

16. John Romey, "Singing the Fronde: Placards, Street Songs, and Performed Politics," *Early Modern French Studies* 41, no. 1 (2019): 52–73.

17. Charles Burney, *An Eighteenth-Century Musical Tour in France and Italy*, ed. Percy A. Scholes (London: Oxford University Press, 1959), 247.

18. Katherine Bergeron, "The Castrato As History," *Cambridge Opera Journal* 8, no. 2 (1996): 167–84; Anne Desler, "'The Little That I Have Done Is Already Gone and Forgotten': Farinelli and Burney Write Music History," *Cambridge Opera Journal* 27, no. 3 (2015): 215; Elisabeth Krimmer, "'Eviva il Coltello'? The Castrato Singer in Eighteenth-Century German Literature and Culture," *PMLA* 120, no. 5 (2005): 1543–59; John Rosselli, "The Castrati as a Professional Group and a Social Phenomenon, 1550–1850," *Acta musicologica* 60 (1988): 143–79; and Alan Sikes, "'Snip Snip Here, Snip Snip There, and a Couple of Tra La Las': The Castrato and the Nature of Sexual Difference," *Studies in Eighteenth-Century Culture* 34, no. 1 (2005): 197–229.

19. Rosemary Sweet, *Cities and the Grand Tour: The British in Italy, c. 1690–1820* (New York: Cambridge University Press, 2012); and Vanessa Agnew, "Hearing Things: Music and Sounds the Traveller Heard and Didn't Hear on the Grand Tour," *Cultural Studies Review* 18, no. 3 (2012): 67–84.

20. Michael Kelly, *Reminiscences*, ed. Roger Fiske (New York: Oxford University Press, 1975), 18–19.

21. Salvatore Di Giacomo, *I quattro antichi conservatorii di musica di Napoli*. Vol. 1, *Il conservatorio di Sant'Onofrio a Capuana e quello di S. Maria della Pietà dei Turchini* (Milan: Sandron, 1924); vol. 2, *Il conservatorio dei Poveri di Gesù Cristo e quello di S. Maria di Loreto* (Milan: Sandron, 1928).

22. John Phillips, "Sade," *French Studies* 68, no. 4 (2014): 526–33.

23. Richard Sherr, "*Ex Concordia Discors*: Popes, Cardinals, Nationalities, Conflict, Deviance, and Irrational Behavior; A Crisis in the Papal Chapel in the Pontificate of Paul IV," *Journal of Musicology* 32, no. 4 (2015): 494–523.

24. Much of the text of the nuncio's letter appears in Aidan McGrath, *A Controversy Concerning Male Impotence* (Rome: Edificia Pontificia Università Gregoriana, 1988), 2. McGrath cites the original located in Archivum Secretum Vaticanum, Fondo Segreteria di Stato, Spagna, vol. 32, fol. 152r. Technically a eunuch was a man without testicles or penis, and a *spadone* was a man with testicles but no penis. But this bull was directed at the castrati made for the purpose of singing.

25. Amerigo Vespucci, *The Letters of Amerigo Vespucci and Other Documents Illustrative of His Career*, trans. Clement R. Markham (London: Printed for the Haylukt Society, 1894), 46.

26. Nicolás Wey Gómez, *The Tropics of Empire: Why Columbus Sailed South to the Indies* (Cambridge, MA: MIT Press, 2008), xiv.

27. Cristóbal Colón, *Textos y documentos completos*, ed. Consuelo Varela (Madrid: Alianza, 1992), 207.

28. Julius E. Olson and Edward Gaylord Bourne, eds., *The Northmen, Columbus, and Cabot, 985–1503: The Voyages of the Northmen* (New York: Charles Scribner's Sons, 1906), 218.

29. Vespucci, *Letters*, 46.

30. Colón, *Textos y documentos completos*, 207. This letter was one of four copies of a letter Columbus wrote describing the new world. It exists in versions addressed to Rafael Sánchez and Luis de Santágal.

31. Thomas Jefferson to John Adams, June 11, 1812, "The Letters of Thomas Jefferson,

1743–1826," American History, http://www.let.rug.nl/usa/presidents/thomas-jefferson/letters-of-thomas-jefferson/jefl214.php.

32. Peter Lowe, *The Whole Course of Chirurgerie* [ . . . ] (London, 1597), chap. 5.

33. Lowe is a rather enigmatic figure and may have been a spy or pirate. Douglas Guthrie, "The Achievement of Peter Lowe, and the Unity of Physician and Surgeon," *Scottish Medical Journal* 10, no. 7 (1965): 261–68; Harold Ellis, "Peter Lowe: A Father Figure of the Royal College of Physicians and Surgeons of Glasgow," *British Journal of Hospital Medicine* 71, no. 8 (2010): 474; I. M. Donaldson, "Peter Lowe: Information and Misinformation about a Scots Surgeon of the Sixteenth Century," *Journal of the Royal College of Physicians of Edinburgh* 44, no. 2 (2014): 170–79.

34. Lucrezia Marinella, *The Nobility and Excellence of Women and the Defects and Vices of Men*, ed. and trans. Anne Dunhill (Chicago: University of Chicago Press, 1999), 77–78.

35. Athanasius Kircher, *Musurgia universalis; Sive, ars magna consoni et dissoni in X. libros digesta* (Rome, 1650), trans. Margaret Murata, in Oliver Strunk and Leo Treitler, eds., *Source Readings in Music History*, rev. ed. (New York: W. W. Norton, 1998), 710.

36. "Et così cominciando da Napoli, si conosce che il Tuono dell'Organo, v'è più grave che questo di Roma." Giovanni Battista Doni, *Annotazioni sopra il compendio de' generi e de' modi della musica* (Roma: Andrea Fei, 1640), 182.

37. "Ho sentito poi discorrere diversamente da i periti di queste cose, circa il tuono di Roma; & attribuirsi da altri la sua gravità alla mollitie, & infingardia de'cantori; da altri alla copia de'castrati, che quando sono provetti in età, non arrivano all'acutezza di voce, che formano I fanciulli interi; e da altri finalmente alla copia maggiore de'bassi profondi, che piú qui, che altrove, si trovano." Doni, *Annotazioni*, 182. For much more on organ pitch, see Bruce Haynes, *A History of Performing Pitch: The Story of "A"* (Lanham, MD: Scarecrow Press, 2002). He cites Patrizio Barbieri's translation of the same passage. See Patrizio Barbieri, "'Chiavette' and Modal Transposition in Italian Practice (c. 1500–1837)," *Recercare* 3 (1991): 5–79.

38. Aníbal Quijano, "Coloniality of Power, Eurocentrism, and Latin America," *Nepantla: Views from the South* 1, no. 3 (2000): 537.

39. Jennifer D. Selwyn, *A Paradise Inhabited by Devils: The Jesuits' Civilizing Mission in Early Modern Naples* (London: Routledge, 2017).

40. On the Gallery of Maps, see Francesca Fiorani, "Post-Tridentine 'Geographia Sacra': The Galleria delle carte geografiche in the Vatican Palace," *Imago mundi* 48, no. 1 (1996): 124–48.

41. The figure of Columbus made his operatic debut at the end of the seventeenth century. See Thomas F. Heck, "Toward a Bibliography of Operas on Columbus: A Quincentennial Checklist," *Notes* 49, no. 2 (1992): 474–97; "The Operatic Christopher Columbus: Three Hundred Years of Musical Mythology," *Annali d'italianistica* 10 (1992): 236–78.

42. George Sandys, *A Relation of a Journey Begun An. Dom. 1610: 4 Bookes, Containing a Description of the Turkish Empire, of Aegypt, of the Holy Land, of the Remote Parts of Italy, and Ilands Adioyning* (London, 1615; repr., New York: Da Capo Press, 1973). See also his *Ovid's Metamorphosis Englished, Mythologiz'd, and Represented in Figures: An Essay to the Translation of Virgil's Aeneis* (Oxford, 1632). On Sandys, see James Ellison, *George Sandys: Travel, Colonialism, and Tolerance in the Seventeenth Century* (Rochester, NY: D. S. Brewer, 2002). Seventeenth- and eighteenth-century libretti referred to this as Asia Minor.

43. George Sandys, *Relation of a Journey*, n.p. (dedicatory material).

44. "Del resto, io lo metto fra i miracoli, che un forestiero abbia qua parlare a a quella foggia, ed a scrivere con più proprietà di noi altre Toscani: e questa è una di quelle cose che

delle volte mi farebbe taroccare, noi altri italiani siamo al di sotto in quasi tutti i generi di letteratura, vedendo per esperienza che le belle arti hanno passato i monti, e son venute a stanziare in quei paesi che altre volte si chiamavano barbari, ed ora sono i più gentili: sicché le scienze, gli studj, l'erudizioni sono allignate, e fanno prova miracolosa in questi terreni oltramontani, ed i nostri, di dove elleno sono state transpiantate, sono sfruttati quasi del tutto." Lorenzo Panciatichi to Lorenzo Magalotti, Paris, January 2, 1671, in Lorenzo Panciatichi, *Scritti vari di Lorenzo Panciatichi*, ed. Cesare Guasti (Florence: Felice Le Monnier, 1856), 266–67.

45. Frederick Hammond, "Orpheus in a New Key: The Barberini and the Rossi-Buti *L'Orfeo*," *Studi musicali* 25, no. 1/2 (1996): 103–25; Murata, "Why the First Opera Given in Paris Wasn't Roman,"; and Freitas, "The Eroticism of Emasculation."

46. Jeffrey Merrick, "The Cardinal and the Queen: Sexual and Political Disorders in the Mazarinades," *French Historical Studies* 18, no. 3 (1994): 667–99.

47. Quoted in Merrick, 668.

48. See Julia Prest, "In Chapel, on Stage, and in the Bedroom: French Responses to the Italian Castrato," *Seventeenth-Century French Studies* 32, no. 2 (2010): 152–64; Sarah Nancy, "The Singing Body in the Tragédie Lyrique of Seventeenth- and Eighteenth-Century France: Voice, Theatre, Speech, Pleasure," in *The Legacy of Opera: Reading Music Theatre as Experience and Performance*, ed. Michael Eigtved and Dominic Symonds (Amsterdam: Rodopi, 2013), 65–78; and Hedy Law, "A Cannon-Shaped Man with an Amphibian Voice: Castrato and Disability in Eighteenth-Century France," in *The Oxford Handbook of Music and Disability Studies*, ed. Blake Howe, Stephanie Jensen-Moulton, Neil Lerner, and Joseph Straus (New York: Oxford University Press, 2016), 329.

49. On Nicolini, see Anne Desler, "From Castrato to Bass: The Late Roles of Nicolò Grimaldi 'Nicolini,'" in *Gender, Age and Musical Creativity*, ed. Catherine Haworth and Lisa Colton (London: Routledge, 2015), 61–80.

50. Thomas McGeary, "Opera and British Nationalism, 1700–1711," *Revue LISA/LISA E-Journal* 4, no. 2 (2006): 5–19.

51. John Dennis, *An Essay on the Opera's after the Italian Manner: Which Are about to Be Establish'd on the English Stage; With Some Reflections on the Damage Which They May Bring to the Publick* (London, 1706), 14.

52. Joseph Addison, *The Spectator* 29 (April 3, 1711), [58].

53. Jill Campbell, *Natural Masques: Gender and Identity in Fielding's Plays and Novels* (Stanford, CA: Stanford University Press, 1995).

54. *The Remarkable Trial of the Queen of Quavers and Her Associates, for Sorcery, Witchcraft, and Enchantments at the Assizes Held in the Moon, for The County of Gelding before the Rt. Hon. Sir Francis Lash, Lord Chief Baron of the Lunar Exchequer* ([London]: Printed for J. Bew, [1777?]), 6–7.

55. Suzanne Aspden, "'An Infinity of Factions': Opera in Eighteenth-Century Britain and the Undoing of Society," *Cambridge Opera Journal* 9, no. 1 (1997): 1–19. On nationalism and Italian opera, see Thomas McGeary, "'Warbling Eunuchs': Opera, Gender, and Sexuality on the London Stage, 1705–1742," *Restoration and Eighteenth-Century Theatre Research* 7, no. 1 (Summer 1992): 1–22; and "Gendering Opera: Italian Opera as the Feminine Other in Britain, 1700–42," *Journal of Musicological Research* 14, no. 1 (1994): 17–34.

56. John Arbuthnot, *The Devil to Pay at St. James's: Or, A Full and True Account of a Most Horrid and Bloody Battle between Madam Faustina and Madam Cuzzoni* [. . .] (London, 1727), 4–5.

57. Lynn Hunt, "Introduction: Obscenity and the Origins of Modernity, 1500–1800," in *The Invention of Pornography, 1500–1800: Obscenity and the Origins of Modernity*, ed. Lynn Hunt (New York: Zone Books, 1996), 9–45.

58. See Anton Luger, "The Origin of Syphilis: Clinical and Epidemiologic Considerations on the Columbian Theory," *Sexually Transmitted Diseases* 20, no. 2 (1993): 110–17; and Eugenia Tognotti, "The Rise and Fall of Syphilis in Renaissance Europe," *Journal of Medical Humanities* 30, no. 2 (2009): 99.

59. See Liza Blake, "Dildos and Accessories: The Functions of Early Modern Strap-Ons," in *Ornamentalism: The Art of Renaissance Accessories*, ed. Bella Mirabella (Ann Arbor: University of Michigan Press, 2011), 130–55; and Michael L. Stapleton, "Nashe and the Poetics of Obscenity: The Choise of Valentines," *Classical and Modern Literature* 12 (1991): 29–48.

60. For a detailed account of the Jesuit project in Naples, including a discussion of the relationship between Naples and the New World in the church imaginary, see Selwyn, *Paradise Inhabited by Devils*.

61. Athanasius Kircher, *Mundus subterraneus in XII Libros digestus*, bk. 1 (Amsterdam: Johan Jansson, 1678).

62. For a discussion of interpretations of Naples in general with reference to Burney, see Melissa Calaresu, "From the Street to Stereotype: Urban Space, Travel and the Picturesque in Late Eighteenth-Century Naples," *Italian Studies* 62, no. 2 (2007): 189–203. On early modern Naples, see John Marino, *Becoming Neapolitan: Citizen Culture in Baroque Naples* (Baltimore: Johns Hopkins University Press, 2011); and Jerry H. Bentley, *Politics and Culture in Renaissance Naples* (Princeton, NJ: Princeton University Press, 2014).

63. Charles Burney, *The Present State of Music in France and Italy* (London, 1773), 370.

64. Benedetto Croce, *Un paradiso abitato da diavoli*, ed. Giuseppe Galasso (Milan: Adelphi, 2006), 11.

65. Burney, *Present State of Music in France*, 321–22.

66. Burney, 323.

67. Charles Burney, *The Present State of Music in Germany, The Netherlands, and United Provinces*, 2nd ed. (London, 1775), 2:3.

68. Jean-Jacques Rousseau, *Complete Dictionary of Music*, trans. William Waring (London, 1779), 56.

69. See Veit Erlmann, *Reason and Resonance: A History of Modern Aurality* (New York: Zone Books, 2010); and Ronald Radano and Tejumola Olaniyan, "Introduction. Hearing Empire—Imperial Listening," in *Audible Empire: Music, Global Politics, Critique*, ed. Ronald Radano and Tejumola Olaniyan (Durham, NC: Duke University Press, 2016), 1–22.

70. Gary Tomlinson, "Musicology, Anthropology, History," *Il saggiatore musicale* 8, no. 1 (2001): 21–37.

71. Gary Tomlinson, "Vico's Songs: Detours at the Origins of (Ethno)Musicology," *Musical Quarterly* 83, no. 3 (1999): 344–77.

72. Charles de Montesquieu, *The Spirit of the Laws*, trans. Thomas Nugent (London: Printed for J. Nourse and P. Vaillant, 1750), 1:427.

73. There is a large literature on climate in Enlightenment thought. See, for example, James Rodger Fleming, *Historical Perspectives on Climate Change* (New York: Oxford University Press, 1998), 11–19.

74. Montesquieu, *Spirit of the Laws*, 1:316.

75. Montesquieu, 1:319.

76. Montesquieu, 2:22–23.

77. Johann Wilhelm von Archenholz, *A Picture of Italy*, trans. Joseph Trapp (London, 1791), 2:200–201.

78. Johann Wilhelm von Archenholz, *Mémoires concernant Christine, Reine de Suède, pour servir à l'éclaircissement de son règne et principalement de sa vie privée et aux évènements de l'histoire de son temps, civile et littéraire: suivi de deux ouvrages de cette savante Princesse, qui n'ont jamais été imprimés* (Amsterdam, 1751), 1:228.

79. Archenholz, *A Picture of Italy*, 2:200.

80. Archenholz, 2:196.

81. The classic study is Johannes Fabian, *Time and the Other: How Anthropology Makes Its Object* (New York: Columbia University Press, 1983).

82. Antonio Gramsci, *The Modern Prince: And Other Writings*, trans. Louis Marks (New York: International, 1957), 108–9.

83. Rosselli, "Castrati as a Professional Group," 174.

84. [Walt Whitman], "The Opera," *Life Illustrated*, November 10, 1855. The 1825 debut of Italian opera in America was a production of Rossini's *Il barbiere di Siviglia* at New York's Park Theater, performed in Italian by a cast consisting mostly of a Spanish family of singers, the Garcías.

85. Kara Keeling, "Electric Feel: Transduction, Errantry, and the Refrain," *Cultural Studies* 28, no. 1 (2014): 51.

86. Keeling, 58.

## CHAPTER EIGHT

1. Johann Wilhelm von Archenholz, *A Picture of Italy*, trans. Joseph Trapp (London, 1791).

2. Angus Heriot repeats the story in *The Castrati in Opera* (London: Secker and Warburg, 1956; repr., New York: Da Capo, 1974).

3. John Christopoulos, "Abortion and the Confessional in Counter-Reformation Italy," *Renaissance Quarterly* 65, no. 2 (2012): 443–84.

4. Heriot, *The Castrati in Opera*, 40.

5. James Q. Davies, "'Veluti in speculum': The Twilight of the Castrato," *Cambridge Opera Journal* 17, no. 3 (2005): 271–301.

6. Lorraine Daston and Katharine Park, *Wonders and the Order of Nature, 1150–750* (New York: Zone Books, 1998); and Lorraine Daston, "Marvelous Facts and Miraculous Evidence in Early Modern Europe," *Critical Inquiry* 18, no. 1 (1991): 93–124.

7. François Raguenet, *A comparison between the French and Italian musick and operas. Translated from the French; with some remarks. To which is added a critical discourse upon operas in England, and a means proposed for their improvement* (London: William Lewis, 1709), 38.

8. Raguenet, 54–55

9. There are many scholarly accounts of Handel's castrato roles. For some classics, see Joke Dame, "Unveiled Voices: Sexual Difference and the Castrato," in *Queering the Pitch: The New Gay and Lesbian Musicology*, ed. Philip Brett, Elizabeth Wood, and Gary C. Thomas, 2nd ed. (New York: Routledge, 2006), 139–54; Anne Desler, "From Castrato to Bass: The Late Roles of Nicolò Grimaldi 'Nicolini,'" in *Gender, Age and Musical Creativity*, ed. Catherine Haworth and Lisa Colton (London: Routledge, 2015), 61–80; and Joseph R. Roach, "Cavaliere Nicolini: London's First Opera Star," *Theatre Journal* 28, no. 2 (1976): 189.

10. This story was repeated in various periodicals. The earliest citation seems to be from Charles Burney, *The Present State of Music in France and Italy* (London, 1771), 205–6.

11. Anne Desler, "'Il novello Orfeo' Farinelli: Vocal Profile, Aesthetics, Rhetoric" (PhD diss., University of Glasgow, 2014).

12. "Il faut être accoutumé à ces voix de castrats pour les goûter. Le timbre en est aussi clair et perçant que celui des enfants de chœur, et beaucoup plus fort; il me paroît qu'ils chantent à l'octave au-dessus de la voix naturelle des femmes. Leurs voix ont presque toujours quelque chose de sec et d'aigre, bien éloigné de la douceur jeune et moelleuse des voix de femmes; mais elles sont brillantes, légères, pleines d'éclat, très-fortes et très-étendues. Les voix de femmes italiennes sont aussi d'un pareil genre, légères et flexibles au dernier point; en un mot, du même caractère que leur musique." Charles de Brosses, *Lettres familières écrites d'Italie en 1739 et 1740*, ed. R. Colomb, 5th ed. (Paris: Garnier Frères, 1885), 2:318–19.

13. For a good summary of work on gender in the early modern period, see Katherine Crawford, "Privilege, Possibility, and Perversion: Rethinking the Study of Early Modern Sexuality," *Journal of Modern History* 78, no. 2 (2006): 412–33.

14. François Raguenet, *Les Monumens de Rome, ou Description des plus beaux ouvrages de peinture, de sculpture et d'architecture qui se voyent à Rome aux Environs* (Amsterdam, 1701), 32–34.

15. Giles Jacob, *Tractatus de Hermaphroditis: Or, A Treatise of Hermaphrodites* (London, 1718), sig. B.

16. See Palmira Fontes da Costa, "Anatomical Expertise and the Hermaphroditic Body," *Spontaneous Generations: A Journal for the History and Philosophy of Science* 1, no. 1 (2007): 78–85; Felicity Nussbaum, *The Limits of the Human: Fictions of Anomaly, Race and Gender in the Long Eighteenth Century* (New York: Cambridge University Press, 2003); and Palmira Fontes da Costa, "The Medical Understanding of Monstrous Births at the Royal Society of London during the First Half of the Eighteenth Century," *History and Philosophy of the Life Sciences* 26, no. 2 (2004): 157–75.

17. On Roman satire, see Irving Lavin, "High and Low before Their Time: Bernini and the Art of Social Satire," in *Modern Art and Popular Culture: Readings in High and Low*, ed. Kirk Varnedoe and Adam Gopnik (New York: Museum of Modern Art, 1990), 19–50.

18. Courtney Quaintance, *Textual Masculinity and the Exchange of Women in Renaissance Venice* (Toronto: University of Toronto Press, 2015), 6.

19. For a specifically Roman version of this trope, see Amy Brosius, "Singers Behaving Badly: Rivalry, Vengeance, and the Singers of Cardinal Antonio Barberini," *Women and Music: A Journal of Gender and Culture* 19, no. 1 (2015): 45–53.

20. Elizabeth S. Cohen, "Moving Words: Everyday Oralities and Social Dynamics in Roman Trials *circa* 1600," in *Voices and Texts in Early Modern Italian Society*, ed. Stefano Dall'Aglio, Brian Richardson, and Massimo Rospocher (London: Routledge, 2017), 69–83.

21. "Che diremo di D. Antonio / Ch'è di corpo assai imperfetto / ha grand naso e poco petto piu superbo del demonio / vol grand bene à Marc Antonio / non so per se questa musica / o per altra inclinatione / Tocca Musa il Colascione." Translated in Amy Brosius, "Courtesan Singers as Courtiers: Power, Political Pawns, and the Arrest of Virtuosa Nina Barcarola," *Journal of the American Musicological Society* 73, no. 2 (2020): 228.

22. Thomas Walter Laqueur, *Making Sex: Body and Gender from the Greeks to Freud* (Cambridge, MA: Harvard University Press, 1992). Laqueur put forth a similar theory in an earlier article; see "Orgasm, Generation, and the Politics of Reproductive Biology," *Representations* 14 (Spring 1986): 1–41.

23. See Katharine Park, "Cadden, Laqueur, and the 'One-Sex Body,'" *Medieval Feminist Forum* 46, no. 1 (2010): 98; and Helen King, *The One-Sex Body on Trial: The Classical and Early Modem Evidence* (Burlington, VT: Ashgate, 2013).

24. Thomas Laqueur, *Making Sex: Body and Gender from the Greeks to Freud* (Cambridge, MA: Harvard University Press, 1992), 82.

25. Valeria Finucci, *The Manly Masquerade: Masculinity, Paternity, and Castration in the Italian Renaissance* (Durham, NC: Duke University Press, 2003), 242.

26. For challenges to Laqueur, see Patricia Parker, "Gender Ideology, Gender Change: The Case of Marie Germain," *Critical Inquiry* 19, no. 2 (1993): 337–64; and Lorraine Daston and Katharine Park, "The Hermaphrodite and the Orders of Nature: Sexual Ambiguity in Early Modern France," in *Premodern Sexualities*, ed. Louise Fradenburg and Carla Freccero (New York: Routledge, 1996): 117–36. See also John Jeffries Martin, "'Et nulle autre me faict plus proprement homme que cette cy': Michel de Montaigne's Embodied Masculinity," *European Review of History: Revue européenne d'histoire* 22, no. 4 (2015): 563–78; and Gary Ferguson, "Early Modern Transitions: From Montaigne to Choisy," *L'esprit créateur* 53, no. 1 (2013): 145–57.

27. Translated in Martin, "Michel de Montaigne's Embodied Masculinity," 568, emphasis mine.

28. Michel de Montaigne, *Complete Essays*, trans. Donald M. Frame (Stanford, CA: Stanford University Press, 1958), 69. For the French, see Montaigne, *Oeuvres complètes*, ed. Albert Thibaudet and Maurice Rat (Paris: Gallimard, 1962), 1118–19.

29. Todd W. Reeser argues that Montaigne was, on the one hand, invested in a proto–Judith Butler idea of the body as materializing through experience and discourse and, on the other, extremely invested in his own penis. See Todd W. Reeser, "Montaigne on Gender," in *The Oxford Handbook of Montaigne*, ed. Philippe Desan (New York: Oxford University Press, 2016), 562–80.

30. Ambroise Paré, *On Monsters and Marvels*, trans. and ed. Janis L. Pallister (Chicago: University of Chicago Press, 1982), 32–33, emphasis added.

31. See, for example, Susan McClary, "Constructions of Gender in Monteverdi's Dramatic Music," in *Feminine Endings: Music, Gender, and Sexuality* (Minneapolis: University of Minnesota Press, 1991), 37; Linda Phyllis Austern, "Nature, Culture, Myth, and Musician in Early Modern England," *Journal of the American Musicological Society* 51, no. 1 (1998): 1–49; and Suzanne G. Cusick, "On Musical Performances of Gender and Sex," in *Audible Traces: Gender, Identity, and Music*, ed. Elaine Barkin and Lydia Hamessley (Los Angeles: Carciofoli, 1999), 25–49.

32. Stephen Greenblatt, *Shakespearean Negotiations: The Circulation of Social Energy in Renaissance England* (Berkeley: University of California Press, 1988), 89.

33. Roger Freitas, "The Eroticism of Emasculation: Confronting the Baroque Body of the Castrato," *Journal of Musicology* 20, no. 2 (2003): 196–249.

34. Naomi André, *Voicing Gender: Castrati, Travesti, and the Second Woman in Early-Nineteenth-Century Italian Opera* (Bloomington: Indiana University Press, 2006), 46.

35. Freya Jarman, "Pitch Fever: The Castrato, the Tenor, and the Question of Masculinity in Nineteenth-Century Opera," in *Masculinity in Opera: Gender, History, and New Musicology*, ed. Philip Purvis (New York: Routledge, 2013), 63.

36. Patricia Simons, for example, argues that the model Laqueur puts forth was a teaching device and that more detailed accounts stressed the male body as more perfect than the

female rather than homologous. See Simons, *The Sex of Men in Premodern Europe: A Cultural History* (New York: Cambridge University Press, 2011).

37. Nancy G. Siraisi, *Avicenna in Renaissance Italy: The Canon and Medical Teaching in Italian Universities after 1500* (Princeton, NJ: Princeton University Press, 2014), 30.

38. Lucrezia Marinella, *La nobiltà et l'eccellenza delle donne, co' diffetti et mancamenti de gli huomini* (Venice: Giovanni Battista Ciotti, 1601). The book appeared a year earlier as *Le nobiltà, et eccellenze delle donne: et i diffetti, e mancamenti de gli huomini. Discorso di Lucretia Marinella. In due parti diviso* (Venice: Giovanni Battista Ciotti, 1600). For a partial English translation with informative introduction, see *The Nobility and Excellence of Women and the Defects and Vices of Men*, ed. and trans. Anne Dunhill (Chicago: University of Chicago Press, 1999). I am deeply grateful to Marguerite Deslauriers for these references and unpublished translations.

39. "Alcuni altri dicono, come fù il buono Aristotile, che le Donne sono men calde de gli huomini, & però sono più imperfette, & meno nobili, di loro: ò che ragione indissolubile, & onnipotente. Nonconsiderò, credo io allhora Aristotile con maturità d'ingegno l'operationi del calore, & quello, ch'importi l'esser più caldo, & men caldo, & quanti effetti buoni, & rei da questo derivano; percioche s'egli havesse ben pensato quante pessime operationi produce il calore, che eccede quello della donna, non havrebbe detto una minima parola. Ma se ne andò alla cieca il cattivello, & però comise mille errori. Non è dubbio alcuno, come scrive Plutarco, che il calore è instrumento dell'anima; ma può esser buono, & ancho poce atto alle sue operationi, ricercandosi in esso una certa mediocrità fra il poco, & il molto: percioche il poco, & manchevole come ne'vecchi è impotentissimo alle operationi. Il molto, & eccedente rende quelle precipitose, & sfrenate. adunque ogni calore non è buono, & atto à servire alle operationi dell'anima, come dice Marsilio Ficino. Ma bene in un certo grado, et proportione conveniente, come quello della donna. Onde non vale la ragione d'Aristo sono i maschi più caldi delle Donne, adunque sono più nobili." Marinella, *La nobiltà et l'eccellenza delle donne*, 119, unpublished English translation by Marguerite Deslauriers.

40. Marguerite Deslauriers, "Marinella and Her Interlocutors: Hot Blood, Hot Words, Hot Deeds," *Philosophical Studies* 174, no. 10 (2017): 2525–37.

41. Sherry Sayed Gadelrab, "Discourses on Sex Differences in Medieval Scholarly Islamic Thought," *Journal of the History of Medicine and Allied Sciences* 66, no. 1 (2010): 40–81.

42. Monica Azzolini, "Exploring Generation. A Context to Leonardo's Anatomies of the Female and Male Bodies," in *Leonardo da Vinci's Anatomical World: Language, Context and "Disegno,"* ed. Alessandro Nova and Domenico Laurenza (Venice: Marsilio, 2011), 79–97.

43. For a catalog of early modern editions of key texts, see Ferdinand Edward Cranz and Paul Oskar Kristeller, eds., *Catalogus translationum et commentariorum: Mediaeval and Renaissance Latin Translations and Commentaries*, vol. 4 (Washington, DC: Catholic University of America Press, 1980). For Hippocrates, see Pearl Kibre, *Hippocrates Latinus: Repertorium of Hippocratic Writings in the Latin Middle Ages*, rev. ed. (New York: Fordham University Press, 1985); and Gilles Maloney and Raymond Savoie, *Cinq cent ans de bibliographie hippocratique: 1473–1982* (Québec: Éditions du Sphinx, 1982). For Galen, see Richard J. Durling, "A Chronological Census of Renaissance Editions and Translations of Galen," *Journal of the Warburg and Courtauld Institutes* 24 (1961): 230–305.

44. See Craig Martin, "Printed Medical Commentaries and Authenticity: The Case of 'De Alimento,'" *Journal of the Washington Academy of Sciences* 90, no. 4 (2004): 17–28; and Guido Giglioni, "'If You Don't Feel Pain, You Must Have Lost Your Mind': The Early Mod-

ern Fortunes of a Hippocratic Aphorism," in *Et Amicorum: Essays on Renaissance Humanism and Philosophy*, ed. Anthony Ossa-Richardson and Margaret Meserve (Boston: Brill, 2018), 313–37.

45. See Thomas Rütten, "Hippocrates and the Construction of 'Progress' in Sixteenth- and Seventeenth-Century Medicine," in *Reinventing Hippocrates: The History of Medicine in Context*, ed. David Cantor (New York: Routledge, 2016), 37–58; and Clément Godbarge, "Hippocrates for Princes: Ippolito de' Medici's *Retratti d'aphorismi*," in *Renaissance Rewritings*, ed. Helmut Pfeiffer, Irene Fantappiè, and Tobias Roth (Berlin: De Gruyter, 2017), 143–56.

46. Charles Estienne, *De dissectione partium corporis humani* (Paris, 1545),

47. Andreas Vesalius, *Observatonum anatomicarum Gabrielis Falloppi examen* (Venice, 1564), 143. Cited in Katherine Park, "The Rediscovery of the Clitoris," in *The Body in Parts: Fantasies of Corporeality in Early Modern Europe*, ed. David Hillman and Carolo Mazzio (London: Routledge, 2013), 187.

48. Shaun Tougher, "Two Views on the Gender Identity of Byzantine Eunuchs," in *Changing Sex and Bending Gender*, ed. Alison Shaw and Shirley Ardener (New York: Berghahn, 2005), 60.

49. Patricia Simons, "Manliness and the Visual Semiotics of Bodily Fluids in Early Modern Culture," *Journal of Medieval and Early Modern Studies* 39, no. 2 (2009): 331–73. Combining the arguments of many scholars, Simons makes the case that the testicles, more than the penis, made a man a man, and that potency meant privilege. See also Simons, *Sex of Men*.

50. Leah DeVun, *The Shape of Sex: Nonbinary Gender from Genesis to the Renaissance* (New York: Columbia University Press, 2021), 164.

51. Translated in Richard Sherr, "Gugliemo Gonzaga and the Castrati," *Renaissance Quarterly* 33, no. 1 (1980): 56.

52. Valeria Finucci, "'There's the Rub': Searching for Sexual Remedies in the New World," *Journal of Medieval and Early Modern Studies* 38, no. 3 (2008): 523–57.

53. Aristotle, *Generation of Animals*, trans. A. L. Peck, Loeb Classical Library 366 (Cambridge, MA: Harvard University Press, 1942), 121.

54. See Angus McLaren, *Impotence: A Cultural History* (Chicago: University of Chicago Press, 2007), for much more detail and context than I can provide here.

55. Juan de la Huarte, *The Examination of Men's Wits*, reprinted and with an introduction and notes by Rocío G. Sumillera (London: Modern Humanities Research Association, 2014), 279–80. "If when nature hath finished to form a man in all perfection, she would convert him into a woman, there needeth nought else to be done save only to turn his instruments of generation inwards. And if she have shaped a woman, and would make a man of her, by taking forth her belly and her cods, it would quickly be performed." Juan de la Huarte, *The Examination of Men's Wits*, trans. Richard Carew (London, 1594), 125. The original Spanish is available in Juan Huarte de San Juan, *Examen de ingenios para las ciencias* (Barcelona: Linkgua, 2011), 269: "Y de tal manera es esto verdad, que si acabando Naturaleza de fabricar un hombre perfecto, le quisiese convertir en mujer, no tenía otro trabajo más que tornarle adentro los instrumentos de la generación; y, si hecha mujer, quisiese volverla en varón, con arrojarle el útero y los testículos fuera, no había más que hacer."

56. Huarte, *Examination of Men's Wits* (2014), 287. "How much it endamageth a man to be deprived of those parts (though so small) there need not many reasons to prove, seeing we see by experience that forthwith the hair and the beard pill away, and the big and shrill voice becometh small, and herewithal a man leeseth his forces and natural heat, and resteth in far

worse and more miserable condition than if he had been a woman." Huarte, *Examination of Men's Wits* (1594). "Cuánto daño haga al hombre privarle de estas partes, aunque pequeñas, no serán menester muchas razones para probarlo. Pues vemos por experiencia que luego se le cae el vello y la barba; y la voz gruesa y abultada se le vuelve delgada; y con esto pierde las fuerzas y el calor natural, y queda de peor condición y más mísera que si fuera mujer." Huarte de San Juan, *Examen de ingenios*, 238.

57. Huarte, *Examination of Men's Wits* (2014), 287. "But the matter most worth the noting is that if a man before his gelding had much wit and hability, so soon as his stones be cut away, he groweth to leese the same, so far-forth as if he had received some notable damage in his very brain. And this is a manifest token that the cods give and reave the temperature from all the other parts of the body, and he will not yield credit hereunto, let him consider (as myself have done oftentimes) that of 1000 such capons who addict themselves to their book, none attaineth to any perfection, and even in music (which is their ordinary profession) we manifestly see how blockish they are, which springeth because music is a work of the imagination, and this power requireth much heat, whereas they are cold and moist." Huarte, *Examination of Men's Wits* (1594). "Pero lo que más conviene notar es que, si antes que capasen al hombre, tenía mucho ingenio y habilidad, después de cortados los testículos lo viene a perder como si en el mismo celebro hubiera recebido alguna notable lesión. Lo cual es evidente argumento que los testículos dan y quitan el temperamento a todas las partes del cuerpo. Y si no, consideremos, como yo muchas veces lo he hecho, que de mil capones que se dan a letras ninguno sale con ellas; y en la música, que es su profesión ordinaria, se echa más claro de ver cuán rudos son; y es la causa que la música es obra de la imaginativa, y esta potencia pide mucho calor, y ellos son fríos y húmidos. Luego cierto está que por el ingenio y habilidad sacaremos el temperamento de los testículos. Y, por tanto, el hombre que se mostrare agudo en las obras de la imaginativa terná calor y sequedad en el tercer grado; y si el hombre no supiere mucho, es señal que con el calor se ha juntado humidad, la cual echa siempre a perder la parte racional; y confirmarse ha más si tiene mucha memoria." Huarte de San Juan, *Examen de ingenios*, 238.

58. Huarte, *Examination of Men's Wits* (2014), 76. "To the end that artificers may attain the perfection requisite for the use of the commonwealth, me thinketh, Catholic royal Majesty, a law should be enacted that no carpenter should exercise himself in any work which appertained to the occupation of a husbandman, nor a tailor to that of an architect, and that the advocate should not minister physic, nor the physician play the advocate, but each one exercise only that art to which he beareth a natural inclination, and let pass the residue. For considering how base and narrowly bounded a man's wit is for one thing and no more, I have always held it for a matter certain that no man can be perfectly seen in two arts without failing in one of them." Huarte, *Examination of Men's Wits* (1594). "A la Majestad del rey don Filipe, nuestro señor. Para que las obras de los artífices tuviesen la perfección que convenía al uso de la república, me pareció, Católica Real Majestad, que se había de esta- blecer una ley: que el carpintero no hiciese obra tocante al oficio del labrador, ni el tejedor del arquitecto, ni el jusrisperito curase, ni el médico abogase; sino que cada uno ejercitase sola aquel arte para la cual tenía talento natural, y dejase las demás. Porque, considerando cuán corto y limitado es el ingenio del hombre para una cosa y no más, tuve siempre entendido que ninguno podía saber dos artes con perfección sin que en la una faltase. Y, porque no errase en elegir la que a su natural estaba mejor, había de haber diputados en la república, hombres de gran prudencia y saber, que en la tierna edad descubriesen a cada uno su ingenio, haciéndole estudiar por fuerza la ciencia que le convenía, y no dejarlo a su elección. De lo cual resultaría en vuestros

estados y señoríos haber los mayores artífices del mundo y las obras de mayor perfección, no más de por juntar el arte con naturaleza." Huarte de San Juan, *Examen de ingenios*, 17.

59. Huarte, *Examination of Men's Wits* (2014), 116.

60. Huarte, *Examination of Men's Wits* (2014), 56.

61. Rocío G. Sumillera, introduction to Juan de la Huarte, *The Examination of Men's Wits* (Cambridge: Modern Humanities Research Association, 2014), 45.

62. The most detailed and most cited discussion of the bull is Aidan McGrath, *A Controversy Concerning Male Impotence* (Rome: Editrice Pontificia Università Gregoriana, 1988). McGrath notes that the text of the bull itself can be found in a number of sources, including Petri Gasparri, ed., *Codicis Iuris Canonici Fontes* (Rome: Typis Polyglottis Vaticanis, 1947), 1:298–99. For more sources, see McGrath, 18n2. Much of the text of the nuncio's letter appears McGrath, 2. McGrath cites the original located in the Archivum Secretum Vaticanum, Fondo Segreteria di Stato, Spagna, vol. 32, fol. 152r.

63. For other discussions of the papal bull, particularly in the context of musicology, see Helen Berry, *The Castrato and His Wife* (New York: Oxford University Press, 2011), 164–92; Katherine Crawford, *Eunuchs and Castrati: Disability and Normativity in Early Modern Europe* (New York: Routledge, 2019), 70–99; Martha Feldman, "Strange Births and Surprising Kin: The Castrato's Tale," in *Italy's Eighteenth Century: Gender and Culture in the Age of the Grand Tour*, ed. Paula Findlen, Wendy Wassyng Roworth, and Catherine M. Sama (Stanford, CA: Stanford University Press, 2009), 175–215; Feldman, *The Castrato: Reflections on Natures and Kinds* (Oakland: University of California Press, 2015), 48; Finucci, *The Manly Masquerade*, 263–71; Mary E. Frandsen, "*Eunuchi conjugium*: The Marriage of a Castrato in Early Modern Germany," *Early Music History* 24 (2005): 53–124; and Giuseppe Gerbino, "The Quest for the Soprano Voice: Castrati in Renaissance Italy," *Studi musicali* 33, no. 2 (2004): 334–39.

64. I owe much of my sense of early modern street life to John M. Hunt, *The Vacant See in Early Modern Rome: A Social History of the Papal Interregnum* (Leiden: Brill, 2016). This incident is described on p. 189.

65. The brief also appears in many discussions of Catholic approaches to fertility and marriage. See Joseph Bajada, *Sexual Impotence and the Contribution of Paolo Zacchia, 1584–1659* (Rome: Editrice Pontificia Università Gregoriana, 1988); Edward Behrend-Martínez, "Manhood and the Neutered Body in Early Modern Spain," *Journal of Social History* 38, no. 4 (2005): 1073–93; Edward H. Nowlan, "Double Vasectomy and Marital Impotence," *Theological Studies* 6, no. 3 (1945): 392–427; and Thomas J. O'Donnell, "A Recent Decision of the Holy See Regarding Impotence and Sterility Implications for Medical Practice," *Linacre Quarterly* 45, no. 1 (1978): 16.

66. "Cum frequenter in istis regionibus Eunuchi quidam, et Spadones, qui utroque teste carent, et ideo certum ac manifestum est eos verum semen emittere non posse; quia impura carnis tentigine atque immundis complexibus cum mulieribus se commiscent, et humorem forsan quemdam similem semini, licet ad generationem et ad matrimonii causam minime aptum, effundunt, matrimonia cum mulieribus praesertim hunc ipsum eorum defectum scientibus contrahere praesumant, idque sibi licere pertinaciter contendant, et super hoc diversae lites et controversiae ad tuum et Ecclesiasticum forum deducantur, requisivit a nobis Fraternitas tua quid de huiusmodi connubiis sit statuendum." This passage is transcribed and summarized in McGrath, *Controversy Concerning Male Impotence*, 6.

67. "Nos igitur attendentes, quod secundum Canonicas sanctiones, et naturae rationem, qui frigidae naturae sunt et impotentes, iidem minime apti ad contrahenda matrimonia rep-

utantur, quodque praedicti Eunuchi, aut Spadones, quas tamquam uxores habere non possunt, easdem habere ut sorores nolunt, quia experientia docet, tam ipsos, dum se potentes ad coeundum iactitant, quam mulieres, quae eis nubunt, non ut caste vivant, sed ut carnaliter invicem coniugantur prava et libidinosa intentione, sub praetextu et in figura matrimonii turpes huiusmodi commixtiones affectare, quae cum peccati et scandali occasionem praebeant et in animarum damnationem tendant, sunt ab Ecclesia Dei prorsus exterminandae." This passage is transcribed and summarized in McGrath, *Controversy Concerning Male Impotence*, 9.

68. "Et insuper considerantes, quod ex Spadonum huiusmodi et Eunuchorum coniugiis nulla utilitas provenit, sed potius tentationum illecebrae et incentiva libidinis oriuntur, eidem Fraternitati tuae per praesentes committimus et mandamus, ut coniugia per dictos, et alios quoscumque Eunuchos et Spadones, utroque teste carentes cum quibusvis mulieribus, defectum praedictum sive ignorantibus, sive etiam scientibus, contrahi prohibeas, eosque ad Matrimonia quomodocumque contrahenda inhabiles auctoritate nostra declares, et tam locorum Ordinariis, ne huiusmodi coniunctiones de cetero fieri quoquomodo permittant, interdicas, quam eos etiam, qui sic de facto Matrimonium contraxerint, separari cures, et Matrimonia ipsa sic de facto contracta, nulla, irrita, et invalida esse decernas." This passage is transcribed and summarized in McGrath, *Controversy Concerning Male Impotence*, 10.

69. Tomas Sánchez, *Disputationum de sancto matrimonii sacramento* (Antwerp: M. Nutii, 1620), 2:364.

70. On Zacchia, see Alessandro Pastore and Giovanni Rossi, eds., *Paolo Zacchia: Alle origini della medicina legale: 1584–1659* (Milan: Franco Angeli, 2008). On medicine and law, see Gianna Pomata, *Contracting a Cure: Patients, Healers, and the Law in Early Modern Bologna* (Baltimore: Johns Hopkins University Press, 1998). For the book's history, see Jacalyn Duffin, "Questioning Medicine in Seventeenth-Century Rome: The Consultations of Paolo Zacchia," *Canadian Bulletin of Medical History* 28, no. 1 (2011): 149–70.

71. Nancy G. Siraisi, *Medieval and Early Renaissance Medicine: An Introduction to Knowledge and Practice* (Chicago: University of Chicago Press, 1990).

72. On the *Consilia* (Consultations), see Duffin, "Questioning Medicine." And see a collaborative translation project, https://jacalynduffin.ca/zacchia/latin-titles/.

73. For a discussion of Zacchia in the context of homosexuality, see George Rousseau, "Policing the Anus: Stuprum and Sodomy: According to Paolo Zacchia's Forensic Medicine," in *The Sciences of Homosexuality in Early Modern Europe*, ed. Kenneth Borris and Georges Rousseau (London and New York: Routledge, 2008), 75–91.

74. On this incident, see Francesco Paolo De Ceglia, "The Woman Who Gave Birth to a Dog: Monstrosity and Bestiality in *Quaestiones Medico-Legales* by Paolo Zacchia," *Medicina nei secoli* 26, no. 1 (2014): 117–44.

75. Paolo Zacchia, *Quaestiones medico-legales*, 4th ed., ed. Johannes Daniel Horstius, vol. 1 (Venice, 1789), 5.3.1.13. Quoted in Bajada, *Sexual Impotence*, 51: "Testes habere distinctam operationem, atque illam quidem insignem, ac perfectissimam seminis scilicet conficiendi."

76. For two very recent assessments of these phenomena, see Martha Feldman and Judith T. Zeitlin, eds., *The Voice as Something More: Essays toward Materiality* (Chicago: University of Chicago Press, 2019); and Nina Sun Eidsheim and Katherine Meizel, eds, *The Oxford Handbook of Voice Studies* (New York: Oxford University Press, 2019).

77. Susan McClary, *Desire and Pleasure in Seventeenth-Century Music* (Berkeley: University of California Press, 2012).

78. Suzanne G. Cusick, "Response: 'This Song Is for You,'" *Journal of the American Musi-*

*cological Society* 66, no. 3 (2013), 862. See also Cusick, "Gendering Modern Music: Thoughts on the Monteverdi-Artusi Controversy," *Journal of the American Musicological Society* 46, no. 1 (1993): 1–25.

79. Kathryn M. Ringrose, *The Perfect Servant: Eunuchs and the Social Construction of Gender in Byzantium* (Chicago: University of Chicago Press, 2007).

80. Katherine Crawford, "Desiring Castrates, or How to Create Disabled Social Subjects," *Journal for Early Modern Cultural Studies* 16, no. 2 (2016): 63.

81. Crawford, 62.

82. Emily Wilbourne, "The Queer History of the Castrato," in *The Oxford Handbook of Music and Queerness*, ed. Fred Everett Maus and Sheila Whiteley (New York: Oxford University Press, 2018).

83. C. Riley Snorton, *Black on Both Sides: A Racial History of Trans Identity* (Minneapolis: University of Minnesota Press, 2017).

84. Carla Freccero, "Queer Times," *South Atlantic Quarterly* 106, no. 3 (2007): 485–94.

85. Valerie Traub, *Thinking Sex with the Early Moderns* (Philadelphia: University of Pennsylvania Press, 2016).

## CHAPTER NINE

1. "Dimandargli perdono dell'errore publico già doi anni sono; comesso; N. S.re lo sentì volontieri, è poi li diede l'assolutione, che tutti intessero vade in pace noli amplius peccare." Quoted and translated in Frederick Hammond, *Music and Spectacle in Baroque Rome: Barberini Patronage under Urban VIII* (New Haven, CT: Yale University Press, 1994), 177.

2. "150 cantori, divisi in sei cori, cantarono musiche dei migliori maestri di cappella romani del momento, quali Stefano Fabbri, Virgilio Mazzocchi, Francesco Foggia e Giacomo Carissimi." Bianca Maria Antolini, "La carriera di cantante e compositore di Loreto Vittori," *Studi musicali* 7 (1978): 160.

3. Fabrizio Antolini "Il 'lieve error di giovanil desio' del cavalier Loreto," in *Spoletium: Revista de arte, storia, cultura* 23, no. 1 (1978): 70–75; Giano Nicio Eritreo (Gian Vittorio Rossi) *Iani Nicii Erythraei Dialogi Septendecim* (Cologne: Idocum Kalcovium, 1645); Thomas D. Dunn, "Introduction," in Loreto Vittori, *La Galatea*, ed. Thomas D. Dunn (Middleton, WI: A-R Editions, 2002), vii–xiii; Hammond, *Music and Spectacle in Baroque Rome*, 175; Jean Lionnet, "André Maugars: Risposta data a un curioso sul sentimento della musica dell'Italia," *Nuova rivista musicale italiana* 19, no. 3/4 (1985): 697.

4. "Interdum Romae, per hyemem, in sacello Patrum Congregationis Oratorii exaudiebatur. Ubi eum ego, nocte quadam, Magdalenae sua deflentis crimina seque ad Christi pedes abjicientis querimoniam canentem audivi: qui, eo ardore animi, ea vi vocis, iis tam mollibus tamque delicatis in cantu flexionibus Magdalenam nostris pene oculis subjiciebat ut, si revixisset, in illa eius, poenitentiae ipsius, imitatione suos veros luctus doloresque agnovisset atque admirata esset." Gian Vittorio Rossi, *Iani Nicii Erithraei Pinacotheca Altera Imaginum, Illvstrium, doctrinae vel ingenii laude, Virorum, qui, auctore superstite, diem suum obierunt.* Amsterdam, 1645, 1:217.

5. "Se posse eam ostenderet, sicut mollissimam ceram, quocunque vellet, contorquere ac flectere." Rossi, 217.

6. "Vocem suavem ac splendidam, et ad quamcunque varietatem et commutationem efficiendam aptissimam, quae, ut in fidibus chordae, ad quemcunque tactum respondeat, acuta, gravis, cita, tarda, magna, tenuis." Rossi, 218.

7. For details of the debate, see Antolini, "La carriera de cantante," 152–53; Antolini, "Il 'lieve error di giovanil desio'"; Claudio Gallico, "La 'Querimonia' di Maddalena di Domenico Mazzocchi e l'interpretazione di Lpreto Vittori [1966]," in *Sopra li fondamenti della verità: Musica italiana fra XV e XVII secolo* (Rome: Bulzoni, 2001): 1000–1013.

8. The most complete history is in Arnaldo Morelli, *Il tempio armonico: Musica nell'Oratorio dei Filippini in Roma (1575–1705)* (Laaber: Laaber-Verlag, 1991). A recent dissertation is also very complete; see Rosemarie Darby, "The Liturgical Music of the Chiesa Nuova, Rome (1575–1644)" (PhD diss., University of Manchester, 2018). On the liturgical origins, see Iain Fenlon, "Church Reform and Devotional Music in Sixteenth-Century Rome: The Influence of Lay Confraternities," in *Forms of Faith in Sixteenth-Century Italy*, ed. Abigail Brundin and Matthew Treherne (Burlington, VT: Ashgate, 2009), 229–46.

9. Elisabeth Le Guin, *Boccherini's Body: An Essay in Carnal Musicology* (Berkeley: University of California Press, 2005).

10. Carla Freccero, "Queer Times," *South Atlantic Quarterly* 106, no. 3 (2007): 485–94.

11. Carolyn Dinshaw, "Chaucer's Queer Touches/A Queer Touches Chaucer," *Exemplaria* 7, no. 1 (1995): 76–77.

12. Virginia Burrus, *The Sex Lives of Saints: An Erotics of Ancient Hagiography* (Philadelphia: University of Pennsylvania Press, 2007), 2.

13. Emily Wilbourne, "The Queer History of the Castrato," in *The Oxford Handbook of Music and Queerness*, ed. Fred Everett Maus and Sheila Whiteley (New York: Oxford University Press, 2018).

14. See Antolini, "La carriera di cantante"; Luigi Fausti, "Un autobiografia poetica di Loreto Vittori," *Atti dell'Accademia spoletina, 1920–22*: 158–68; and Carl August Rau, *Loreto Vittori: Beiträge zur historisch-kritischen Würdigung seines Lebens, Wirkens und Schaffens* (Munich, 1916).

15. "Che la Giovine non in mano del marito, mà in un Convento si riponesse per assicurarsi, che non potesse far mala vita già ché le relationi della sua prava Inclinatione e del consenso dato antecedentemente al Ratto persuadevono, che ella non fosse per vivere castamente." From Spada's report as transcribed in Antolini, "Il 'lieve error di giovanil desio,'" 71.

16. Amy Brosius, "Singers Behaving Badly: Rivalry, Vengeance, and the Singers of Cardinal Antonio Barberini," *Women and Music: A Journal of Gender and Culture* 19, no. 1 (2015): 45–53.

17. Valeria Finucci, *The Manly Masquerade: Masculinity, Paternity, and Castration in the Italian Renaissance* (Durham, NC: Duke University Press, 2003), 261.

18. John Hunt, "Carriage, Violence and Masculinity in Early Modern Rome," *I Tatti Studies in the Italian Renaissance* 17, no. 1 (2014): 175–96.

19. When, in the next century, the castrato Giusto Ferdinando Tenducci married Dorothea Maunsell, she was kidnapped back by her family, and he was accused of abduction and rape. See Helen Berry, *The Castrato and His Wife* (New York: Oxford University Press, 2011).

20. On the academy, see Laura Alemanno, "L'Accademia degli umoristi," *Roma moderna e contemporanea* 3 (1995): 97–120; and "Le 'Rime degli Accademici Umoristi,'" in *Letteratura italiana e utopia*, ed. Alberto Asor Rosa (Roma: Riuniti, 1996): 2:275–90; Francis Gravit, "The Accademia degli Umoristi and Its French Relationships," *Papers of the Michigan Academy of Science, Arts, and Letters* 20 (1934): 505–21; and Piera Russo, "L'Accademia degli umoristi. Fondazione, strutture e leggi: Il primo decennio di attività," *Esperienze Ietterarie: Rivista trimestrale di critica e di cultura* 4, no. 4 (1979): 47–61.

21. See Leicester Bradner, "The Latin Drama of the Renaissance (1340–1640)," *Studies*

*in the Renaissance* 4 (1957): 31–54; David Marsh, "Latin Poetry," in *The Oxford Handbook of Neo-Latin*, ed. Stefan Tilg and Sarah Knight (Oxford: Oxford University Press, 2015), 395; and Elena Tamburini, "Dietro la scena: Comici, cantanti e letterati nell'Accademia romana degli Umoristi," *Studi secenteschi* 50 (2009): 89–112.

22. "Mense Iunio Superiore, cum noctu a cœna, vitandi æstus causa, urbem obambularem, ante ædes fœminæ primariæ, in cuius tutela lateo, honoris illius gratia, cantiunculam cecini: sed ea completa, admonitus sum a pedissequis eiusdem, ea cantione graviter fuisse offensum scrutarium quendam, qui sub illis ædibus, una cum uxore ac liberis, habitabat, cupidum in primis existimationis bonæ; quod eximia pulchritudine domi filiam haberet, paulo a viro relictam." Gian Vittorio Rossi, *Iani Nicii Erythraei Dialogi septendecim*, dialogue 1 (Cologne: Kalcovius, 1645), 7–8. Thanks to Amy Brosius for pointing me to this abduction and source in 2007.

23. Martha Feldman, *The Castrato: Reflections on Natures and Kinds* (Oakland: University of California Press, 2015), 50.

24. Feldman, *The Castrato*, 293n29.

25. Dinko Fabris, *Music in Seventeenth-Century Naples: Francesco Provenzale (1624–1704)* (Burlington, VT: Ashgate, 2007), 23.

26. "Il loro canto in effetto non mi attrae [ . . . ] Ma, in somma, assai più volentieri vo dove sento cantar bene, e dalle buone musiche più volte mi ricordo di aver sentito eccitarsi in me spiriti di divozione e di compunzione e fino desiderio dell'altra vita e delle cose celesti." Pietro della Valle in Angelo Solerti, ed., *Le origini del melodramma: Testimonian ze dei contemporanei* (Turin: Fratelli Bocca, 1903), 175–6.

27. For more on Teresa and her virility, see Alison Weber, *Teresa of Avila and the Rhetoric of Femininity* (Princeton, NJ: Princeton University Press, 1990), 16–20.

28. Despite his reservations, Augustine did ultimately endorse singing in church. He follows his statement that song moves him to "greater religious fervor" than speech and kindles his "ardent flame of piety" with a cautionary note: "But I ought not to allow my mind to be paralyzed by the gratification of my senses, which often leads it astray." Translated in Richard Taruskin and Piero Weiss, eds., *Music in the Western World* (New York: Schirmer, 1984), 32.

29. The church even had something to say about "secular art," as in Cardinal Robert Bellarmine's public insistence in 1603 that *Il pastor fido* was more harmful to Catholic morals than Protestantism itself. See Nicolas J. Perella, *The Critical Fortune of Battista Guarini's "Il Pastor Fido"* (Florence: Olschki, 1973), 28–29.

30. The Index of Forbidden Books outlawed "books that professedly deal with, narrate or teach things lascivious or obscene." The mandate by the Council of Trent in 1563 required that images remain devoid of "seductive charm" and set off a systematic censorship of pagan imagery and nudity. Translated in Robert Klein and Henri Zerner, *Italian Art 1500–1600: Sources and Documents* (Evanston, IL: Northwestern University Press, 1966), 121.

31. Richard Sherr, "*Ex Concordia Discors*: Popes, Cardinals, Nationalities, Conflict, Deviance, and Irrational Behavior; A Crisis in the Papal Chapel in the Pontificate of Paul IV," *Journal of Musicology* 32, no. 4 (2015): 494–523.

32. Gioseffo Zarlino, "Gli Instituzioni armoniche," translated in Oliver Strunk, ed., *Source Readings in Music History: The Renaissance* (New York: W. W. Norton, 1965), 59. On connections to Cicero and Bembo, see Martha Feldman, *City Culture and the Madrigal at Venice* (Berkeley: University of California Press, 1995).

33. Translated in Strunk, *The Renaissance*, 108.

34. "Decimus Tonus, flebilis, amoroeus, mollis, hominem molli coversationi deditum

exprimit rebus spiritualibus et pietatis operibus, si remittatur, aptus est si intendatur in lasciviam ac mollitiem animi, amoresque mundanos facile rapit." Athanasius Kircher, *Musurgia universalis; Sive, ars magna consoni et dissoni in X. libros digesta* (Rome, 1650), 2:576.

35. Craig A. Monson, "The Council of Trent Revisited," *Journal of the American Musicological Society* 55, no. 1 (2002): 9.

36. Translated in Guido Giglioni, "'Bolognan Boys Are Beautiful, Tasteful and Mostly Fine Musicians': Cardano on Male Same-Sex Love and Music," in *The Sciences of Homosexuality in Early Modern Europe*, ed. Kenneth Borris and George Rousseau (New York: Routledge, 2008), 214–15. For more on *Proxeneta*, see Giglioni, "*Musicus puer*: A Note on Cardano's Household and the Dangers of Music," *Bruniana & Campanelliana* 11, no. 1 (2005): 83–88. See also Anthony Grafton, "Cardano's 'Proxeneta': Prudence for Professors," *Bruniana & Campanelliana* 7, no. 2 (2001): 363–80.

37. Eric Bianchi, "Was Man Made for the Sabbath? Site, Space and Identity in Jesuits' Musical Life," in *The Grand Theater of the World: Music, Space, and the Performance of Identity in Early Modern Rome*, ed. Valeria De Lucca and Christine Jeanneret (New York: Routledge, 2019), 72–87.

38. "Quasi vero, si tanti referebat ad Creatoris laudes modulate concinendas, ipse, qui omnia perfecta condidit, non etiam huic indigentiae providisset: an fortasse nos homunciones Dei providentiam incusamus, & quae ipse bene operatus est, emendare praesumimus? Quod si tanta tenet cupido mollitudinis vocum mulcentium, cur non feminas potius admittimus, naturae suam integritatem relinquentes, quam viros effeminamus, velut Neronem in tot Sporis imitati?" This was written in 1648 and first published in Cologne in 1653. It was reprinted as Leone Allacci, ed., *Symmikta, sive Opuscula Graeca et Latina, vetustiora & recentiora*, ed. (Venice, 1733), 78.

39. "Ut nihil denique dicam de praeclaris supremae vocis phonascis, qui in Collegio Germanico reperiuntur et vocis suavitate et artis peritia excellentissimis. His igitur aut similibus egregiis Musicis instructus et industrius quispiam choragus, accedente praesertim eleganti, ac ingeniosa compositione, posset uti mihi certo persuadeo, eos ipsos admirandos effectus, affectusque in auditorium animis excitare, quos tantopere depraedicant, ac posteritati commendarunt veterum olim historiarum scriptores." Kircher, *Musurgia universalis* (1650), 598.

40. Translated in Bonnie Blackburn and Edward Lowinsky, "Luigi Zenobi and His Letter on the Perfect Musician," *Studi musicali* 22 (1993): 65.

41. Blackburn and Lowinsky, 66.

42. On Chigi patronage, see Valeria De Lucca, "L'Alcasta and the Emergence of Collective Patronage in Mid-Seventeenth-Century Rome," *Journal of Musicology* 28, no. 2 (2011): 195–230; and *"Dalle sponde del tebro alle rive dell'adria": Maria Mancini and Lorenzo Onofrio Colonna's Patronage of Music and Theater between Rome and Venice (1659–1675)* (Princeton, NJ: Princeton University Press, 2009); and Colleen Reardon, *A Sociable Moment: Opera and Festive Culture in Baroque Siena* (New York: Oxford University Press, 2016).

43. Translated in Robert F. Hayburn, *Papal Legislation on Sacred Music, 95 A.D. to 1977 A.D.* (Collegeville, MN: Liturgical Press, 1979), 76, 78.

44. "Mentre pur l'avevano [la voce] come persone che per l'età non avevano giudizio, anche cantavano senza gusto e senza grazia, come cose appunto imparate a mente, che alle volte a sentirli mi davano certe strappate di corda insopportabili. I soprani di oggi, persone di giudizio, d'età, di sentimento e di perizia nell'arte esquisita cantano le loro cose con grazia, con gusto, con vero garbo; vestendosi degli affetti rapiscono a sentirli." Pietro della Valle in Solerti, *Le origini del melodramma*, 163–64.

45. "C'est une expression merveilleuse, mais franchement beaucoup trop vive pour une église." Charles de Brosses, *Le président de Brosses en Italie: Lettres familières écrites d'Italie en 1739 et 1740* (Paris: Didier, 1858), 78–79.

46. Teresa of Avila, *The Collected Works of St. Teresa of Avila*, trans. Kieran Kavanaugh and Otilio Rodriguez, 2nd ed. (Washington, DC: ICS, 1987), 1:161.

47. Warren Kirkendale offers an in-depth description in *Emilio de' Cavalieri "Gentiluomo Romano": His Life and Letters, His Role as Superintendent of All the Arts at the Medici Court, and His Musical Compositions* (Florence: L. S. Olschki, 2001), 233–95.

48. "Acciò vediate voi stesso esser vero quanto vi dico, mi condusse al cembalo e cantò alcuni pezzi di quella rappresentatione et in particolare quell loco del Corpo, che lo moveva tanto, e mi piacque in maniera ch'io lo pregai a farmene parte, il che molto cortesemente face, e me lo copiò di sua mano, et io lo imparai alla mente, et andavo spesso a casa sua per sentirlo cantare da lui." Quoted in Morelli, *Il tempio armonico*, 179.

49. "E che egli vi si trovò presente quel giorno, che si rappresentò tre volte senza potersi mai satiare, e mi disse in particolare che sentendo la parte del Tempo, si sentì entrare adosso un timore e spavento grande, et alla parte del Corpo, rappresentata dal medesimo che faceva il Tempo, quando stato alquanto in dubbio, che cosa doveva fare o seguire Dio e 'l Mondo, si risolveva di seguire Iddio, che gli uscirno da gl'occhi in grandissima abbondanza le lacrime e sentiì destarsi nel core un pentimento grande e dolore dei suoi peccati, né questo fu per allora solamente, ma di poi sempre che la cantava, talché ogni olta che si voleva communicare, per eccitare in sé la divotione, cantava quella parte, e prorompeva in un fiume di pianto. Lodava ancora in estremo la parte del'Anima, che oltre esser stata rappresentata divinamente da quell putto, diceva nella musica esser un artifitio inestimabile, che esprimeva gli affetti di dolore e di dolcezza con certe seste false, che tiravano alla settima, che rapivano l'anima; in soma concludeva in quell genere non potersi fare cosa più bella, né più perfetta." Quoted in Morelli, *Il tempio armonico*, 179.

50. "FALSIRENA da Arsete consigliata al bene; ma da Idonia persuasa al male, e l'Anima consigliata dalla Ragione; ma persuasa dalla Concupiscenza. E come Falsirena a Idonia facilmente cede, cosi mostra, ch'ogni Affetto e dal Senso agevolmente superato. E se finalmente a duro Scoglio e legata la malvagia Falsirena, si deve anco intendere, che la Pena al fine e seguace della Colpa." Domenico Mazzocchi, *La catena d'Adone* (Venice: Vicenti, 1626; repr., Bologna: Forni, 1969), 126.

51. See Margaret Murata, "'Colpe mie venite a piangere': The Penitential Cantata in Baroque Rome," *Listening to Early Modern Catholicism: Perspectives from Musicology*, ed. Daniele V. Filippi and Michael Noone (Boston: Brill, 2017), 204–32. Ignatius of Loyola, *Ignatius of Loyola: The Spiritual Exercises and Selected Works*, ed. George E. Ganss (New York: Paulist Press, 1991), sections 65–69, pp. 46–47. Bruna Filippi, "The Orator's Performance: Gesture, Word, and Theatre at the Collegio Romano," in *The Jesuits 2: Cultures, Sciences, and the Arts, 1540–1773*, ed. Thomas Frank Kennedy, Steven J. Harris, Gauvin Alexander Bailey, and John W. O'Malley (Toronto: University of Toronto Press, 2006), 512–29; Robert L. Kendrick, "Martyrdom in Seventeenth-Century Italian Music," in *From Rome to Eternity: Catholicism and the Arts in Italy, ca. 1550–1650*, ed. Pamela M. Jones and Thomas Worcester (Leiden: Brill, 2002), 121–41.

52. "Christo dal cielo in terra, vedendo questo il detto Re con gran furia si buttò a basso in ginocchione a li piedi di Christo, che dopo haverli parlato risalì in cielo, et si serrò la scena, et il detto Re rimontò sopra il suo carro, et passò via per mezzo della scena." The full *avviso* from 1640 is transcribed in Antolini, "La carriera di cantante," 187.

53. "La gloria del Paradiso non si può esplicare con parole. È poco il dire quel giubilo, quel gaudio, quella dolcezza, quella soavità, quel contento de' Beati. Non si comprende a bastanza, e particolarmente dalle persone semplici. Per innamorarne maggiormente gli huomini, per commovere maggiormente gli animi, si dipingono i Chori degl'Angeli in atto di cantare, e di sonare varij stromenti Musicali. Ma perché la pittura è muta, e non porge all'orecchie melodia alcuna, si aggiunge il canto delle voci, & il suono degli stromenti, per rappresentare più vivamente all'udito soave armonia." Grazioso Uberti, *Contrasto musico: Opera dilettevole* (Rome: Grignani, 1630; facsimile ed., Lucca: LIM, 1991), 110–11.

54. Translation modified from Richard Eric Engelhart, "Domenico Mazzocchi's 'Dialoghi e Sonetti' and 'Madrigali a cinque voci' (1638): A Modern Edition with Biographical Commentary and New Archival Documents" (PhD diss., Kent State University, 1987), 631.

55. Emily Wilbourne, "A Question of Character: Artemisia Gentileschi and Virginia Ramponi Andreini," *Italian Studies* 71, no. 3 (2016): 335–55. The Fedeli troupe performed a drama called *La Maddalena* in 1617 in honor of the wedding of Ferdinando Gonzaga to Caterina de' Medici

56. Ellen Rosand, "The Descending Tetrachord: An Emblem of Lament," *Musical Quarterly* 65, no. 3 (1979): 346–59, http://www.jstor.org/stable/741489.

57. "Aer concitet naturalem humorem objecto, & harmonicis motibus prorsus similem." Athanasius Kircher, *Musurgia universalis* (1650), 1:552.

58. See Caroline Walker Bynum, *Fragmentation and Redemption: Essays on Gender and the Human Body in Medieval Religion* (New York: Zone Books, 1992).

59. Quoted in Paolo Fabbri, *Monteverdi*, trans. Tim Carter (New York: Cambridge University Press, 1994), 86.

60. Ellen Rosand, "Barbara Strozzi, *virtuosissima cantatrice*: The Composer's Voice," *Journal of the American Musicological Society* 31, no. 2 (1978): 279.

61. "Ut Auditores sæpe contineri nescii in clamores gemitus, suspiria exoticos corporum motus erumpentes quanto interiorum affectuum aestu incitarẽntur signis extrinsecis clare exprimerent." Kircher, *Musurgia universalis* (1650), 1:546.

62. Jeffrey Levenberg, "Worth the Price of the 'Musurgia universalis': Athanasius Kircher on the Secret of the 'Metabolic Style,'" *Recercare* 28, no. 1/2 (2016): 43–88; Patrizio Barbieri and Ken Hurry, "Pietro della Valle: The 'Esthèr' Oratorio (1639) and Other Experiments in the 'Stylus Metabolicus.' With New Documents on Triharmonic Instruments," *Recercare* 19, no. 1/2 (2007): 73–124.

63. "Porro mutatio utraque magnam emphasin habet, notabilesque alterationes in auditoribus efficit, potestque infinities variari, & quibuslibet affectibus exprimendis appositissima est." Kircher, *Musurgia universalis*, 1:672.

64. See Eva Linfield, "Modulatory Techniques in Seventeenth-Century Music: Schütz, a Case in Point," *Music Analysis* 12, no. 2 (1993): 197–214.

65. Feldman, *The Castrato*, 113–19.

66. Giulio Caccini, *Le nuove musiche*, ed. H. Wiley Hitchcock (Madison: A-R Editions, 1970), 13.

67. "Ivi si faccia sollevatione, o (come si suol dire volgarmente) messa di voce, che è l'andar crescendo à poco à poco la voce di fiato insieme, e di tuono, & è specie della metà del sopradetto X (come si pratica ne gli Enarmonici)." Domenico Mazzocchi, *Dialoghi e sonetti* (Rome, 1638), 180.

68. "Ma quando si haverà da crescer la voce solamente di fiato, e di spirito, e non di tuono, sarà segnato con la lettera C, come si è fatto in alcuni Madrigali, & all'hora si osserverà,

che si come la tenuta di voce si deve prima dolcemente ingrandire, così anche doppo successivamente si debba à poco à poco andare smorzando, e tanto pianeggiarla, sino che si riduca all'insensibile, o al nulla; cavato da una Cisterna, che cosi rispondeva a certe voci. Ma questi, & altri avvisi solo il buon concerto, e la discretione del buon Cantante potrà bene aggiustarli." Mazzocchi, 180.

69. Mazzocchi, 180

70. Juvenal, *Satires*, trans. Susanna Morton Braund, Loeb Classical Library 91 (Cambridge, MA: Harvard University Press, 2004), 417. "Facere hoc non possis quinque diebus, continuis, quia sunt talis quoque taedia vitae magna: voluptates commendat rarior usus." Decimus Iunius Iuvenalis, *Saturae*, 11.1, lines 206–8, in Classical Latin Texts: A Resources Prepared by the Packard Humanities Institute, https://latin.packhum.org/loc/1276/1/10/9447–9458@1#10.

71. "Si deve anco ingegnare il discreto Cantore di proferire distinto, e scolpito, accio, se sia possibile, non si perda nè pure una nota di quello, che canta: nella qual parte è singolare il Sig. Cavaliere Loreto; perciocchè oltre il dono naturale, è stato allevato in Firenze, dove si fa particolare professione di bella pronunzia, come già in Atene, così nel cantare, come nel favellare." Doni, *De' trattati di musica*, ed. A. F. Gori (Florence, 1763), 2:134.

72. "Acre in canendo judicium, atque artificem rationem: quo sit ut suum, cuique animi motui, vocis genus adhibeat." Rossi, *Pinacotheca Altera Imaginum, Illvstrium*, 218. Cited in Claudio Gallico, "La 'Querimonia' di Maddalena di Domenico Mazzocchi e l'interpretazione di Loreto Vittori [1966]," in *Sopra li fondamenti della verità: Musica italiana fra XV e XVII secolo* (Rome: Bulzoni, 2001): 145.

73. "Ille vero, a natura atque ab optimorum magistrorum praeceptis edoctus, si iracunda perturbati vox & oratio cantu sit exprimenda, vocis genus sumit acutum, incitatum, crebro incidens; si miseratio ac maeror explicandus, flexibile, plenum, interruptum, flebili voce; si metus referendus, demissum, haesitans, abjectum; si truculenti vis hominis tradenda, contentum, vehemens, imminens; si exhilarati voluptas animi anteponenda, effusum, lene, tenerum, hilaratum ac remissum; si aegritudine aliqua oppressi molestia significanda, sine commiseratione grave quiddam, & uno pressu ac sono obductum adhibet. Haec, a L. Crasso aliisque summis doctoribus relata, felicius, in illius modulate canentis voce, quam in eorum libris, impressa cernuntur. Quod autem ita scienter in cantu, ad quemcunque velit animi affectum exprimendum, se formet ac fingat, a natura consecutus est."

74. "At vero, illo canere incipiente, omnes alacres & erecti consistunt, sermones, quos instituerant, dirimunt, animam continent, ne, crebro respirantes, sibi quodammodo obstrepere videantur; tum, cantu, eiusdem mirabiliter deliniti, nullis ab eo fastidiis abalienantur, nulla aurium satietate defatigati discedunt, sed ab eo, tanquam capti alligatique, continentur." Rossi, 218.

75. On Mazzocchi's lament, see Claudio Gallico, "La 'Querimonia' di Maddalena di Domenico Mazzocchi e l'interpretazione di Loreto Vittori," *Collectanea Historiae Musicae* 4 (1966): 133–47; Levenberg, "Worth the price of the 'Musurgia universalis'" Christine Jeanneret, "Gender Ambivalence and the Expression of Passions in the Performances of Early Roman Cantatas by Castrati and Female Singers," in *The Emotional Power of Music: Multidisciplinary Perspectives on Musical Arousal, Expression, and Social Control*, ed. Tom Cochrane, Bernardino Fantini, and Klaus R. Scherer (Oxford: Oxford University Press, 2013), 85–101; and Murata, "'Colpe mie venite a piangere,'" 204–32.

76. Elizabeth Freeman, "Time Binds, or, Erotohistoriography," *Social Text* 23, no. 3/4 (2005): 66. For a discussion of temporal drag, see Freeman, "Packing History, Count(er)ing Generations" *New Literary History* 31, no. 4 (Autumn 2000): 727–44.

77. Freeman, "Packing History," 728.

78. Giovanni Battista Marino, *La galeria* (Venice, 1620), 78.

79. Marino, 79.

80. Marino, 82.

81. "Accordianci ancora noi à questa pietosa Musica, se non col canto, col pianto, se non con le voci, almeno con le lagrime, con queste acque l'acque di que' fonti canori imitando, che naturalmente rispondono al suono. Questa questa era la tua Musica, ò Madalena. . . . *Auribus percipe lachrymas meas.* No pregava, che le mirasse, ma che le sentisse. Indi soggiungeva. *Exaudivit Dominus vocem fletus mei* Dice, ch'à Dio era piaciuta la Musica delle sue lagrime." Giovanni Battista Marino, *Le dicerie sacre* (Venice, 1674), 306.

82. "At neminem eorum, qui aderant, arbitror fuiffe tam leni animo tamque remiffo, qui non ad eos motus se perduci sentiret, ad quos ab illo impellebatur; nimirum, ad fletum ad iram ad odium peccatorum, nefcio alios; me quidem, fcio, acriter vehementerque in delicta mea exarfiffe, cum ille Magdalenæ perfonæ actor, præteritæ illius vitæ crimina exfecraretur, propter quæ in tantam Dei atque hominum offenfionem in curriffet; fenfi, mihi ubertim lacrimas ab oculis ire, cum ille flentis peccatricis gemitus, voce ad miferabilem fonum inflexa, repræsentaret; senf, me ad incredibilem admirationem efferri, cum vocem a graviffimo ad acutiffimum fonum gradatim impellens, eandemque ab acutiffimo ad graviffimum per varios anfractus volubilitate incredibili colligens, se poffe eam oftenderet sicut molliffimam ceram, quocunque vellet, contorquere ac flectere." Rossi, *Pinacotheca Altera Imaginum, Illvstrium*, 217.

83. Freeman, "Time Binds," 60.

84. For more on Barberini relics, see David Jaffé, "The Barberini Circle: Some Exchanges between Peiresc, Rubens, and Their Contemporaries," *Journal of the History of Collections* 1, no. 2 (1989): 119–47. On the chapel at S. Andrea della Valle, see Peter Rietbergen, *Power and Religion in Baroque Rome: Barberini Cultural Policies* (Boston: Brill, 2006), 61–94. For a theoretical take on relics, see Alice E. Sanger, "Sensuality, Sacred Remains and Devotion in Baroque Rome," in *Sense and the Senses in Early Modern Art and Cultural Practice*, ed. Alice E. Sanger and Siv Tove Kulbrandstad Walker (Burlington, VT: Ashgate, 2012), 199–215.

85. Heather K. Love, *Feeling Backward: Loss and the Politics of Queer History* (Cambridge, MA: Harvard University Press, 2007), 10.

## CHAPTER TEN

1. "V. Ecc.a resti per servita d'havere ch'io crederei, che fosse men male veder havergli figliuoli, et che havessero solumente habilità di farsi credendo io che la si sarebbero meglio che qua." Negri to Guglielmo Gonzaga, transcribed in Richard Sherr, "Guglielmo Gonzaga and the Castrati," *Renaissance Quarterly* 33, no. 1 (1980): 52–53, my translation.

2. Donna Haraway, "A Cyborg Manifesto: Science, Technology, and Socialist-Feminism in the Late Twentieth Century," in *Simians, Cyborgs and Women: The Reinvention of Nature* (New York: Routledge, 1991), 149.

3. Sandra Cavallo, *Artisans of the Body in Early Modern Italy: Identities, Families and Masculinities* (Manchester: Manchester University Press, 2007).

4. John Gagné, "Emotional Attachments: Iron Hands, Their Makers, and Their Wearers, 1450–1600," in *Feeling Things: Objects and Emotions through History*, ed. Stephanie Downes, Sally Holloway, and Sarah Randles (Oxford: Oxford University Press, 2018), 134–35.

5. Julius Casserius, "The Three Books of the First Treatise on the Anatomical History of the Larynx, the Organ of Voice," trans. Malcolm H. Hast and E. B. Holtsmark, supplement, *Acta Oto-Laryngologica* 68, S261 (1969): 9.

6. David Abulafia, "The First Atlantic Slaves, 1350–1520: Conquest, Slavery and the Opening of the Atlantic," in *Western Fictions, Black Realities: Meanings of Blackness and Modernities*, eds. Isabel Soto and Violet Showers Johnson (East Lansing: Michigan State University Press, 2012), 107.

7. Shaun Tougher, "Eunuchs in the East, Men in the West? Dis/unity, Gender and Orientalism in the Fourth Century," in *East and West in the Roman Empire of the Fourth Century: An End to Unity?*, ed. Roald Dijkstra, Sanne van Poppel, and Daniëlle Slootjes (Boston: Brill, 2015), 147–63.

8. Claudian, *Against Eutropius*, trans. Maurice Platnauer, Loeb Classical Library 135 (Cambridge, MA: Harvard University Press, 1922), para. 22.

9. Claudian, para. 3.

10. Aristotle, *Generation of Animals*, trans. A. L. Peck, Loeb Classical Library 366 (Cambridge, MA: Harvard University Press, 1942), 521.

11. Galen, *On Semen*, ed. and trans. Phillip DeLacy, Corpus medicorum Graecorum, vol. 5, pt. 3, fasc. 1. (Berlin: Akademie, 1992), 1.15.34–38. For more, see Dale B. Martin, "Contradictions of Masculinity: Ascetic Inseminators and Menstruating Men in Greco-Roman Culture," in *Generation and Degeneration: Tropes of Reproduction in Literature and History from Antiquity through Early Modern Europe*, ed. Valeria Finucci and Kevin Brownlee (Durham, NC: Duke University Press, 2001), 81–108.

12. Juvenal, *The Sixteen Satires*, ed. and trans. Peter Green (Baltimore: Penguin, 1974), 55–56,

13. Quintilian, *Institutio oratoria*, trans. H. E. Butler (New York: G. P. Putnam's Sons, 1921), 2:307, 309.

14. Joy Connolly, "Virile Tongues: Rhetoric and Masculinity," in *A Companion to Roman Rhetoric*, ed. William Dominik and Jon Hall (Malden, MA: Blackwell, 2007), 83–97.

15. Quintilian, *Institutio oratoria*, 2:309.

16. Paul of Aegina, *The Seven Books of Paulus Aegineta*, trans. Francis Adams (London: Printed for the Sydenham Society, 1846), 2:379–80.

17. There were seventy editions in various languages over the next four centuries, and it was widely read through the nineteenth century.

18. Leah DeVun, "Erecting Sex: Hermaphrodites and the Medieval Science of Surgery," *Osiris* 30, no. 1 (2015): 20.

19. Cynthia Klestinec, "Translating Learned Surgery," *Journal of the History of Medicine and Allied Sciences* 72, no. 1 (2017): 34–50.

20. Giovanni Andrea della Croce, translated in Cynthia Klestinec, "Translating Learned Surgery," *Journal of the History of Medicine and Allied Sciences* 72, no. 1 (2017): 42.

21. In Italian, the monumental book was published as Giovanni Andrea Della Croce, *Cirugia universale e perfetta di tutte le parti pertinenti all'ottimo Chirurgo* (Venice: Ziletti, 1574, 1583).

22. "Cresce ancor la carne trà le tuniche, lequali, i Greci chiamano Oscheon gli Arabi Barichem, & i Latini Scrotum, di donde nasce una certa durezza carnosa, detta da Greci Sarcocelum, dalli Arabi Burum, & da Latini Hernia carnosa; è del tutto fuor di natura, la onde ha bisogno d'esser rimossa: nondimeno se sarà congiunta col testicolo, overo rivolta intorno, facciasi castratura, alla qual'opera si ricercano quattro cose, il castratore, il rasoio, il filo, & li cauterij." Croce, *Cirugia universale*, bk. 7, 21.

23. Richard Sherr, "*Ex Concordia Discors*: Popes, Cardinals, Nationalities, Conflict, Deviance, and Irrational Behavior; A Crisis in the Papal Chapel in the Pontificate of Paul IV,"

*Journal of Musicology* 32, no. 4 (2015): 494–523; "Gugliemo Gonzaga and the Castrati," *Renaissance Quarterly* 33, no. 1 (1980): 33–56; "The 'Spanish Nation' in the Papal Chapel, 1492–1521," *Early Music* 20, no. 4 (1992): 601–10; Iain Fenlon, "Monteverdi's Mantuan *Orfeo*: Some New Documentation," *Early Music* 12, no. 2 (May 1984): 163–72; and Giuseppe Gerbino, "The Quest for the Soprano Voice: Castrati in Renaissance Italy," *Studi musicali* 33, no. 2 (2004): 303–57.

24. Susan Parisi has documented the career of Paolo Faccone, who recruited talented musicians from 1586 to 1615. In 1613 he wrote to Mantua of a boy who desired to have himself castrated but delicately avoided the subject of castration. Susan Parisi, "Acquiring Musicians and Instruments in the Early Baroque: Observations from Mantua," *Journal of Musicology* 14, no. 2 (1996): 117–50.

25. Valeria Finucci, *The Prince's Body: Vincenzo Gonzaga and Renaissance Medicine* (Cambridge, MA: Harvard University Press, 2015), 84–87.

26. "E quanto alla relatione c'ha alla medicina, ritiene in se qual che segno d'honore, ma per il soggetto medicabile, è più presto vile, e negletto, che altro, perché all'ultimo un Castradore, non è altro, che un Medico da testicoli, anzi più tosto un Barbiero, il quale pien di rigore non sa sanar la piaga se non impiaga: Di questa professione sono communemente i Norsini come anco da Norsia vengono quelli che acconciano le braccia rotte, & quei che fanno Brachieri detti latinamente fasciæ ò cerotti nelle parti virili d'un'altra specie di medicina molto differente. & perche questo mestiero si risolve in poca cosa, cioè, nel taglio d'una borsa solamente mentre che l'huomo è legato, e tenuto a modo d'una bestia." Tommaso Garzoni, *La piazza universale di tutte le professioni del mondo* (Venice, 1610), 363.

27. David Gentilcore, *Medical Charlatanism in Early Modern Italy* (Oxford: Oxford University Press, 2006), 136–88; Fiona Davidson, "Lithotomists, Cataract-Curers, and Hernia-Carvers: The Surgical School of Preci," *Journal of Medical Biography* 24, no. 4 (2016): 440–52; F. Guiggi and G. M. Giustozzi, "History of Surgery in Preci Better Known as Surgeons of Norcia," *Minerva chirurgica* 54, no. 12 (1999): 917–26.

28. Eric Bianchi has written extensively about this; see "Was Man Made for the Sabbath? Site, Space and Identity in Jesuits' Musical Life," in *The Grand Theater of the World: Music, Space, and the Performance of Identity in Early Modern Rome*, ed. Valeria De Lucca and Christine Jeanneret (New York: Routledge, 2019), 72–87; and "Scholars, Friends, Plagiarists: The Musician as Author in the Seventeenth Century," *Journal of the American Musicological Society* 70, no. 1 (2017): 61–128.

29. "Hinc fit, quod excellentia vocis in ordine ad cantum sit magnifacienda; tum quia est perfectio boni corporalis; quod pluribus membris praefertur; tum quia ratione ipsius persona redditur honorabilis etiam apud Principes; tum quia est sufficiens, ut unde acquiratur nobilis sustenatio vitae: & plus commodi ab ipsa percipitur, quam ex eo, quod quis sit habilis ad generandum. Unde maius bonum censendum est quod sit excellentia vocis; quam non esse castratum. Poterit igitur rationabiliter, & secundum ius naturale excellentia vocis praeferri virilibus: & proinde abhiberi castratio, ne pereat excellentia vocis." Zaccaria Pasqualigo, *Decisiones morales iuxta principia theologica* (Verona, 1641), 439. Pietro Redondi spells out the important political and ideological implications of this debate; see *Galileo Heretic*, trans. Raymond Rosenthal (Princeton, NJ: Princeton University Press, 1987), 251–54.

30. Michael John Gorman writes extensively about this in *The Scientific Counter-Revolution: The Jesuits and the Invention of Modern Science* (New York: Bloomsbury, 2020).

31. Camillo Maffei, *Scala naturale, overo fantasia dolcissima di Gio. Camillo Maffei da Solofra intorno alle cose occulte e desiderate nella filosofia* (Venice: Gio. Varisco, 1581).

32. Maffei, translated in Carol MacClintock, *Readings in the History of Music in Performance* (Bloomington: Indiana University Press, 1979), 47.

33. Maffei, translated in MacClintock, 39.

34. Maffei, translated in MacClintock, 44.

35. Maffei, translated in MacClintock, 45.

36. "Le Scuole di Roma obligano i Discepoli ad impiegare ogni giorno un'ora nel cantar cose difficili e malagevoli, per l'acquisto della esperienza; un'altra, nell'esercitio del Trillo; un'altra in quello de' Passaggi; un'altra negli studi delle Lettere; & un'altra negli ammaestramenti & eserciti del Canto, e sotto l'udito del Maestro, e davanti ad uno Specchio, per assuefarsi a non far moto alcuno inconveniente, ne di vita, ne di fronte, ne di ciglia, ne di bocca. E tutti questi erano gl'impieghi della mattina." Giovanni Andrea Angelini Bontempi, *Historia musica* (1695; repr. Geneva: Minkoff, 1976), 169.

37. "Erano l'andar spesse volte e cantare e sentire la risposta da un'Eco fuori della Porta Angelica, verso Monte Mario, per farsi giudice da se stessi de' propri accenti, l'andare a cantar quasi in tutte le Musiche che si faceuano nelle Chiese di Roma; e l'osservare le maniere del Canto di tanti Cantori insigni che fiorivano nel Pontificato di Urbano Ottavo; l'esercitarsi sopra quelle, & il renderne le ragioni al Maestro, quando si ritornava a Casa: il quale poi per maggiormente imprimerle nella mente de' Discepoli, vi faceva sopra i necessari discorsi, e ne dava i necessari avvertimenti." Bontempi, 170.

38. Mancini, translated in Julianne Baird, "An 18th-Century Controversy about the Trill: Mancini v. Manfredini," *Early Music* 15, no. 1 (1987): 39.

39. On Royer, see Gorman, *Scientific Counter-Revolution*, 216–18. Anthony Grafton also discusses Royer in the context of technology and machines; see *Magic and Technology in Early Modern Europe* (Washington, DC: Smithsonian Institution Libraries, 2005). See also Ingrid D. Rowland, "Athanasius Kircher's Guardian Angel," in *Conversations with Angels: Essays towards a History of Spiritual Communication, 1100–1700*, ed. Joad Raymond (London: Palgrave Macmillan, 2011), 250–70.

40. "Qui e stomacho suo deprimit duodecim, quatordecimve diversorum colorum aquas odoriferas, liquores perfectissimos, vinum adustum quod inceditur, oleum saxi quod sine ellychnio comburitur, lactucas, & flores omnis generis, integris & recentissimis foliis. Fontem etiam exhibet projiciendo aquam ex ore in altum per spatium duorum Miserere." Gaspar Schott, *Mechanica hydraulico-pneumatica cum figuris Aeneis, et privilegio sacrae Cesare Majestatis* (Frankfort: Herbipoli, 1657), 311–12.

41. On the traffic of musicians between Italy and Poland, see Barbara Przybyszewska-Jarmińska, "The Careers of Italian Musicians Employed by the Polish Vasa Kings (1587–1668)," *Musicology Today* 6 (2009): 26–43.

42. "E non creda il Musico Lettore, esser favoloso il presente raccontamento, per contener cose impossibili a praticarsi: al Ferri non solo erano possibili ma facilissime." Bontempi, *Historia musica*, 110.

43. "Per non haver come gli altri la nel Diafragma, tanto piu vigoroso degli altri haveva le viscere spiritali, e naturali. Et havendo petto intero e grande, è da persuadersi ancora ch'avesse il polmone fongoso e raro, & il temperamento piu frigido; ond'era che piu degli altri poteva contener lo spirito, e far sentir i miracoli dell'Arte; co' quali si rendeva maggiore della maraviglia, & quasi eccedeva il possibile della humanità." Bontempi, 110.

44. For more, see John Rosselli, "The Castrati as a Professional Group and a Social Phenomenon, 1550–1850," *Acta Musicologica* 60, no. 2 (1988): 174; and Valeria Finucci, *The Manly*

*Masquerade: Masculinity, Paternity, and Castration in the Italian Renaissance* (Durham, NC: Duke University Press, 2003), 245.

45. "Ampissima è l'efficacia del Canto regolato dalle vere norme dell'uno o dell'altro Modo moderno; poiché, sicome la virtù o lo spirito naturale dove egli è potentissimo subito ammollisce, e liquefà gli alimenti durissimi, e di austeri gli rende dolci, e genera fuori di sè col producimento dello spirito seminale la propagine: così la virtù vitale & animale dove ella è col proprio spirito efficacissima, agita per via del Canto col movimento di se stessa il proprio corpo, e con l'effusione muove il corpo vicino; e con una certa proprietà stellata, la quale concepisce e dalla forma di se stessa, e dalla eletta opportunità del tempo, dispone tanto il suo, quanto l'alieno. Questa agitatione non nasce che dalla forza e virtù del numero Harmonico, diffuso nella diversità degli' intervalli de' Modi; i quali sono così potenti, che se con l'osservare quei si usano in questa e quella Regione, e da quali stelle sia questa o quella dominata, vi s'indirizzassero le Cantilene appropriate, come composte secondo la convenienza di quegli Astri dominanti, si sarebbe quasi una forma comune, & in quella ne sottonascerebbe ancora qualche virtù celeste. Ma essendone difficile la cognitione, la quale senza il sussidio della Divina sorte, non si potrebbe conseguir da qualsivoglia humana diligenza, si può nella Dottrina de' Modi tralasciar la consideratione delle Stelle, e prender quella degli Affetti. Nascono i loro movimenti dalla diversità de' Numeri Harmonici diversamente ne' Modi constituiti, e dalla diversità delle complessioni degli Uditori che gli ascoltano." Bontempi, *Historia musica*, 239–40.

46. Bruno Latour, "An Attempt at a 'Compositionist Manifesto,'" *New Literary History* 41, no. 3 (Summer 2010): 480.

47. Rosi Braidotti, "Posthuman Humanities," *European Educational Research Journal* 12, no. 1 (2013): 1–19.

48. Donna Haraway, "A Manifesto for Cyborgs: Science, Technology, and Socialist Feminism in the 1980s," in *Feminism/Postmodernism*, ed. Linda J. Nicholson (New York: Routledge, 1990), 190–233.

49. Donna J. Haraway, *Staying with the Trouble: Making Kin in the Chthulucene* (Durham, NC: Duke University Press, 2016).

50. Tim R. Birkhead and Robert Montgomerie, "Three Centuries of Sperm Research," in *Sperm Biology: An Evolutionary Perspective*, ed. Tim R. Birkhead, Dave J. Hosken, and Scott Pitnick (Burlington, MA: Academic Press, 2009), 1–42.

51. Jill Burke, "Nakedness and Other Peoples: Rethinking the Italian Renaissance Nude," *Art History* 36, no. 4 (2013): 714–39.

## EPILOGUE

1. Valerie Traub, *Thinking Sex with the Early Moderns* (Philadelphia: University of Pennsylvania Press, 2016), 3.

2. Jacques Derrida, "Archive Fever: A Freudian Impression," trans. Eric Prenowitz, *Diacritics* 25, no. 2 (Summer 1995): 27.

3. Jessica Gabriel Peritz, *The Lyric Myth of Voice: Civilizing Song in Enlightenment Italy* (Oakland: University of California Press, 2022).

4. Baldassare Ferri and Biancamaria Brumana, *"Il pianto de' cigni in morte della fenice de' musici": Poesie per Baldassare Ferri e nuove ipotesi sulla carriera del cantante* (Perugia: Deputazione di storia patria per l'Umbria, 2010). This modern volume includes transcriptions of the 1680 volume.

5. "Per richiamarlo con sì erudite voci a più stabil vita." Ferri and Brumana, 18.

6. "Si duole il poeta che non possano, come i volti; effigiarsi ancora le voci; bramando lasciare al mondo un ritratto immortale dell'incomparabile ed angelica musica del signor Baldasarre Ferri." Ferri and Brumana, 38.

7. Jacques Rancière, *The Names of History: On the Poetics of Knowledge*, trans. Hassan Melehy (Minneapolis: University of Minnesota Press, 1994), 63.

8. Bontempi, translated in Feldman, "The Castrato as a Rhetorical Figure," in *Rhetoric and Drama*, ed. Daniel Scott Mayfield (Berlin: De Gruyter, 2017), 73–74. See also Feldman, *The Castrato: Reflections on Natures and Kinds* (Oakland: University of California Press, 2015), 112, 281.

9. Recently, the nascent field of critical organology has argued for a blending of the classification of musical instruments with broader questions of technology and has pushed against paradigms that consider instruments as things, objects that stand passive as human subjects act on them. See John Tresch and Emily I. Dolan, "Toward a New Organology: Instruments of Music and Science," *Osiris* 28, no. 1 (2013): 278–98; Eliot Bates, "The Social Life of Musical Instruments," *Ethnomusicology* 56, no. 3 (2012): 363–95; and Deirdre Loughridge, "Technologies of the Invisible: Optical Instruments and Musical Romanticism" (PhD diss., University of Pennsylvania, 2011).

10. Genesis 4:21 (New Revised Standard Version).

11. "Ma che parliamo della propria esperienza, noi che fra i professori dell'Arte del Canto occupiamo appena l'ultimo luogo? E qual maggior esperienza per autenticar questa verità si può trovare che il canto divino del Cavalier Baldassarre Ferri nostro Compatriota? Ciò che non ha spiegato con la voce un si sublime cantore non pensi alcuno di poterlo mostrare." Bontempi, trans. Feldman, "Castrato as a Rhetorical Figure," 110.

12. Anne Desler, "'The Little That I Have Done Is Already Gone and Forgotten': Farinelli and Burney Write Music History," *Cambridge Opera Journal* 27, no. 3 (2015): 230.

13. "Senza contrasto si può chiamare il Baldassare Ferri del nostro secolo." Giambattista Mancini, *Pensieri, e riflessioni pratiche sopra il canto figurato* (Vienna, 1774), 105.

14. "Possedeva la messa di voce a tanta perfezzione, che, a senso comune, e mio, fù quella, che lo rese per fama immortale nel canto." Mancini, 104.

15. Translated in Jean-Jacques Rousseau, *Essay on the Origin of Languages and Writings Related to Music*, trans. and ed. John T. Scott (Hanover, NH: University Press of New England, 1998), 485.

16. "Nelle carte musicali non apparisce vestigio del *da capo* se non verso la fine del secolo scorso. Il primo ad introdurlo sembra essere stato il cantore Baldassarre Ferri perugino, come si può argomentare dalla prefazione d'una raccolta de poesie a lui dedicata, ove nello stile ampolloso di quel secolo si dice, parlando di non so quale cantilena: Che il popolo sopraffato da vostri sovrumani concenti, guardandovi qual novello portentoso Orfeo della età nostra, vi sentì replicar più volte sulle nostre scene rimbombanti coi vostri applausi, ed inaffiata coi torrenti dell' armonia vostra dolcissima." Stefano Arteaga, *Le rivoluzioni del teatro musicale italiano dalla sua origine fino al presente* (Bologna, 1785), 2:71.

17. Jean-Jacques Rousseau, *Complete Dictionary of Music*, trans. William Waring (London, 1779), 56. On the twilight of the castrato, see Martha Feldman, "Denaturing the Castrato," *Opera Quarterly* 24, no. 3/4 (2008): 178–99; see also James Q. Davies, "'Veluti in Speculum': The Twilight of the Castrato," *Cambridge Opera Journal* 17, no. 3 (2005): 271–301.

18. Giancarlo Conestabile, *Notizie biografiche di Baldassarre Ferri, musico celebratissimo* (Perugia: Bartelli, 1846), 9–10.

19. "Muore l'esecutore con la sua arte particolare." In Conestabile, 9.

20. Derrida, "Archive Fever," 27.

21. Thanks to Martha Feldman and Jessica Peritz for thinking through the ideas of remains and archaeology.

22. Peritz, *Lyric Myth of Voice.*

23. Vernon Lee, *Studies of the Eighteenth Century in Italy* (London: T. F. Unwin, 1978), xxi.

24. Lee, xxiii.

25. Lee, xlviii.

26. Vernon Lee, *Selected Letters of Vernon Lee 1856–1935*, vol. 1, *1865–1884*, ed. Amanda Gagel (New York: Routledge, 2017), 20.

27. Lee, *Studies of the Eighteenth Century in Italy*, 172.

28. For a discussion of Lee in similar terms, see Anthony Teets, "Singing Things: The Castrato in Vernon Lee's Biography of a 'Culture-Ghost,'" *The Sibyl: A Journal of Vernon Lee Studies*, http://thesibylblog.com/singing-things-the-castrato-in-vernon-lees-biography-of-a-culture-ghost-by-anthony-teets/.

29. Vernon Lee, "An Eighteenth-Century Singer: An Imaginary Portrait," *Fortnightly Review* 50, no. 300 (December 1, 1891): 853.

30. Vernon Lee, *Hauntings and Other Fantastic Tales*, ed. Catherine Maxwell and Patricia Pulham (Peterborough: Broadview Press, 2006), 154.

31. Lee, *Studies of the Eighteenth Century in Italy*, 182.

32. Lee, 186.

33. Peritz, *Lyric Myth of Voice.*

34. Lee, *Studies of the Eighteenth Century in Italy*, 186.

35. Vernon Lee, "Old Italian Gardens," in *Limbo and Other Essays to Which Is Now Added Ariadne in Mantua* (New York: John Lane, 1908), 131–32.

36. Lee, *Studies of the Eighteenth Century in Italy*, 188.

37. Alberto Zanatta et al., "Occupational Markers and Pathology of the Castrato Singer Gaspare Pacchierotti (1740–1821)," *Scientific Reports* 6, 28463 (2016): 1–9, https://doi.org/10.1038/srep28463.

38. Cecilia Bartoli, "The Great Pretender: My Tribute to Farinelli, the Greatest Castrato," *Guardian*, January 23, 2020, https://www.theguardian.com/music/2020/jan/23/cecilia-bartoli-my-farinelli-album-countertenor-songs.

39. Derrida, "Archive Fever," 22.

40. Derrida, 16–17.

# BIBLIOGRAPHY


Abbate, Carolyn. *In Search of Opera*. Princeton, NJ: Princeton University Press, 2001.

———. "Outside Ravel's Tomb." *Journal of the American Musicological Society* 52, no. 3 (1999): 465–530.

———. "Sound Object Lessons." *Journal of the American Musicological Society* 69, no. 3 (2016): 793–829.

———. *Unsung Voices: Opera and Musical Narrative in the Nineteenth Century*. Princeton, NJ: Princeton University Press, 1991.

Abulafia, David. "The First Atlantic Slaves, 1350–1520: Conquest, Slavery and the Opening of the Atlantic." In *Western Fictions, Black Realities: Meanings of Blackness and Modernities*, edited by Isabel Soto and Violet Showers Johnson, 107–28. East Lansing: Michigan State University Press, 2012.

Ademollo, Alessandro. *I teatri di Roma nel secolo decimosettimo: Memorie sincrone, inedite, o non conosciute, di fatti ed artisti teatrali, "librettisti," commediografi e musicisti, cronologicamente ordinate per servire alla storia del teatro italiano*. Bologna: L. Pasqualucci, 1888. Facsimile edition. Bologna: Forni, 1969.

———. *La bell'Adriana: Ed altre virtuose del suo tempo alla corte di Mantova*. Città di Castello: S. Lapi, 1888.

Adler, Guido. "The Scope, Method, and Aim of Musicology." Translated by Erica Mugglestone. *Yearbook for Traditional Music* 13 (1981): 1–21.

Aercke, Kristiaan. *Gods of Play: Baroque Festive Performances as Rhetorical Discourse*. Albany: State University of New York Press, 1994.

Agazzari, Agostino. *Del sonare sopra il basso con tutti li stromenti e dell'uso loro nel conserto*. Siena: Domenico Falcini, 1607.

Agnew, Vanessa. *Enlightenment Orpheus: The Power of Music in Other Worlds*. New York: Oxford University Press, 2008.

———. "Hearing Things: Music and Sounds the Traveller Heard and Didn't Hear on the Grand Tour." *Cultural Studies Review* 18, no. 3 (2012): 67–84.

Albert the Great. *Questions Concerning Aristotle's* On Animals. Translated by Irven M. Resnick and Kenneth F. Kitchell Jr. Washington, DC: Catholic University of America Press, 2008.

Alemanno, Laura. "L'Accademia degli umoristi." *Roma moderna e contemporanea* 3 (1995): 97–120.

———. "Le 'Rime degli Accademici Umoristi.'" In *Letteratura italiana e utopia*, edited by Alberto Asor Rosa, 2:275–90. Rome: Riuniti, 1996.

Alessandrini, Anthony C., ed. *Frantz Fanon: Critical Perspectives*. New York: Routledge, 1999.

Allacci, Leone. *Symmikta, sive Opuscula Graeca et Latina, vetustiora & recentiora*. Venice, 1733.

Allegri, Luigi. "Introduzione allo studio degli apparati a Parma nel Settecento." In *La Parma in festa: Spettacolarità e teatro nel Ducato di Parma nel Settecento*, edited by Luigi Allegri and Renato di Benedetto, 13–34. Modena: Mucchi, 1987.

Álvarez, Ángel Medina. "Los atributos del Capón: Imagen histórica de los cantores castrados en España." *Música oral del sur: Revista internacional* 3 (1998): 9–29.

Ammianus, Marcellinus. *The Surviving Books of the History of Ammianus Marcellinus*. Vol. 1 of *Ammianus Marcellinus*, translated by John C. Rolfe. Loeb Classical Library 300. Cambridge, MA: Harvard University Press, 1935. https://penelope.uchicago.edu/Thayer/E/Roman/Texts/Ammian/home.html.

André, Naomi. *Voicing Gender: Castrati, Travesti, and the Second Woman in Early-Nineteenth-Century Italian Opera*. Bloomington: Indiana University Press, 2006.

Antolini, Bianca Maria. "La carriera di cantante e compositore di Loreto Vittori." *Studi musicali* 7 (1978): 141–88.

———. *Loreto Vittori: Musico spoletino*. Spoleto: Edizioni dell'Accademia Spoletino, 1984.

Antolini, Fabrizio. "Il 'lieve error di giovanil desio' del cavalier Loreto." *Spoletium: Revista de arte, storia, cultura* 23, no. 1 (1978): 70–75.

Arbuthnot, John. *The Devil to Pay at St. James's: Or, A Full and True Account of a Most Horrid and Bloody Battle between Madam Faustina and Madam Cuzzoni* [ . . . ] . London, 1727.

Archenholz, Johann Wilhelm von. *Mémoires concernant Christine, Reine de Suède, pour servir à l'éclaircissement de son règne et principalement de sa vie privée et aux évènements de l'histoire de son temps, civile et littéraire: suivi de deux ouvrages de cette savante Princesse, qui n'ont jamais été imprimés*. 4 vols. Amsterdam, 1751–1760.

———. *A Picture of Italy*. Translated by Joseph Trapp. 2 vols. London, 1791.

Aristotle. *Generation of Animals*. Translated by A. L. Peck. Loeb Classical Library 366. Cambridge, MA: Harvard University Press, 1942.

———. *On the Soul*. Translated by J. A. Smith. The Internet Classics Archive. Daniel C. Stevenson, Web Atomics, 2009. http://classics.mit.edu/Aristotle/soul.html.

———. *Politics*. Vol. 21 of *Aristotle in 23 Volumes*, translated by H. Rackham. Cambridge, MA: Harvard University Press, 1944. http://www.perseus.tufts.edu/hopper/text?doc=Perseus%3atext%3a1999.01.0058.

Arteaga, Stefano. *Le rivoluzioni del teatro musicale italiano dalla sua origine fino al presente*. 3 vols. Bologna, 1785.

Ascione, Francesco. *Le lodi e grandezze della Aguglia e Fontana di piazza Navona*. Rome: Francesco Cavalli, 1651.

Aspden, Suzanne. "'An Infinity of Factions': Opera in Eighteenth-Century Britain and the Undoing of Society." *Cambridge Opera Journal* 9, no. 1 (1997): 1–19.

Atkinson, James. *Medical Bibliography: A. and B*. London: John Churchill, 1834.

Austern, Linda Phyllis. "Nature, Culture, Myth, and Musician in Early Modern England." *Journal of the American Musicological Society* 51, no. 1 (1998): 1–49.

Averett, Matthew Knox. "'Redditus orbis erat': The Political Rhetoric of Bernini's Fountains in Piazza Barberini." *Sixteenth Century Journal* 45, no. 1 (2014): 3–24.

Azzolini, Monica. "Exploring Generation. A Context to Leonardo's Anatomies of the Female and Male Bodies." In *Leonardo da Vinci's Anatomical World: Language, Context and "Disegno,"* edited by Alessandro Nova and Domenico Laurenza, 79–97. Venice: Marsilio, 2011.

Bacon, Francis. *The Major Works*. Edited by Brian Vickers. Oxford: Oxford University Press, 2008.

———. *The New Organon* [*Novum organum*]. In *The Works of Francis Bacon*, edited and translated by James Spedding, Robert Leslie Ellis, and Douglas Denon Heath, 4:39–248. London: Longmans, 1858.

———. "Orpheus; or Philosophy." In *The Works of Francis Bacon, Baron of Verulam, Viscount St. Alban, and Lord High Chancellor of England*, edited and translated by James Spedding, Robert Ellis, and Douglas Heath, 6:720–22. London: Longmans, 1890.

———. *Sylva Sylvarum, or, A Natural History, in Ten Centuries* [ . . . ]. The Ninth and Last Edition. London, 1670.

———. *The Works of Lord Bacon: Philosophical Works*. London: Henry G. Bohn, 1855.

Baird, Julianne. "An 18th-Century Controversy about the Trill: Mancini v. Manfredini." *Early Music* 15, no. 1 (1987): 36–45.

Bajada, Joseph. *Sexual Impotence and the Contribution of Paolo Zacchia, 1584–1659*. Rome: Editrice Pontificia Università Gregoriana, 1988.

Baldinucci, Filippo. *The Life of Bernini*. University Park: Pennsylvania State University Press, 2007.

Banchieri, Adriano. *Conclusions for Playing the Organ (1609)*. Translated by Lee R. Garrett. Colorado Springs: Colorado College Music Press, 1982.

Barbieri, Patrizio. "'Chiavette' and Modal Transposition in Italian Practice (c. 1500–1837)." *Recercare* 3 (1991): 5–79.

———. "Giambattista Della Porta's 'Singing' Hydraulis and Other Expressive Devices for the Organ, c. 1560–1860." *Journal of the American Musical Instrument Society* 32 (2006): 145–67.

———. "Michele Todini's *Galleria Armonica*: Its Hitherto Unknown History." *Early Music* 30, no. 4 (November 2002): 565–82.

———. "The New Water Organ of the Villa d'Este, Tivoli." *Organ Yearbook* 33 (2004): 33–41.

———. "Organi e automati musicali idraulici de Villa d'Este a Tivoli." *L'organo* 24 (1986): 3–61.

———. "The Speaking Trumpet: Developments of della Porta's 'Ear Spectacles' (1589–1967)." *Studi musicali* 33, no. 1 (2004): 205–47.

———. *The World of the Castrati: The History of an Extraordinary Operatic Phenomenon*. Translated by Margaret Crosland. London: Souvenir, 1996.

Barbieri, Patrizio, and Ken Hurry. "Pietro della Valle: The 'Esthèr' Oratorio (1639) and Other Experiments in the 'Stylus Metabolicus.' With New Documents on Triharmonic Instruments." *Recercare* 19, no. 1/2 (2007): 73–124.

Barolsky, Paul. "Bernini and Ovid." *Source: Notes in the History of Art* 16, no. 1 (1996): 29–31.

Barthes, Roland. "The Grain of the Voice." In *Image, Music, Text*, edited and translated by Stephen Heath, 179–89. New York: Hill and Wang, 1977.

———. *The Responsibility of Forms: Critical Essays on Music, Art, and Representation*. Translated by Richard Howard. Berkeley: University of California Press, 1991.

———. *S/Z*. Translated by Richard Miller. New York: Farrar, Straus and Giroux, 1974.

Bartoli, Cecilia. "The Great Pretender: My Tribute to Farinelli, the Greatest Castrato." *Guardian*. January 23, 2020. https://www.theguardian.com/music/2020/jan/23/cecilia-bartoli-my-farinelli-album-countertenor-songs.

Bates, Eliot. "The Social Life of Musical Instruments." *Ethnomusicology* 56, no. 3 (2012): 363–95.

Beagon, Mary. "Situating Nature's Wonders in Pliny's *Natural History*." In "*Vita vigilia est*: Essays in Honour of Barbara Levick." Supplement, *Bulletin of the Institute of Classical Studies*, no. 100 (2007): 19–40.

Beck, Thomas E. "Gardens as a 'Third Nature': The Ancient Roots of a Renaissance Idea." *Studies in the History of Gardens & Designed Landscapes* 22, no. 4 (2002): 327–34.

Beecher, D. A. "Gianlorenzo Bernini's *The Impresario*: The Artist as the Supreme Trickster." *University of Toronto Quarterly* 53, no. 3 (1984): 236–47.

Behrend-Martínez, Edward. "Manhood and the Neutered Body in Early Modern Spain." *Journal of Social History* 38, no. 4 (2005): 1073–93.

Benjamin, Walter. *Selected Writings*. Vol. 1, *1913–1926*, edited by Marcus Bullock and Michael W. Jennings. Cambridge, MA: Harvard University Press, 1996.

Bennett, Jane. *Vibrant Matter: A Political Ecology of Things*. Durham, NC: Duke University Press, 2010.

Bentley, Jerry H. *Politics and Culture in Renaissance Naples*. Princeton, NJ: Princeton University Press, 2014.

Bergeron, Katherine. "The Castrato as History." *Cambridge Opera Journal* 8, no. 2 (1996): 167–84.

Bernini, Gian Lorenzo. *Fontana di Trevi: Commedia inedita*. Edited by Cesare D'Onofrio. Rome: Staderini, 1963.

———. *The Impresario (Untitled)*. Translated by Donald Beecher and Massimo Ciavolella. Ottawa, ON: Dovehouse Editions Canada, 1995.

Bernstein, Jane. *Print Culture and Music in Sixteenth-Century Venice*. New York: Oxford University Press, 2001.

———. "Print Culture, Music, and Early Modern Catholicism in Rome." In *Listening to Early Modern Catholicism: Perspectives from Musicology*, edited by Daniele V. Filippi and Michael J. Noone, 112–28. Boston: Brill, 2017.

Berry, Helen. *The Castrato and His Wife*. New York: Oxford University Press, 2011.

Beth, Wendy, ed. *The Automaton in English Renaissance Literature*. Burlington, VT: Ashgate, 2011.

Biagioli, Mario. *Galileo, Courtier: The Practice of Science in the Culture of Absolutism*. Chicago: University of Chicago Press, 1993.

Bianchi, Eric. "Scholars, Friends, Plagiarists: The Musician as Author in the Seventeenth Century." *Journal of the American Musicological Society* 70, no. 1 (2017): 61–128.

———. "Was Man Made for the Sabbath? Site, Space and Identity in Jesuits' Musical Life." In *The Grand Theater of the World: Music, Space, and the Performance of Identity in Early Modern Rome*, edited by Valeria De Lucca and Christine Jeanneret, 72–87. New York: Routledge, 2019.

Bianconi, Lorenzo. *Music in the Seventeenth Century*. Translated by David Bryant. New York: Cambridge University Press, 1987.

Bisaccioni, Maiolino. *Il cannocchiale per La finta pazza drama dello Strozzi. Dilineato da M.B. C. di G.* Venice, 1641.

Blackburn, Bonnie J., and Edward Lowinsky. "Luigi Zenobi and His Letter on the Perfect Musician." *Studi musicali* 22 (1993): 61–114.

Blackmer, Corinne E., and Patricia Juliana Smith, eds. *En travesti: Women, Gender Subversion, Opera*. New York: Columbia University Press, 1995.

Blake, Liza. "Dildos and Accessories: The Functions of Early Modern Strap-Ons." In *Ornamentalism: The Art of Renaissance Accessories*, edited by Bella Mirabella, 130–55. Ann Arbor: University of Michigan Press, 2011.

Bloechl, Olivia A. *Native American Song at the Frontiers of Early Modern Music*. New York: Cambridge University Press, 2008.

———. *Opera and the Political Imaginary in Old Regime France*. Chicago: University of Chicago Press, 2017.

Bontempi, Giovanni Andrea Angelini. *Historia musica*. Geneva: Minkoff, 1976. First published 1695 by L. Constantin (Perugia).

Boriaud, Jean-Yves. "La Poésie et le théâtre latins au Collegio Romano d'après les manuscrits du Fondo Gesuitico de la Bibliothèque Nationale Vittorio Emanuele II." *Mélanges de l'École française de Rome, Italie et Méditerranée* 102, no. 1 (1990): 77–96.

Borlik, Todd Andrew. "'More than Art': Clockwork Automata, the Extemporizing Actor, and the Brazen Head in Friar Bacon and Friar Bungay." In *The Automaton in English Renaissance Literature*, edited by Wendy Beth Hyman, 129–44. Burlington, VT: Ashgate, 2011.

Bottegari, Cosimo. *Il libro de canto e liuto / The Song and Lute Book*. Edited by Dinko Fabris and John Griffiths. Bologna: Arnaldo Forni, 2006.

Bottrigari, Ercole. *Il desiderio*. Venice: Ricciardo Amadino, 1594.

Bradner, Leicester. "The Latin Drama of the Renaissance (1340–1640)." *Studies in the Renaissance* 4 (1957): 31–54.

Braidotti, Rosi. *Nomadic Theory: The Portable Rosi Braidotti*. New York: Columbia University Press, 2011.

———. *The Posthuman*. Cambridge: Polity, 2013.

———. "Posthuman Humanities." *European Educational Research Journal* 12, no. 1 (2013): 1–19.

Bridgman, Nanie. "Giovanni Camillo Maffei et sa lettre sur le chant." *Revue de Musicologie* 38, no. 113 (1956): 3–34.

Brosius, Amy. "Courtesan Singers as Courtiers: Power, Political Pawns, and the Arrest of Virtuosa Nina Barcarola." *Journal of the American Musicological Society* 73, no. 2 (2020): 207–67.

———. "'Il Suon, Lo Sguardo, Il Canto': The Function of Portraits of Mid-Seventeenth-Century Virtuose in Rome." *Italian Studies* 63, no. 1 (2008): 17–39.

———. "Singers Behaving Badly: Rivalry, Vengeance, and the Singers of Cardinal Antonio Barberini." *Women and Music: A Journal of Gender and Culture* 19, no. 1 (2015): 45–53.

Brown, Patricia Fortini. "Seduction and Spirituality: The Ambiguous Roles of Music in Venetian Art." In *The Music Room in Early Modern France and Italy: Sound, Space, and Object*, edited by Deborah Howard and Laura Moretti, 19–36. Oxford: Oxford University Press, 2012.

Brummett, Palmira. "Placing the Ottomans in the Mediterranean World: The Question of Notables and Households." *Osmanlı Araştırmaları* 36 (2010): 77–96.

Burke, Jill. "Nakedness and Other Peoples: Rethinking the Italian Renaissance Nude." *Art History* 36, no. 4 (2013): 714–39.

Burke, Peter. "Varieties of Performance in Seventeenth-Century Italy." In *Performativity and Performance in Baroque Rome*, edited by Peter Gillgren and Mårten Snickare, 29–38. London: Routledge, 2017.

Burney, Charles. *An Eighteenth-Century Musical Tour in France and Italy*. Edited by Percy A. Scholes. London: Oxford University Press, 1959.

———. *A General History of Music: From the Earliest Ages to the Present Period*. Critical and historical notes by Frank Mercer. 2 vols. New York: Dover, 1957.

———. *The Present State of Music in France and Italy*. London, 1773.

———. *The Present State of Music in Germany, The Netherlands, and United Provinces*. 2nd ed. 2 vols. London, 1775.

Burroughs, Charles. "Opacity and Transparence: Networks and Enclaves in the Rome of Sixtus V." *RES: Anthropology and Aesthetics* 41, no. 1 (2002): 56–71.

Burrus, Virginia. *The Sex Lives of Saints: An Erotics of Ancient Hagiography*. Philadelphia: University of Pennsylvania Press, 2007.

Butler, Shane. *The Ancient Phonograph*. New York: Zone Books, 2015.

Bynum, Caroline Walker. *Fragmentation and Redemption: Essays on Gender and the Human Body in Medieval Religion*. New York: Zone Books, 1992.

Caballero, Carlo. "'A Wicked Voice': On Vernon Lee, Wagner, and the Effects of Music." *Victorian Studies* 35, no. 4 (Summer 1992): 385–408.

Caccini, Giulio. *Le nuove musiche*. Edited by H. Wiley Hitchcock. Madison, WI: A-R Editions, 1970.

Calaresu, Melissa. "From the Street to Stereotype: Urban Space, Travel and the Picturesque in Late Eighteenth-Century Naples." *Italian Studies* 62, no. 2 (2007): 189–203.

Calcagno, Mauro. *From Madrigal to Opera: Monteverdi's Staging of the Self*. Berkeley: University of California Press, 2012.

Campbell, Jill. *Natural Masques: Gender and Identity in Fielding's Plays and Novels*. Stanford, CA: Stanford University Press, 1995.

Canal, Pietro. "Della musica in Mantova: Notizie tratte principalmente dall'Archivio Gonzaga." *Memorie del Reale istituto Veneto di scienze lettere ed arti* 21 (1879): 655–744.

Carlino, Andrea. *Books of the Body: Anatomical Ritual and Renaissance Learning*. Chicago: University of Chicago Press, 1999.

Carter, Tim. *Monteverdi's Musical Theatre*. New Haven, CT: Yale University Press, 2002.

———. "Musical Sources." In *The Cambridge Companion to Monteverdi*, edited by John Whenham and Richard Wistreich, 20–30. New York: Cambridge University Press, 2007.

———. "*Possente Spirto*: On Taming the Power of Music." *Early Music* 21, no. 4 (1993): 517–24.

———. "Singing Orfeo: On the Performers of Monteverdi's First Opera." *Recercare* 11 (1999): 75–118.

Casserius, Julius. *De vocis auditusque organis historia anatomica*. Ferrara, 1601.

———. "The Three Books of the First Treatise on the Anatomical History of the Larynx, the Organ of Voice." Translated by Malcolm H. Hast and E. B. Holtsmark. Supplement, *Acta Oto-Laryngologica* 68, S261 (1969): 9–15.

Cather, Willa. *The Song of the Lark*. Boston: Houghton Mifflin, 1915.

Cavalieri, Emilio de'. *Rappresentatione di anima, et di corpo* (1600). Edited by Murray C. Bradshaw. Middleton, WI: American Institute of Musicology, 2007.

Cavallo, Sandra. *Artisans of the Body in Early Modern Italy: Identities, Families and Masculinities*. Manchester: Manchester University Press, 2007.

Cavarero, Adriana. *For More Than One Voice: Toward a Philosophy of Vocal Expression*. Translated by Paul A. Kottman. Stanford, CA: Stanford University Press, 2005.

Chafe, Eric T. *Monteverdi's Tonal Language*. New York: Schirmer Books, 1992.

Chakrabarty, Dipesh. "The Climate of History: Four Theses." *Critical Inquiry* 35, no. 2 (2009): 197–222.

Chian, Howard. *After Eunuchs: Science, Medicine, and the Transformation of Sex in Modern China*. New York: Columbia University Press, 2018.

Christiansen, Keith. "Music and Painting in Cardinal Del Monte's Household." *Metropolitan Museum Journal* 26 (1991): 213–27.

Christopoulos, John. "Abortion and the Confessional in Counter-Reformation Italy." *Renaissance Quarterly* 65, no. 2 (2012): 443–84.

Chua, Daniel K. L. *Absolute Music and the Construction of Meaning*. New York: Cambridge University Press, 1999.

———. "Untimely Reflections on Operatic Echoes: How Sound Travels in Monteverdi's *L'Orfeo* and Beethoven's *Fidelio* with a Short Instrumental Interlude." *Opera Quarterly* 21, no. 4 (2005): 573–96.

Claudian. *Against Eutropius*. Translated by Maurice Platnauer. Loeb Classical Library 135. Cambridge, MA: Harvard University Press, 1922.

Clementi, Filippo. *Il carnevale romano nelle cronache contemporanee*. Rome, 1899.

Clement of Alexandria. *Exhortation to the Greeks. The Rich Man's Salvation. To the Newly Baptized*. Translated by G. W. Butterworth. Loeb Classical Library 92. Cambridge, MA: Harvard University Press, 1919.

Cohen, Elizabeth S. "Moving Words: Everyday Oralities and Social Dynamics in Roman Trials *circa* 1600." In *Voices and Texts in Early Modern Italian Society*, edited by Stefano Dall'Aglio, Brian Richardson, and Massimo Rospocher, 69–83. London: Routledge, 2017.

———. "Open City: An Introduction to Gender in Early Modern Rome." *I Tatti Studies in the Italian Renaissance* 17, no. 1 (2014): 35–54.

Coleman, Kathleen M. "Fatal Charades: Roman Executions Staged as Mythological Enactments." *Journal of Roman Studies* 80 (1990): 44–73.

Collins, Jeffrey, and Anthony Grafton. "Obelisks and Empires of the Mind." *American Scholar* 71, no. 1 (2002): 123–27.

Collier, William. "New Light on Bernini's Neptune and Triton." *Journal of the Warburg and Courtauld Institutes* 31(1968): 438–40.

Colón, Cristóbal. *Textos y documentos completos*. Edited by Consuelo Varela. Madrid: Alianza, 1992.

Columbus, Christopher. *The Four Voyages of Christopher Columbus*. Edited and translated by J. M. Cohen. New York: Penguin, 2004.

Conestabile, Giancarlo. *Notizie biografiche di Baldassarre Ferri, musico celebratissimo*. Perugia: Bartelli, 1846.

Connolly, Joy. "Virile Tongues: Rhetoric and Masculinity." In *A Companion to Roman Rhetoric*, edited by William Dominik and Jon Hall, 83–97. Malden, MA: Blackwell, 2007.

Connors, Joseph. "Alliance and Enmity in Roman Baroque Urbanism." *Römisches Jahrbuch der Bibliotheca Hertziana* 25 (1989): 207–94.

Cope, Jackson I. "Bernini and Roman *Commedie Ridicolose*." *PMLA* 102, no. 2 (1987): 177–86.

Costa, Margherita. *La tromba di Parnaso*. Paris: Cramoisy, 1647.

Cottom, Daniel. "The Work of Art in the Age of Mechanical Digestion." *Representations* 66 (1999): 52–74.

Cox, Christoph, and Daniel Warner. "Introduction: Music and the New Audio Culture." In *Audio Culture: Readings in Modern Music*, edited by Christoph Cox and Daniel Warner, xiii–xvii. New York: Continuum, 2004.

Cranz, Ferdinand Edward, and Paul Oskar Kristeller, eds. *Catalogus translationum et commentariorum: Mediaeval and Renaissance Latin Translations and Commentaries*. Vol. 4. Washington, DC: Catholic University of America Press, 1980.

Crawford, Katherine. "Desiring Castrates, or How to Create Disabled Social Subjects." *Journal for Early Modern Cultural Studies* 16, no. 2 (2016): 59–90.

———. *Eunuchs and Castrati: Disability and Normativity in Early Modern Europe*. New York: Routledge, 2018.

———. "Privilege, Possibility, and Perversion: Rethinking the Study of Early Modern Sexuality." *Journal of Modern History* 78, no. 2 (2006): 412–33.

Cremades, Fernando Checa, and Laura Fernández-González. *Festival Culture in the World of the Spanish Habsburgs*. New York: Routledge, 2016.

Croce, Benedetto. *Un paradiso abitato da diavoli*. Edited by Giuseppe Galasso. Milan: Adelphi, 2006.

Croce, Giovanni Andrea della. *Cirugia universale e perfetta di tutte le parti pertinenti all'ottimo Chirurgo*. Venice: Ziletti, 1574.

Crutchfield, Will. "Giovanni Battista Velluti e lo sviluppo della melodia romantica." *Bollettino del Centro Rossiniano di Studi* 54 (2014): 9–83.

Cuneo, Pia F., ed. *Animals and Early Modern Identity*. Burlington, VT: Ashgate, 2014.

Cunningham, Andrew. *The Anatomical Renaissance: The Resurrection of the Anatomical Projects of the Ancients*. New York: Routledge, 2016.

Cusick, Suzanne G. "Gendering Modern Music: Thoughts on the Monteverdi-Artusi Controversy." *Journal of the American Musicological Society* 46, no. 1 (1993): 1–25.

———. "On Musical Performances of Gender and Sex." In *Audible Traces: Gender, Identity, and Music*, edited by Elaine Barkin and Lydia Hamessley, 25–49. Los Angeles: Carciofoli, 1999.

———. "Response: 'This Song Is for You.'" *Journal of the American Musicological Society* 66, no. 3 (2013): 861–66.

———. "'There Was Not One Lady Who Failed to Shed a Tear': Arianna's Lament and the Construction of Modern Womanhood." *Early Music* 22, no. 1 (1994): 21–44.

Cypess, Rebecca. *Curious and Modern Inventions: Instrumental Music as Discovery in Galileo's Italy*. Chicago: University of Chicago Press, 2016.

———. "Instrumental Music and 'Conversazione' in Early Seicento Venice: Biagio Marini's 'Affetti Musicali' (1617)." *Music & Letters* 93, no. 4 (2013): 453–78.

D'Accone, Frank. "Cardinal Chigi and Music Redux." In *Music Observed: Studies in Memory of William C. Holmes*, edited by Colleen Reardon and Susan Helen Parisi, 65–100. Warren, MI: Harmonie Park Press, 2004.

da Costa, Palmira Fontes. "Anatomical Expertise and the Hermaphroditic Body." *Spontaneous Generations: A Journal for the History and Philosophy of Science* 1, no. 1 (2007): 78–85.

———. "The Medical Understanding of Monstrous Births at the Royal Society of London during the First Half of the Eighteenth Century." *History and Philosophy of the Life Sciences* 26, no. 2 (2004): 157–75.

Dame, Joke. "Unveiled Voices: Sexual Difference and the Castrato." In *Queering the Pitch: The New Gay and Lesbian Musicology*, edited by Philip Brett, Elizabeth Wood, and Gary C. Thomas, 2nd ed., 139–54. New York: Routledge, 2006.

d'Ancillon, Charles. *Eunuchism Display'd. Describing All the Different Sorts of Eunuchs* [ . . . ]. London, 1718. Originally published as *Traité des eunuques, dans lequel on explique toutes les différentes sortes d'eunuque* [ . . . ]. Sweden: Marknadsdomstolen, 1707.

Darby, Rosemarie. "The Liturgical Music of the Chiesa Nuova, Rome (1575–1644)." PhD diss. University of Manchester, 2018.

Darwin, Charles. *The Descent of Man, and Selection in Relation to Sex*. 2nd ed. London: John Murray, 1875.

Daston, Lorraine. "Intelligences: Angelic, Animal, Human." In *Thinking with Animals: New Perspectives on Anthropomorphism*, edited by Lorraine Daston and Gregg Mitman, 37–58. New York: Columbia University Press, 2005.

———. "Introduction: Speechless." In *Things That Talk: Object Lessons from Art and Science*, edited by Lorraine Daston, 9–26. New York: Zone Books, 2004.

———. "Marvelous Facts and Miraculous Evidence in Early Modern Europe." *Critical Inquiry* 18, no. 1 (1991): 93–124.

Daston, Lorraine, and Katharine Park. "The Hermaphrodite and the Orders of Nature: Sexual Ambiguity in Early Modern France." In *Premodern Sexualities*, edited by Louise Fradenburg and Carla Freccero, 117–36. New York: Routledge, 1996.

———. *Wonders and the Order of Nature, 1150–1750*. New York: Zone Books, 1998.

Davies, James Q. *Romantic Anatomies of Performance*. Berkeley: University of California Press, 2014.

———. "'Veluti in speculum': The Twilight of the Castrato." *Cambridge Opera Journal* 17, no. 3 (2005): 271–301.

Dean, Winton. "The Castrati in Opera by Angus Heriot." *Music & Letters* 38, no. 1 (1957): 82–84.

de Brosses, Charles. *Le président de Brosses en Italie: Lettres familières écrites d'Italie en 1739 et 1740*. Paris: Didier, 1858.

———. *Le président de Brosses en Italie: Lettres familières écrites d'Italie en 1739 et 1740*. Edited by M. R. Colomb. 2nd ed. 2 vols. Paris: È. Perrin, 1885.

De Ceglia, Francesco Paolo. "The Woman Who Gave Birth to a Dog: Monstrosity and Bestiality in *Quaestiones medico-legales* by Paolo Zacchia." *Medicina nei secoli* 26, no. 1 (2014): 117–44.

Deleuze, Gilles. *The Fold: Leibniz and the Baroque*. Translated by Tom Conley. Minneapolis: University of Minnesota Press, 1993.

Deleuze, Gilles, and Félix Guattari. *A Thousand Plateaus: Capitalism and Schizophrenia*. Translated by Brian Massumi. Minneapolis: University of Minnesota Press, 1987.

Dell'Antonio, Andrew. *Listening as Spiritual Practice in Early Modern Italy*. Berkeley: University of California Press, 2011.

———. *Syntax, Form, and Genre in Sonatas and Canzonas, 1621–1635*. Lucca: Libreria Musicale Italiana, 1997.

Della Porta, Giambattista. *Natural Magick [ . . . ] Wherein Are Set Forth All the Riches and Delights of the Natural Sciences*. London, 1658.

Dell'Arco, Maurizio Fagiolo, and Silvia Carandini. *La festa barocca*. Rome: De Luca, 1997.

———. *L'effimero Barocco: Strutture della festa nella Roma del '600*. 2 vols. Rome: Bulzoni, 1978.

De Lucca, Valeria. *"Dalle sponde del tebro alle rive dell'adria": Maria Mancini and Lorenzo Onofrio Colonna's Patronage of Music and Theater between Rome and Venice (1659–1675)*. Princeton, NJ: Princeton University Press, 2009.

———. "L'Alcasta and the Emergence of Collective Patronage in Mid-Seventeenth-Century Rome." *Journal of Musicology* 28, no. 2 (2011): 195–230.

———. *The Politics of Princely Entertainment: Music and Spectacle in the Lives of Lorenzo Onofrio and Maria Mancini Colonna*. New York: Oxford University Press, 2020.

De Lucca, Valeria, and Christine Jeanneret, eds. *The Grand Theater of the World: Music, Space, and the Performance of Identity in Early Modern Rome*. New York: Routledge, 2019.

Dennis, John. *An Essay on the Opera's after the Italian Manner: Which Are about to Be Establish'd on the English Stage; With Some Reflections on the Damage Which They May Bring to the Publick*. London, 1706.

Derrida, Jacques. "Archive Fever: A Freudian Impression." Translated by Eric Prenowitz. *Diacritics* 25, no. 2 (Summer 1995): 9–63.

———. *Of Grammatology*. Translated by Gayatri Chakravorty Spivak. Baltimore: Johns Hopkins University Press, 1976.

Descartes, René. *The Philosophical Writings of Descartes*. Translated by John Cottingham, Robert Stoothoff, and Dugald Murdoch. 3 vols. New York: Cambridge University Press, 1985.

Deslauriers, Marguerite. "Marinella and Her Interlocutors: Hot Blood, Hot Words, Hot Deeds." *Philosophical Studies* 174, no. 10 (2017): 2525–37.

Desler, Anne. "From Castrato to Bass: The Late Roles of Nicolò Grimaldi 'Nicolini.'" In *Gender, Age and Musical Creativity*, edited by Catherine Haworth and Lisa Colton, 61–80. London: Routledge, 2015.

———. "'Il novello Orfeo' Farinelli: Vocal Profile, Aesthetics, Rhetoric." PhD diss. University of Glasgow, 2014.

———. "'The Little That I Have Done Is Already Gone and Forgotten': Farinelli and Burney Write Music History." *Cambridge Opera Journal* 27, no. 3 (2015): 215–38.

DeVun, Leah. "Erecting Sex: Hermaphrodites and the Medieval Science of Surgery." *Osiris* 30, no. 1 (2015): 17–37.

———. *The Shape of Sex: Nonbinary Gender from Genesis to the Renaissance*. New York: Columbia University Press, 2021.

Diderot, Denis. *Oeuvres complètes de Diderot, revues sur les éditions originales*. Edited by J. Assézat. Vol. 2. Paris: Garnier Frères, 1875.

———. *Rameau's Nephew and Other Works*. Translated by Jacques Barzun and Ralph H. Bowen. Indianapolis, IN: Hackett, 2001.

Di Giacomo, Salvatore. *I quattro antichi Conservatorii di musica di Napoli*. Vol. 1, *Il conservatorio di Sant'Onofrio a Capuana e quello di S. Maria della Pietà dei Turchini*. Milan: Sandron, 1924.

———. *I quattro antichi Conservatorii di musica di Napoli*. Vol. 2, *Il conservatorio dei Poveri di Gesù Cristo e quello di S. Maria di Loreto*. Milan: Sandron, 1928.

di Gioya, Antonio Bernal. *Copiosissimo discorso della fontana, e guglia eretta in Piazza Navona per ordine della Santità di Nostro Signore Innocentio X*. Rome, 1651.

Dinshaw, Carolyn. "Chaucer's Queer Touches / A Queer Touches Chaucer." *Exemplaria* 7, no. 1 (1995): 75–92.

Dolan, Emily I. "E. T. A. Hoffmann and the Ethereal Technologies of 'Nature Music.'" *Eighteenth Century Music* 5, no. 1 (2008): 7–26.

———. "Toward a Musicology of Interfaces." *Keyboard Perspectives* 5 (2012): 1–12.

Dolar, Mladen. *A Voice and Nothing More*. Cambridge: MIT Press, 2006.

Dominicis, Giulia de. "I teatri di Roma nell'età di Pio VI." *Archivio della Società romana della storia patria* 46 (1923): 49–243.

———. "The Roman Theatres in the Age of Pius VI." *Theatre History Studies* 21 (2001): 81–86.

Donadio, Rachel. "A Museum Display of Galileo Has a Saintly Feel." *New York Times*, July 23, 2010. https://www.nytimes.com/2010/07/23/world/europe/23galileo.html.

Donaldson, I. M. "Peter Lowe: Information and Misinformation about a Scots Surgeon of the Sixteenth Century." *Journal of the Royal College of Physicians of Edinburgh* 44, no. 2 (2014): 170–79.

Doni, Anton Francesco. *Le ville del Doni*. Bologna, 1566.

Doni, Giovanni Battista. *Annotazioni sopra il compendio de' generi e de' modi della musica.* Rome: Andrea Fei, 1640.

———. *De' trattati di musica.* Edited by A. F. Gori. 2 vols. Florence, 1763.

D'Onofrio, Cesare. *Le fontane di Roma.* Rome: Romana Società, 1986.

Drake, Stillman. "Music and Philosophy in Early Modern Science." In *Music and Science in the Age of Galileo*, edited by Victor Coelho, 3–16. Dordrecht: Kluwer Academic, 1992.

Duffin, Jacalyn. "Questioning Medicine in Seventeenth-Century Rome: The Consultations of Paolo Zacchia." *Canadian Bulletin of Medical History* 28, no. 1 (2011): 149–70.

Duffin, Ross W. "Cipriano de Rore, Giovanni Battista Benedetti, and the Just Tuning Conundrum." *Journal of the Alamire Foundation* 9, no. 2 (2017): 239–66.

———. "Just Intonation in Renaissance Theory and Practice." *Music Theory Online* 12 (2006). http://www.mtosmt.org/issues/mto.06.12.3/mto.06.12.3.duffin.html.

Durling, Richard J. "A Chronological Census of Renaissance Editions and Translations of Galen." *Journal of the Warburg and Courtauld Institutes* 24 (1961): 230–305.

Dussel, Enrique. "The Invention of the Americas: Eclipse of 'the Other' and the Myth of Modernity." Translated by Michael D. Barber. New York: Continuum, 1995.

Dyer, Frank Lewis, and Thomas Commerford Martin. *Edison, His Life and Inventions.* New York: Harper and Brothers, 1910.

Eamon, William. *Science and the Secrets of Nature: Books of Secrets in Medieval and Early Modern Culture.* Princeton, NJ: Princeton University Press, 1994.

Eidsheim, Nina Sun. *Sensing Sound: Singing and Listening as Vibrational Practice.* Durham, NC: Duke University Press, 2015.

Eidsheim, Nina Sun, and Katherine Meizel, eds. *The Oxford Handbook of Voice Studies.* New York: Oxford University Press, 2019.

Ellis, Harold. "Peter Lowe: A Father Figure of the Royal College of Physicians and Surgeons of Glasgow." *British Journal of Hospital Medicine* 71, no. 8 (2010): 474.

Ellison, James. *George Sandys: Travel, Colonialism, and Tolerance in the Seventeenth Century.* Rochester, NY: D. S. Brewer, 2002.

Elmi, Giuseppe. *Relazione de' fuochi artificiati e feste fatte in Roma per la Coronazione del novello Cesare Leopoldo Primo dall'Eminentissimo, e Reverendissimo Sig. Cardinal Colonna Protettore del S. Romano Imperio* [ . . . ]. Rome: Francesco Cavalli, 1658.

Engelhart, Richard Eric. "Domenico Mazzocchi's 'Dialoghi e Sonetti' and 'Madrigali a cinque voci' (1638): A Modern Edition with Biographical Commentary and New Archival Documents." PhD diss., Kent State University, 1987.

Enterline, Lynn. *The Rhetoric of the Body from Ovid to Shakespeare.* New York: Cambridge University Press, 2000.

Erlmann, Veit. *Reason and Resonance: A History of Modern Aurality.* New York: Zone Books, 2010.

Fabbri, Paolo. *Monteverdi.* Translated by Tim Carter. New York: Cambridge University Press, 1994.

Fabbri, Paolo, and and Angelo Pompilio, eds. *Il Corago: Overo alcune osservazioni per metter bene in scena le composizioni drammatiche.* Firenze: Olschki, 1983.

Fabian, Johannes. *Time and the Other: How Anthropology Makes Its Object.* New York: Columbia University Press, 1983.

Fabricius ab Aquapendente, Hieronymus. *De brutorum loquela.* Venice, 1599.

Fabris, Dinko. *Music in Seventeenth-Century Naples: Francesco Provenzale (1624–1704).* Burlington, VT: Ashgate, 2007.

Fahrner, Robert and William Kleb. "The Theatrical Activity of Gianlorenzo Bernini." *Educational Theatre Journal* 25, no. 1 (1973): 5–14.

Fausti, Luigi. "Un'autobiografia poetica di Loreto Vittori." *Atti dell'Accademia spoletina, 1920–22* (1922): 158–68.

Favaro, Antonio. "Galileo astrologo secondo documenti editi e inediti: Studi e ricerche." *Mente e cuore: Periodico mensile di scienze, letteratura e cose scolastiche* 8, no. 3 (1881): 99–108.

Feldman, Martha. "Castrato Acts." In *The Oxford Handbook of Opera*, edited by Helen M. Greenwald, 395–418. Oxford: Oxford University Press, 2014.

———. "The Castrato as a Rhetorical Figure." In *Rhetoric and Drama*, edited by Daniel Scott Mayfield, 71–96. Berlin: De Gruyter, 2017.

———. *The Castrato: Reflections on Natures and Kinds*. Oakland: University of California Press, 2015.

———. *City Culture and the Madrigal at Venice*. Berkeley: University of California Press, 1995.

———, ed. "Colloquy: Why Voice Now?" *Journal of the American Musicological Society* 68, no. 3 (2015): 653–85.

———. "Denaturing the Castrato." *Opera Quarterly* 24, no. 3/4 (2008): 178–99.

———. *Opera and Sovereignty: Transforming Myths in Eighteenth-Century Italy*. Chicago: University of Chicago Press, 2007.

———. "Red Hot Voice." In *The Castrato: Reflections on Natures and Kinds*, 79–132. Oakland: University of California Press, 2015.

———. "Strange Births and Surprising Kin: The Castrato's Tale." In *Italy's Eighteenth Century: Gender and Culture in the Age of the Grand Tour*, edited by Paula Findlen, Wendy Wassyng Roworth, and Catherine M. Sama, 175–215. Stanford, CA: Stanford University Press, 2009.

Feldman, Martha, and Judith T. Zeitlin, eds. *The Voice as Something More: Essays toward Materiality*. Chicago: University of Chicago Press, 2019.

Fenlon, Iain. "Church Reform and Devotional Music in Sixteenth-Century Rome: The Influence of Lay Confraternities." In *Forms of Faith in Sixteenth-Century Italy*, edited by Abigail Brundin and Matthew Treherne, 229–46. Burlington, VT: Ashgate, 2009.

———. "Monteverdi's Mantuan *Orfeo*: Some New Documentation." *Early Music* 12, no. 2 (May 1984): 163–72.

Ferguson, Gary. "Early Modern Transitions: From Montaigne to Choisy." *L'esprit créateur* 53, no. 1 (2013): 145–57.

Ferrein, Antoine. *Mémoire de l'Académie royale des sciences*. Paris, 1754.

Ferri, Baldassare and Biancamaria Brumana. *"Il pianto de' cigni in morte della fenice de' musici": Poesie per Baldassare Ferri e nuove ipotesi sulla carriera del cantante*. Perugia: Deputazione di storia patria per l'Umbria, 2010.

Filippi, Bruna. "The Orator's Performance: Gesture, Word, and Image in Theatre at the Collegio Romano." In *The Jesuits 2: Cultures, Sciences, and the Arts, 1540–1773*, edited by Thomas Frank Kennedy, Steven J. Harris, Gauvin Alexander Bailey, and John W. O'Malley, 512–29. Toronto: University of Toronto Press, 2006.

Findlen, Paula, ed. *Athanasius Kircher: The Last Man Who Knew Everything*. New York: Routledge, 2004.

———. *Possessing Nature: Museums, Collecting, and Scientific Culture in Early Modern Italy*. Berkeley: University of California Press, 1994.

Findlen, Paula, Michelle Fontaine, and Duane J. Osheim, eds. *Beyond Florence: The Contours of Medieval and Early Modern Italy*. Stanford, CA: Stanford University Press, 2003.

Finucci, Valeria. *The Manly Masquerade: Masculinity, Paternity, and Castration in the Italian Renaissance*. Durham, NC: Duke University Press, 2003.

———. *The Prince's Body: Vincenzo Gonzaga and Renaissance Medicine*. Cambridge, MA: Harvard University Press, 2015.

———. "'There's the Rub': Searching for Sexual Remedies in the New World." *Journal of Medieval and Early Modern Studies* 38, no. 3 (2008): 523–57.

Fiorani, Francesca. "Post-Tridentine 'Geographia Sacra': The Galleria delle carte geografiche in the Vatican Palace." *Imago mundi* 48, no. 1 (1996): 124–48.

Fleming, James Rodger. *Historical Perspectives on Climate Change*. New York: Oxford University Press, 1998.

Fontana, Carlo, Pietro Santi Bartoli, and Teresa Del Po. *Risposta del Signor Carlo Fontana alla lettera dell'Illustrissimo Signor Ottavio Castiglioni*. Rome: Angelo Bernabò, 1668.

Ford, Terence. "Andrea Sacchi's 'Apollo Crowning the Singer Marc Antonio Pasqualini.'" *Early Music* 12, no. 1 (1984): 79–84.

Foreman, Edward. *The Art of Bel Canto in the Italian Baroque: A Study of the Original Sources*. Minneapolis: Pro Musica Press, 2006.

Foreman, Edward Vaught. "A Comparison of Selected Italian Vocal Tutors of the Period circa 1550 to 1800." DMA thesis, University of Illinois, 1969.

Foucault, Michel. *The History of Sexuality: An Introduction*. Translated by Robert Hurley. New York: Vintage, 1990.

Franchi, Saverio. "Osservazioni sulla scenografia dei melodrammi romani nella prima metà del seicento." In *Musica e immagine: Tra iconografia e mondo dell'opera: Studi in onore di Massimo Bogianckino*, edited by Biancamaria Brumana and Galliano Ciliberti, 151–75. Florence: Olschki, 1993.

Frandsen, Mary E. "*Eunuchi conjugium*: The Marriage of a Castrato in Early Modern Germany." *Early Music History* 24 (2005): 53–124.

Frank, Felicia Miller. *The Mechanical Song: Women, Voice, and the Artificial in Nineteenth-Century French Narrative*. Stanford, CA: Stanford University Press, 1995.

Fraschetti, Stanislao. *Il Bernini, la sua vita, la sua opera, il suo tempo*. Milan: Hoepli, 1900.

Freccero, Carla. "Queer Times." *South Atlantic Quarterly* 106, no. 3 (2007): 485–94.

Freeman, Elizabeth. "Packing History, Count(er)ing Generations." *New Literary History* 31, no. 4 (Autumn 2000): 727–44.

———. "Time Binds, or, Erotohistoriography." *Social Text* 23, no. 3/4 (2005): 57–68.

Freitas, Roger. "The Eroticism of Emasculation: Confronting the Baroque Body of the Castrato." *Journal of Musicology* 20, no. 2 (2003): 196–249.

———. *Portrait of a Castrato: Politics, Patronage, and Music in the Life of Atto Melani*. Cambridge: Cambridge University Press, 2009.

———. "Sex without Sex: An Erotic Image of the Castrato Singer." In *Italy's Eighteenth Century: Gender and Culture in the Age of the Grand Tour*, edited by Paula Findlen, Wendy Wassyng Roworth, and Catherine M. Sama, 203–15. Stanford, CA: Stanford University Press, 2009.

———. "Singing and Playing: The Italian Cantata and the Rage for Wit," *Music & Letters* 82, no. 4 (2001): 509–42.

Freud, Sigmund. *The Standard Edition of the Complete Psychological Works of Sigmund Freud*. Edited by James Strachey et al. 24 vols. London: Hogarth Press, 1953–1974.

Fudge, Erica, ed. *Renaissance Beasts: Of Animals, Humans, and Other Wonderful Creatures*. Urbana: University of Illinois Press, 2004.

Fudge, Erica, Ruth Gilbert, and Susan Wiseman, eds. *At the Borders of the Human: Beasts, Bodies and Natural Philosophy in the Early Modern Period*. New York: Macmillan; St. Martin's, 1999.

Gadelrab, Sherry Sayed. "Discourses on Sex Differences in Medieval Scholarly Islamic Thought." *Journal of the History of Medicine and Allied Sciences* 66, no. 1 (2010): 40–81.

Gagné, John. "Emotional Attachments: Iron Hands, Their Makers, and Their Wearers, 1450–1600." In *Feeling Things: Objects and Emotions through History*, edited by Stephanie Downes, Sally Holloway, and Sarah Randles, 133–53. Oxford: Oxford University Press, 2018.

Galen. *On Prognosis*. Translated and edited by Vivian Nutton. Corpus medicorum Graecorum, vol. 5, pt. 8, fasc. 1. Berlin: Akademie, 1979).

———. *On Semen*. Edited and translated by Phillip DeLacy. Corpus medicorum Graecorum, vol. 5, pt. 3, fasc. 1. Berlin: Akademie, 1992.

———. *On the Usefulness of the Parts of the Body*. Translated by Margaret Tallmadge May. 2 vols. Ithaca, NY: Cornell University Press, 1968.

Galilei, Galileo. *Dialogues Concerning Two New Sciences by Galileo Galilei*. Translated by Henry Crew and Alfonso de Salvio. New York: Macmillan, 1914.

———. *Discoveries and Opinions of Galileo*. Edited and translated by Stillman Drake. Garden City, NY: Doubleday Anchor Books, 1957.

———. *Opere di Galileo Galilei*. Edited by Antonio Favaro. 20 vols. Florence: Le Monnier, 1890.

———. *Sidereus nuncius, or The Sidereal Messenger*. Edited and translated by Albert van Helden. Chicago: University of Chicago Press, 1989.

Galilei, Galileo, and Christoph Scheiner. *On Sunspots*. Translated and introduced by Eileen Reeves and Albert Van Helden. Chicago: University of Chicago Press, 2010.

Galilei, Vincenzo. *Fronimo Dialogo / Di Vincentio Galilei Fiorentino / Nel Quale Si Contengono Le Vere*. Venice: Girolamo Scotto, 1568.

Gallico, Claudio. "La 'Querimonia' di Maddalena di Domenico Mazzocchi e l' interpretazione di Loreto Vittori." *Collectanea Historiae Musicae* 4 (1966): 133–47.

———. *Sopra li fondamenti della verità: Musica italiana fra XV e XVII secolo*. Rome: Bulzoni, 2001.

Galluzzi, Paolo. *The Italian Renaissance of Machines*. Cambridge, MA: Harvard University Press, 2020.

Gamba, Antonio. "Il primo teatro anatomico stabile in Padova non fu quello di Fabrici d'Acquapendente." *Atti e memorie dell'Accademia patavina si scienze, lettere ed arti*, pt. 3, 99 (1986–1987): 157–61.

Garzoni, Tommaso. *La piazza universale di tutte le professioni del mondo*. Venice, 1610.

Gasparri, Petri, ed. *Codicis Iuris Canonici Fontes*. 9 vols. Rome: Typis Polyglottis Vaticanis, 1947.

Gattei, Stefano. "Galileo's Legacy: A Critical Edition and Translation of the Manuscript of Vincenzo Viviani's *Grati Animi Monumenta*." *British Journal for the History of Science* 50, no. 2 (2017): 181–228.

Gay, Peter, ed. *The Freud Reader*. New York: W. W. Norton, 1989.

Gensini, Stefano. "Il *De brutorum loquela* di Girolamo Fabrici d'Acquapendente." *Bruniana & Campanelliana* 17, no. 1 (2011): 163–74.

Gerbino, Giuseppe. "The Quest for the Soprano Voice: Castrati in Renaissance Italy." *Studi musicali* 33, no. 2 (2004): 303–57.

Gharipour, Mohammad, ed. *Gardens of Renaissance Europe and the Islamic Empires: Encounters and Confluences*. University Park: Pennsylvania State University Press, 2017.

Gibson, Thomas. "The Prostheses of Ambroise Paré." *British Journal of Plastic Surgery* 8 (1955): 3–8.

Gibson, William. *Neuromancer*. New York: Berkley Publishing Group, 1989.

Gigli, Giacinto. *Diario di Roma*. Edited by Manlio Barberito. 2 vols. Rome: Colombo, 1994.

Giglioni, Guido. "'Bolognan Boys Are Beautiful, Tasteful and Mostly Fine Musicians': Cardano on Male Same-Sex Love and Music." In *The Sciences of Homosexuality in Early Modern Europe*, edited by Kenneth Borris and George Rousseau, 201–20. New York: Routledge, 2008.

———. "'If You Don't Feel Pain, You Must Have Lost Your Mind': The Early Modern Fortunes of a Hippocratic Aphorism." In *Et Amicorum: Essays on Renaissance Humanism and Philosophy*, edited by Anthony Ossa-Richardson and Margaret Meserve, 313–37. Boston: Brill, 2018.

———. "*Musicus puer*: A Note on Cardano's Household and the Dangers of Music." *Bruniana & Campanelliana* 11, no. 1 (2005): 83–88.

Godbarge, Clément. "Hippocrates for Princes: Ippolito de' Medici's *Retratti d'aphorismi*." In *Renaissance Rewritings*, edited by Helmut Pfeiffer, Irene Fantappiè, and Tobias Roth, 143–56. Berlin: De Gruyter, 2017.

Gómez, Nicolás Wey. *The Tropics of Empire: Why Columbus Sailed South to the Indies*. Cambridge, MA: MIT Press, 2008.

Gordon, Bonnie. *Monteverdi's Unruly Women: The Power of Song in Early Modern Italy*. New York: Cambridge University Press, 2004.

———. "Orfeo's Machines." *Opera Quarterly* 24, no. 3/4 (2008): 200–22.

Gorman, Michael John. "Between the Demonic and the Miraculous: Athanasius Kircher and the Baroque Culture of Machines." In *The Great Art of Knowing: The Baroque Encyclopedia of Athanasius Kircher*, edited by Daniel Stolzenberg, 1000–1012. Stanford, CA: Stanford University Libraries, 2001.

———. *The Scientific Counter-Revolution: The Jesuits and the Invention of Modern Science*. New York: Bloomsbury, 2020.

Gorman, Michael John, and Nick Wilding. "Technica Curiosa: The Mechanical Marvels of Kaspar Schott (1608–1666)." In *La technica curiosa di Kaspar Schott*, edited by Maurizio Sonnino, 253–79. Rome: Edizione dell'Elefante, 2000.

Gouk, Penelope M. "Acoustics in the Early Royal Society 1660–1680." *Notes and Records of the Royal Society of London* 36, no. 2 (1982): 155–75.

———. *Music, Science, and Natural Magic in Seventeenth-Century England*. New Haven, CT: Yale University Press, 1999.

———. "Music in Francis Bacon's Natural Philosophy." In *Number to Sound: The Musical Way to the Scientific Revolution*, edited by Paolo Gozza, 135–150. Dordrecht: Kluwer Academic, 2000.

Grafton, Anthony. "Cardano's 'Proxeneta': Prudence for Professors." *Bruniana & Campanelliana* 7, no. 2 (2001): 363–80.

———. *Magic and Technology in Early Modern Europe*. Washington, DC: Smithsonian Institution Libraries, 2005.

Gramsci, Antonio. *The Modern Prince: And Other Writings*. Translated by Louis Marks. New York: International, 1957.

Gravit, Francis. "The Accademia degli Umoristi and Its French Relationships." *Papers of the Michigan Academy of Science, Arts, and Letters* 20 (1934): 505–21.

Green, Jerry. "Melody and Rhythm at Plato's *Symposium* 187d2." *Classical Philology* 110, no. 2 (April 2015): 152–58.

Greenblatt, Stephen. *Shakespearean Negotiations: The Circulation of Social Energy in Renaissance England*. Berkeley: University of California Press, 1988.

Greene, Robert. *Friar Bacon and Friar Bungay*. Lincoln: University of Nebraska Press, 1963.

Guarracino, Serena. "Voices from the South: Music, Castration, and the Displacement of the Eye." In *Anglo-Southern Relations: From Deculturation to Transculturation*, edited by Luigi Cazzato, 40–51. Lecce: Negroamaro: 2012.

Gurunluoglu, Raffi, and Aslin Gurunluoglu. "Paul of Aegina: Landmark in Surgical Progress." *World Journal of Surgery* 27, no. 1 (2003): 18–25.

Guthrie, Douglas. "The Achievement of Peter Lowe, and the Unity of Physician and Surgeon." *Scottish Medical Journal* 10, no. 7 (1965): 261–68.

Haar, James. "Notes on the 'Dialogo della Musica' of Antonfrancesco Doni." *Music & Letters* 47, no. 3 (1966): 198–224.

Habel, Dorothy Metzger. *"When All of Rome Was under Construction": The Building Process in Baroque Rome*. University Park: Pennsylvania State University Press, 2013.

Hadlock, Heather. *Mad Loves: Women and Music in Offenbach's* Les Contes d'Hoffmann. Princeton, NJ: Princeton University Press, 2000.

Halberstam, Jack [Judith]. *In a Queer Time and Place: Transgender Bodies, Subcultural Lives*. New York: New York University Press, 2005

Hall, Crystal. "Galileo, Poetry, and Patronage: Giulio Strozzi's *Venetia edificata* and the Place of Galileo in Seventeenth-Century Italian Poetry." *Renaissance Quarterly* 66, no. 4 (2014): 1296–1331.

Hammond, Frederick. "The Artistic Patronage of the Barberini and the Galileo Affair." In *Music and Science in the Age of Galileo*, edited by Victor Coelho, 67–89. Dordrecht: Springer, 1992.

———"Bernini and the 'Fiera di Farfa.'" In *Gianlorenzo Bernini: New Aspects of His Art and Thought. A Commemorative Volume*, edited by Irving Lavin, 115–78. University Park: Pennsylvania State University Press, 1985.

———. "A Celestial Siren and Her Music." *Early Music* 30, no. 4 (2002): 637–39.

———. "The Creation of a Roman Festival: Barberini Celebrations for Christina of Sweden." In *Life and the Arts in the Baroque Palaces of Rome: Ambiente Barocco*, edited by Stephanie Walker and Frederick Hammond, 53–69. New Haven, CT: Yale University Press, 1999.

———. "More on Music in Casa Barberini." *Studi musicali* 14 (1985): 235–61.

———. *Music and Spectacle in Baroque Rome: Barberini Patronage under Urban VIII*. New Haven, CT: Yale University Press, 1994.

———. "Orpheus in a New Key: The Barberini and the Rossi-Buti *L'Orfeo*." *Studi musicali* 25, no. 1/2 (1996): 103–25.

Hammond, Susan Lewis. "'Chi soffre speri' and the Influence of the *Commedia dell'arte* on the Development of Roman Opera." MA thesis, University of Arizona, 1995.

Hanafi, Zakiya. *The Monster in the Machine: Magic, Medicine, and the Marvelous in the Time of the Scientific Revolution*. Durham, NC: Duke University Press, 2000.

Hanning, Barbara Russano. "The Ending of *L'Orfeo*: Father, Son, and Rinuccini." *Journal of Seventeenth-Century Music* 9, no. 1 (2003). http://sscm-jscm.org/v9/no1/hanning.html.

———. "Powerless Spirit: Echo on the Musical Stage of the Late Renaissance." In *Word,*

*Image, and Song*, vol. 1, *Essays on Early Modern Italy*, edited by Rebecca Cypess, Beth L. Glixon, and Nathan Link, 193–218. Rochester, NY: University of Rochester Press, 2013.

Haraway, Donna. "A Manifesto for Cyborgs: Science, Technology, and Socialist Feminism in the 1980s." In *Feminism/Postmodernism*, edited by Linda J. Nicholson, 190–233. New York: Routledge, 1990.

———. *Simians, Cyborgs and Women: The Reinvention of Nature*. New York: Routledge, 1991.

———. *Staying with the Trouble: Making Kin in the Chthulucene*. Durham, NC: Duke University Press, 2016.

Harris, Ellen T. "Twentieth-Century Farinelli." *Musical Quarterly* 81, no. 2 (1997): 180–89.

Harris, Jonathan Gil, and Natasha Korda, eds. *Staged Properties in Early Modern English Drama*. Cambridge: Cambridge University Press, 2006.

Hayburn, Robert F. *Papal Legislation on Sacred Music, 95 A.D. to 1977 A.D.* Collegeville, MN: Liturgical Press, 1979.

Hayles, N. Katherine. *Unthought: The Power of the Cognitive Nonconscious*. Chicago: University of Chicago Press, 2017.

Haynes, Bruce. *A History of Performing Pitch: The Story of "A."* Lanham, MD: Scarecrow Press, 2002.

Heck, Thomas F. "The Operatic Christopher Columbus: Three Hundred Years of Musical Mythology." *Annali d'italianistica* 10 (1992): 236–78.

———. "Toward a Bibliography of Operas on Columbus: A Quincentennial Checklist." *Notes* 49, no. 2 (1992): 474–97.

Heidegger, Martin. *On the Way to Language*. Translated by Peter D. Hertz. New York: Harper Collins, 1971.

———. *The Question Concerning Technology and Other Essays*. New York: Harper, 1977.

Heller, Wendy. "Tacitus Incognito: Opera as History." *Journal of the American Musicological Society* 52, no. 1 (1999): 39–97.

Heriot, Angus. *The Castrati in Opera*. London: Secker and Warburg, 1956. Reprint, New York: Da Capo, 1974.

Hewett, Ivan. "Farinelli and the Guilty Appeal of the Castrato." *Telegraph*, October 12, 2015. http://www.telegraph.co.uk/music/classical-music/farinelli-and-the-guilty-appeal-of-the-castrato/.

Heyde, Herbert. "Todini's Golden Harpsichord Revisited." *Journal of the American Musical Instrument Society* 39 (2013): 5–61.

Hill, John Walter. *Roman Monody, Cantata, and Opera from the Circles around Cardinal Montalto*. Oxford: Clarendon Press, 1977.

Hinks, Peter P., ed. *David Walker's Appeal to the Coloured Citizens of the World*. University Park: Pennsylvania State University Press, 2000.

Hirt, Katherine. *When Machines Play Chopin: Musical Spirit and Automation in Nineteenth-Century German Literature*. New York: Walter de Gruyter, 2010.

Höfert, Almut, Matthew Mesley, and Serena Tolino, eds. *Celibate and Childless Men in Power: Ruling Eunuchs and Bishops in the Pre-modern World*. New York: Routledge, 2017.

Hoffman, Alice. "Luxury, Sex, and Music." *New York Times*. October 10, 1982. https://www.nytimes.com/1982/10/10/books/luxury-sex-and-music.html.

Holman, Peter. "*Col nobilissimo esercitio della vivuola*: Monteverdi's String Writing." *Early Music* 21, no. 4 (1993): 576–90.

Holsinger, Bruce W. *Music, Body, and Desire in Medieval Culture: Hildegard of Bingen to Chaucer*. Stanford, CA: Stanford University Press, 2001.

Hornbostel, Erich M. von, and Curt Sachs. "Classification of Musical Instruments." Translated by Anthony Baines and Klaus P. Wachsmann. *Galpin Society Journal* 14 (March 1961): 3–29.

Howard, Patricia. *The Modern Castrato: Gaetano Guadagni and the Coming of a New Operatic Age*. New York: Oxford University Press, 2014.

Huarte, Juan de la. *The Examination of Men's Wits*. Translated by Richard Carew. London, 1594.

———. *The Examination of Men's Wits*. Reprinted with introduction and notes by Rocío G. Sumillera. London: Modern Humanities Research Association, 2014.

Hunt, John Dixon. "Ovid in the Garden." *AA Files*, no. 3 (1983): 3–11.

Hunt, John M. "Carriages, Violence and Masculinity in Early Modern Rome." *I Tatti Studies in the Italian Renaissance* 17, no. 1 (Spring 2014): 175–96.

———. "Ritual Time and Popular Expectations of Papal Rule in Early Modern Rome." *Explorations in Renaissance Culture* 45, no. 1 (2019): 29–49.

———. *The Vacant See in Early Modern Rome: A Social History of the Papal Interregnum*. Boston: Brill, 2016.

Hunt, Lynn. "Introduction: Obscenity and the Origins of Modernity, 1500–1800." In *The Invention of Pornography, 1500–1800: Obscenity and the Origins of Modernity*, edited by Lynn Hunt, 9–45. New York: Zone Books, 1996.

Isoke, Saidah K. "Dream If You Can: A Mother and Daughter's Reflections on Prince, Self-Realization, and Black Womanhood." *Journal of African American Studies* 21, no. 3 (2017): 524–27.

Jacob, Giles. *Tractatus de Hermaphroditis: Or, A Treatise of Hermaphrodites*. London, 1718.

Jaffé, David. "The Barberini Circle: Some Exchanges between Peiresc, Rubens, and Their Contemporaries." *Journal of the History of Collections* 1, no. 2 (1989): 119–47.

Jarman, Freya. "Pitch Fever: The Castrato, the Tenor, and the Question of Masculinity in Nineteenth-Century Opera." In *Masculinity in Opera: Gender, History, and New Musicology*, edited by Philip Purvis, 51–66. New York: Routledge, 2013.

Jeanneret, Christine. "Gender Ambivalence and the Expression of Passions in the Performances of Early Roman Cantatas by Castrati and Female Singers." In *The Emotional Power of Music: Multidisciplinary Perspectives on Musical Arousal, Expression, and Social Control*, edited by Tom Cochrane, Bernardino Fantini, and Klaus R. Scherer, 85–101. Oxford: Oxford University Press, 2013.

Joncus, Berta. "One God, So Many Farinellis: Mythologising the Star Castrato." *Journal for Eighteenth-Century Studies* 28, no. 3 (2005): 437–96.

Judd, Cristle Collins. *Reading Renaissance Music Theory: Hearing with the Eyes*. Cambridge Studies in Music Theory and Analysis 14. Cambridge: Cambridge University Press, 2000.

Juvenal. *Satires*. Translated by Susanna Morton Braund. Loeb Classical Library 91. Cambridge, MA: Harvard University Press, 2004.

———. *The Sixteen Satires*. Edited and translated by Peter Green. Baltimore: Penguin, 1974.

Kaiser, Simone M., and Matteo Valleriani. "The Organ of the Villa d'Este in Tivoli and the Standards of Pneumatic Engineering in the Renaissance." In *Gardens, Knowledge and the Sciences in the Early Modern Period*, edited by Hubertus Fischer, Volker R. Remmert, and Joachim Wolschke-Bulmahn, 77–102. Cham, Switzerland: Springer International, 2016.

Kane, Brian. *Sound Unseen: Acousmatic Sound in Theory and Practice*. New York: Oxford University Press, 2014.

Kaplan, Edwin L., George I. Salti, Manuela Roncella, Noreen Fulton, and Mark Kadowaki.

"History of the Recurrent Laryngeal Nerve: From Galen to Lahey." *World Journal of Surgery* 33, no. 3 (2009): 386–93.

Katz, Mark. *Capturing Sound: How Technology Has Changed Music*. Berkeley: University of California Press, 2004.

Keeling, Kara. "Electric Feel: Transduction, Errantry, and the Refrain." *Cultural Studies* 28, no. 1 (2014): 49–83.

Kelly, Michael. *Reminiscences of Michael Kelly* [ . . . ]. Vol. 1. London: Colburn, 1826. Published as *Reminiscences* and edited by Roger Fiske. New York: Oxford University Press, 1975.

Kelly, Thomas Forrest. *First Nights: Five Musical Premieres*. New Haven, CT: Yale University Press, 2000.

Kemp, Martin, ed. *Leonardo on Painting: An Anthology of Writings by Leonardo da Vinci with a Selection of Documents Relating to His Career as an Artist*. Translated by Martin Kemp and Margaret Walker. New Haven, CT: Yale University Press, 1989.

Kendrick, Robert L. *Celestial Sirens: Nuns and Their Music in Early Modern Milan*. New York: Clarendon Press, 1996.

———. "Martyrdom in Seventeenth-Century Italian Music." In *From Rome to Eternity: Catholicism and the Arts in Italy, ca. 1550–1650*, edited by Pamela M. Jones and Thomas Worcester, 121–41. Leiden: Brill, 2002.

Kibre, Pearl. *Hippocrates Latinus: Repertorium of Hippocratic Writings in the Latin Middle Ages*. Rev. ed. New York: Fordham University Press, 1985.

King, Helen. *The One-Sex Body on Trial: The Classical and Early Modem Evidence*. Burlington, VT: Ashgate, 2013.

Kircher, Athanasius. *Mundus subterraneus in XII Libros digestus*. Bk. 1. Amsterdam: Johan Jansson, 1678.

———. *Musurgia universalis; Sive, ars magna consoni et dissoni in X. libros digesta*. 2 vols. Rome, 1650. Published as *Musurgia universalis* and edited by Ulf Scharlau. New York: Georg Olms, 1970.

———. *Obeliscus Pamphilius, hoc est, Interpretatio nova & hucusque intentata obelisci hieroglyphici quem non ita pridem ex veteri hippodromo Antonini Caracallae caesaris, in Agonale Forum transtulit, integritati restituit, & in Urbis aeternae ornamentum erexit Innocentius X Pont. Max*. Rome: Grignani, 1650.

———. *Phonurgia nova*. Campidonae: Dreherr, 1673. Reprint, New York: Broude Brothers, 1966.

Kirkendale, Warren. *Emilio de'Cavalieri "Gentiluomo Romano": His Life and Letters, His Role as Superintendent of All the Arts at the Medici Court, and His Musical Compositions*. Florence: L. S. Olschki, 2001.

Kittler, Friedrich. *Gramophone, Film, Typewriter*. Stanford, CA: Stanford University Press, 1999.

Klein, Robert, and Henri Zerner. *Italian Art 1500–1600: Sources and Documents*. Evanston, IL: Northwestern University Press, 1966.

Klestinec, Cynthia. *Theaters of Anatomy: Students, Teachers, and Traditions of Dissection in Renaissance Venice*. Baltimore: Johns Hopkins University Press, 2011.

———. "Translating Learned Surgery." *Journal of the History of Medicine and Allied Sciences* 72, no. 1 (2017): 34–50.

Krimmer, Elisabeth. "'Eviva il Coltello'? The Castrato Singer in Eighteenth-Century German Literature and Culture." *PMLA* 120, no. 5 (2005): 1543–59.

Kuefler, Mathew S. "Castration and Eunuchism in the Middle Ages." In *Handbook of Medie-*

*val Sexuality*, edited by Vern L. Bullough and James A. Brundage, 279–306. New York: Garland, 1996.

Landi, Stefano. *Il S. Alessio: dramma musicale Alessio*. Rome: Paolo Masotti, 1634.

Lang, Paul Henry. *George Frideric Handel*. New York: W. W. Norton, 1966.

———. "Performance Practice and the Voice." In *Musicology and Performance*, edited by Alfred Mann and George J. Buelow, 185–98. New Haven, CT: Yale University Press, 1997.

Laqueur, Thomas. *Making Sex: Body and Gender from the Greeks to Freud*. Cambridge, MA: Harvard University Press, 1992.

———. "Orgasm, Generation, and the Politics of Reproductive Biology." *Representations* 14 (Spring 1986): 1–41.

Larson, J. H. "The Conservation of a Marble Group of Neptune and Triton by Gian Lorenzo Bernini." Supplement, *Studies in Conservation* 31, no. S1 (1986): 22–26.

Lasso, Orlando di. *Libro di villanelle, moresche, et altre canzoni*. Paris, 1581.

Latour, Bruno. "An Attempt at a 'Compositionist Manifesto.'" *New Literary History* 41, no. 3 (Summer 2010): 471–90.

Lavin, Irving. "Bernini and the Art of Social Satire." *History of European Ideas* 4, no. 4 (1983): 365–420.

———. *Bernini and the Unity of the Visual Arts*. 2 vols. New York: Oxford University Press, 1980.

———. "High and Low before Their Time: Bernini and the Art of Social Satire." In *Modern Art and Popular Culture: Readings in High and Low*, edited by Kirk Varnedoe and Adam Gopnik, 19–50. New York: Museum of Modern Art, 1990.

Law, Hedy. "A Cannon-Shaped Man with an Amphibian Voice: Castrato and Disability in Eighteenth-Century France." In *The Oxford Handbook of Music and Disability Studies*, edited by Blake Howe, Stephanie Jensen-Moulton, Neil Lerner, and Joseph Straus, 329–44. New York: Oxford University Press, 2016.

Lazzaro, Claudia, and Ralph Lieberman. *The Italian Renaissance Garden: From the Conventions of Planting, Design, and Ornament to the Grand Gardens of Sixteenth-Century Central Italy*. New Haven, CT: Yale University Press, 1990.

Lee, Vernon [Violet Piaget]. *Althea: A Second Book of Dialogues on Aspirations and Duties*. London: Osgood, McIlvaine, 1894.

———. "An Eighteenth-Century Singer: An Imaginary Portrait." *Fortnightly Review* 50, no. 300 (December 1, 1891): 842–80.

———. *Hauntings and Other Fantastic Tales*. Edited by Catherine Maxwell and Patricia Pulham. Peterborough: Broadview Press, 2006.

———. *Limbo and Other Essays to Which Is Now Added Ariadne in Mantua*. New York: John Lane, 1908.

———. "Old Italian Gardens." In *In Praise of Old Gardens*, 29–60. Portland, ME: Thomas B. Mosher, 1912.

———. *Selected Letters of Vernon Lee 1856–1935*. Vol. 1, *1865–1885*. Edited by Amanda Gagel. New York: Routledge, 2017.

———. *Studies of the Eighteenth Century in Italy*. London: T. F. Unwin, 1887. Reprint, New York, Da Capo, 1978.

Le Guin, Elisabeth. *Boccherini's Body: An Essay in Carnal Musicology*. Berkeley: University of California Press, 2005.

Lehoux, Daryn. *What Did the Romans Know? An Inquiry into Science and Worldmaking*. Chicago: University of Chicago Press, 2012.

Leopold, Silke. *Monteverdi: Music in Transition*. Translated by Anne Smith. Oxford: Clarendon Press, 1991.

*Lettera di raguaglio scritta da N. N. ad un suo amico toccante la cavalcata solenne fatta per il ricevimento della Maestà della Reginadi Svetia in Roma*. Rome: Coligni, 1655.

Levenberg, Jeffrey. "Worth the Price of the 'Musurgia universalis': Athanasius Kircher on the Secret of the 'Metabolic Style.'" *Recercare* 28, no. 1/2 (2016): 43–88.

Lewis, Simon L., and Mark A. Maslin. "Defining the Anthropocene." *Nature* 519, no. 7542 (2015): 171–80.

Lightbown, Ronald W. "Nicolas Audebert and the Villa d'Este." *Journal of the Warburg and Courtauld Institutes* 27 (1964): 164–90.

Linfield, Eva. "Modulatory Techniques in Seventeenth-Century Music: Schütz, a Case in Point." *Music Analysis* 12, no. 2 (1993): 197–214.

Lionnet, Jean. "André Maugars: Risposta data a un curioso sul sentimento della musica dell'Italia." *Nuova rivista musicale italiana* 19, no. 3/4 (1985): 681–707.

———. "Les activités musicales de Flavio Chigi cardinal neveu d'Alexandre VII." *Studi Musicali* 9 (1980): 287–302.

Lipking, Lawrence. *What Galileo Saw: Imagining the Scientific Revolution*. Ithaca, NY: Cornell University Press, 2014.

Lomazzo, Giovanni Paolo. *Trattato dell'arte della pittura, scoltura, et architettura*. Milan: Paolo Gottardo Pontio, stampatore Regio. A instantia di Pietro Tini, 1585. Translated as *A Tracte Containing the Artes of Curious Paintinge, Carvinge, and Buildinge* by Richard Haydock. Oxford: Joseph Barnes, 1598.

Loughridge, Deirdre. "Technologies of the Invisible: Optical Instruments and Musical Romanticism." PhD diss., University of Pennsylvania, 2011.

Love, Heather K. *Feeling Backward: Loss and the Politics of Queer History*. Cambridge, MA: Harvard University Press, 2007.

Lowe, Peter. *The Whole Course of Chirurgerie* [ . . . ]. London, 1597.

Lucretius. *On the Nature of Things*. Translated by William Ellery Leonard. 6 bks. http://classics.mit.edu/Carus/nature_things.html.

Luger, Anton. "The Origin of Syphilis: Clinical and Epidemiologic Considerations on the Columbian Theory." *Sexually Transmitted Diseases* 20, no. 2 (1993): 110–17.

Lydiatt, Daniel D., and Gregory S. Bucher. "The Historical Latin and Etymology of Selected Anatomical Terms of the Larynx." *Clinical Anatomy* 23, no. 2 (2010): 131–44.

MacClintock, Carol. *The Bottegari Lutebook*. Wellesley, MA: Wellesley College, 1965.

———. "Giustiniani's 'Discorso Sopra la Musica.'" *Musica Disciplina* 15 (1961): 209–25.

———. *Readings in the History of Music in Performance*. Bloomington: Indiana University Press, 1979.

MacDougall, Elisabeth B. *Fountains, Statues, and Flowers: Studies in Italian Gardens of the Sixteenth and Seventeenth Centuries*. Washington, DC: Dumbarton Oaks Research Library and Collection, 1994.

Maloney, Gilles, and Raymond Savoie. *Cinq cent ans de bibliographie hippocratique: 1473–1982*. Quebec: Éditions du Sphinx, 1982.

Mamczarz, Irene. "La trattatistica dei Gesuiti e la pratica teatrale al Collegio Romano: Maciej Sarbiewski, Jean Dubreuil e Andrea Pozzo." In *I Gesuiti e i primordi del teatro barocco in Europa*, edited by Maria Chiabò and Federico Doglio, 349–87. Rome: Centro studi sul teatro medioevale e Rinascimentale, 1995.

Mancini, Giambattista. *Pensieri, e riflessioni pratiche sopra il canto figurato*. Vienna, 1774.

Manzini, Luigi. *Applausi festivi fatti a Roma per l'elezzione di Ferdinando III al Regno de' Romani dal Ser.mo Princ. Maurizio Card. Di Savoia, descritti al Ser.mo Francesco d'Este Duca di Modena da D. Luigi Manzini*. Rome: Facciotti, 1637.

Marcigliano, Alessandro. "The Development of the Fireworks Display and Its Contribution in Dramatic Art in Renaissance Ferrara." *Theatre Research International* 14 (1989):1–12.

Marder, Tod A. "Vitruvius and the Architectural Treatise in Early Modern Europe." In *Companions to the History of Architecture*, vol. 1, *Renaissance and Baroque Architecture*, edited by Alina Payne, 1–31. Hoboken, NJ: Wiley-Blackwell, 2017.

Marinella, Lucrezia. *The Nobility and Excellence of Women and the Defects and Vices of Men*. Edited and translated by Anne Dunhill. Chicago: University of Chicago Press, 1999.

———. *La nobiltà et l'eccellenza delle donne, co' difetti et mancamenti degli huomini*. Venice: Giovanni Battista Ciotti, 1601.

Marino, Giovanni Battista. *Adonis*. Translated by Harold Martin Priest. Ithaca, NY: Cornell University Press, 1967.

———. *Dicerie sacre e la strage degli innocenti*. Edited by Giovanni Pozzi. Turin: G. Einaudi, 1960.

———. *La galeria*. Venice, 1620.

———. *Le dicerie sacre*. Venice, 1674.

Marino, John. *Becoming Neapolitan: Citizen Culture in Baroque Naples*. Baltimore: Johns Hopkins University Press, 2011.

Marsh, David. "Latin Poetry." In *The Oxford Handbook of Neo-Latin*, edited by Stefan Tilg and Sarah Knight, 395–410. Oxford: Oxford University Press, 2015.

Martin, Craig. "Printed Medical Commentaries and Authenticity: The Case of 'De Alimento.'" *Journal of the Washington Academy of Sciences* 90, no. 4 (2004): 17–28.

Martin, Dale B. "Contradictions of Masculinity: Ascetic Inseminators and Menstruating Men in Greco-Roman Culture." In *Generation and Degeneration: Tropes of Reproduction in Literature and History from Antiquity through Early Modern Europe*, edited by Valeria Finucci and Kevin Brownlee, 81–108. Durham, NC: Duke University Press, 2001.

Martin, John Jeffries. "'Et nulle autre me faict plus proprement homme que cette cy': Michel de Montaigne's Embodied Masculinity." *European Review of History: Revue européenne d'histoire* 22, no. 4 (2015): 563–78.

Mascardi, Vitale. *Festa fatta in Roma alli 25 di Febraio MDCXXXI e data in luce da Vitale Mascardi*. Rome: Vitale Mascardi, 1635.

Mazzocchi, Domenico. *Dialoghi e sonetti*. Rome, 1638.

———. *La catena d'Adone*. Venice: Vicenti, 1626. Reprint, Bologna: Forni, 1969.

McCaffrey, Anne. *The Ship Who Sang*. New York: Ballantine Books, 1969.

McClary, Susan. *Desire and Pleasure in Seventeenth-Century Music*. Berkeley: University of California Press, 2012.

———. *Feminine Endings: Music, Gender, and Sexuality*. Minneapolis: University of Minnesota Press, 1991.

———. "Making Waves: Opening Keynote for the Twentieth Anniversary of the Feminist Theory and Music Conference." *Women and Music: A Journal of Gender and Culture* 16, no. 1 (2012): 86–96.

———. *Modal Subjectivities: Self-Fashioning in the Italian Madrigal*. Berkeley: University of California Press, 2004.

McGeary, Thomas. "Gendering Opera: Italian Opera as the Feminine Other in Britain, 1700–42." *Journal of Musicological Research* 14, no. 1 (1994): 17–34.

———. "Opera and British Nationalism, 1700–1711." *Revue LISA/LISA E-Journal* 4, no. 2 (2006): 5–19.

———. "'Warbling Eunuchs': Opera, Gender, and Sexuality on the London Stage, 1705–1742." *Restoration and Eighteenth-Century Theatre Research* 7, no. 1 (Summer 1992): 1–22.

McGrath, Aidan. *A Controversy Concerning Male Impotence*. Rome: Editrice Pontificia Università Gregoriana, 1988.

McLaren, Angus. *Impotence: A Cultural History*. Chicago: University of Chicago Press, 2007.

Meinardus, Otto. "The Upper Egyptian Practice of the Making of Eunuchs in the XVIIIth and XIXth Century." *Zeitschrift für Ethnologie* 94 (1969): 47–58.

Meli, Emilio. *La Fontana Panfilia comedia del Sig. D. Emilio Meli*. Rome: Francesco Moneta, 1652.

Melicow, Meyer M., and Stanford Pulrang. "Castrati Choir and Opera Singers." *Urology* 3, no. 5 (1974): 663–70.

Menke, Bettine. "'However One Calls into the Forest . . .': Echoes of Translation." Translated by Robert J. Kiss. In *Walter Benjamin and Romanticism*, edited by Beatrice Hanssen and Andrew Benjamin, 83–97. New York: Continuum, 2002.

Merrick, Jeffrey. "The Cardinal and the Queen: Sexual and Political Disorders in the Mazarinades." *French Historical Studies* 18, no. 3 (1994): 667–99.

Mersenne, Marin. *Harmonie universelle*. Paris: S. Cramoisy, 1636–1637. Edited by François Lesure. 3 vols. Paris: Centre national de la recherche scientifique, 1965.

Mignolo, Walter. *The Darker Side of the Renaissance: Literacy, Territoriality, and Colonization*. Ann Arbor: University of Michigan Press, 2003.

Moe, Nelson. *The View from Vesuvius: Italian Culture and the Southern Question*. Berkeley: University of California Press, 2002.

Monson, Craig A. "The Council of Trent Revisited." *Journal of the American Musicological Society* 55, no. 1 (2002): 1–37.

Montaigne, Michel de. *Complete Essays*. Translated by Donald M. Frame. Stanford, CA: Stanford University Press, 1958.

———. *The Journal of Montaigne's Travels in Italy*. Translated and edited by W. G. Waters. 3 vols. London: John Murray, 1903.

———. *Montaigne's Travel Journal*. Translated by Donald M. Frame. San Francisco: North Point Press, 1983.

———. *Oeuvres complètes*. Edited by Albert Thibaudet and Maurice Rat. Paris: Gallimard, 1962.

Montesquieu, Charles de. *The Spirit of the Laws*. Translated by Thomas Nugent. 2 vols. London: Printed for J. Nourse and P. Vaillant, 1750.

Monteverdi, Claudio. *La favola d'Orfeo rappresentata in musica il carnevale dell'anno MDCVII nell'Accademia degli Invaghiti di Mantova*. Mantua: Francesco Osanna, 1607.

———. *The Letters of Claudio Monteverdi*. Translated by Denis Stevens. Cambridge: Cambridge University Press, 1980.

———. *L'Orfeo: Favola in musica: SV 318*. Edited by Claudio Gallico. New York: Eulenburg, 2004.

———. *Madrigals, Book VIII (Madrigali guerrieri et amorosi)*. New York: Dover, 1991.

Moran, Neil. "Byzantine Castrati." *Plainsong & Medieval Music* 11, no. 2 (2002): 99–112.

Morelli, Arnaldo. *Il tempio armonico: Musica nell'Oratorio dei Filippini in Roma (1575–1705)*. Laaber: Laaber-Verlag, 1991.

Morrissey, Jake. *The Genius in the Design: Bernini, Borromini, and the Rivalry That Transformed Rome*. London: Duckworth, 2005.

Moseley, Roger. "Digital Analogies: The Keyboard as Field of Musical Play." *Journal of the American Musicological Society* 68, no. 1 (2015): 151–228.

Mundy, Rachel. *Animal Musicalities: Birds, Beasts and Evolutionary Listening*. Middletown, CT: Wesleyan University Press, 2018.

Murata, Margaret. "'Colpe mie venite a piangere': The Penitential Cantata in Baroque Rome." In *Listening to Early Modern Catholicism: Perspectives from Musicology*, edited by Daniele V. Filippi and Michael Noone, 204–32. Boston: Brill, 2017.

———. "Further Remarks on Pasqualini and the Music of MAP." *Analecta Musicologica* 19 (1979): 125–45.

———. "Image and Eloquence: Secular Song." In *The Cambridge History of Seventeenth-Century Music*, edited by Tim Carter and John Butt, 378–425. Cambridge: Cambridge University Press, 2005.

———. *Operas for the Papal Court 1631–1668*. Ann Arbor, MI: UMI Research Press, 1981.

———. "Pasqualini, Marc'Antonio." In *Dizionario biografico degli italiani*, vol. 81. Rome: Treccani, 2014. https://www.treccani.it/enciclopedia/marcantonio-pasqualini_(Dizionario-Biografico).

———. "Why the First Opera Given in Paris Wasn't Roman." *Cambridge Opera Journal* 7, no. 2 (1995): 87–105.

Muri, Allison. *The Enlightenment Cyborg: A History of Communications and Control in the Human Machine, 1660–1830*. Toronto: University of Toronto Press, 2007.

Nancy, Sarah. "The Singing Body in the Tragédie Lyrique of Seventeenth- and Eighteenth-Century France: Voice, Theatre, Speech, Pleasure." In *The Legacy of Opera: Reading Music Theatre as Experience and Performance*, edited by Michael Eigtved and Dominic Symonds, 65–78. Amsterdam: Rodopi, 2013.

Newcomb, Anthony. *The Madrigal at Ferrara, 1579–1597*. 2 vols. Princeton, NJ: Princeton University Press, 1980.

———. "Notions of Notation around 1600." *Il saggiatore musicale* 22, no. 1 (2015): 5–31.

Nocerino, Francesco. "Il tiorbino fra Napoli e Roma: Notizie e documenti su uno strumento di produzione cembalaria." *Recercare* 12 (2000): 95–109.

Nooter, Sarah. "The Prosthetic Voice in Ancient Greece." In *The Voice as Something More: Essays towards Materiality*, edited by Martha Feldman and Judith Zeitlin, 277–94. Chicago: University of Chicago Press, 2019.

Novak, David, and Matt Sakakeeny. Introduction to *Keywords in Sound*, edited by David Novak and Matt Sakakeeny, 1–11. Durham, NC: Duke University Press, 2015.

Nowlan, Edward H. "Double Vasectomy and Marital Impotence." *Theological Studies* 6, no. 3 (1945): 392–427.

Nussbaum, Felicity. *The Limits of the Human: Fictions of Anomaly, Race and Gender in the Long Eighteenth Century*. New York: Cambridge University Press, 2003.

Nussdorfer, Laurie. *Civic Politics in the Rome of Urban VIII*. Princeton, NJ: Princeton University Press, 2019.

———. "Masculine Hierarchies in Roman Ecclesiastical Households." *European Review of History: Revue européenne d'histoire* 22, no. 4 (2015): 620–42.

———. "Men at Home in Baroque Rome." *I Tatti Studies in the Italian Renaissance* 17, no. 1 (2014): 103–29.

———. "The Politics of Space in Early Modern Rome." *Memoirs of the American Academy in Rome* 42 (1997): 161–86.

———. "Print and Pageantry in Baroque Rome." *Sixteenth Century Journal* 29, no. 2 (1998): 439–64.

———. "The Vacant See: Ritual and Protest in Early Modern Rome." *Sixteenth Century Journal* (1987): 173–89.

O'Brien, Grant. "The Tiorbino: An Unrecognised Instrument Type Built by Harpsichord Makers with Possible Evidence for a Surviving Instrument." *Galpin Society Journal* 58 (2005): 184–208.

O'Donnell, Thomas J. "A Recent Decision of the Holy See Regarding Impotence and Sterility Implications for Medical Practice." *Linacre Quarterly* 45, no. 1 (1978): 15–21.

Olson, Julius E., and Edward Gaylord Bourne, eds. *The Northmen, Columbus, and Cabot, 985–1503: The Voyages of the Northmen*. New York: Charles Scribner's Sons, 1906.

Olson, Todd P. "'Long Live the Knife': Andrea Sacchi's Portrait of Marcantonio Pasqualini." *Art History* 27, no. 5 (2004): 697–722.

Ovid. *Metamorphoses*. Translated by Anthony S. Kline. http://ovid.lib.virginia.edu/trans/Metamorph8.htm#482327661.

———. *Metamorphoses*. Translated by Brookes More. Boston: Cornhill, 1922. http://data.perseus.org/citations/urn:cts:latinLit:phi0959.phi006.perseus-eng1:3.337–3.434.

———. *Metamorphoses*. Translated by Frank Justus Miller. 2 vols. Cambridge, MA: Harvard University Press, 1951.

Palisca, Claude V. *Studies in the History of Italian Music and Music Theory*. Oxford: Clarendon Press, 1994.

Panciatichi, Lorenzo. *Scritti vari di Lorenzo Panciatichi*. Edited by Cesare Guasti. Florence: Felice Le Monnier, 1856.

Panofsky, Erwin. "Galileo as a Critic of the Arts: Aesthetic Attitude and Scientific Thought." *Isis* 47, no. 1 (1956): 3–15.

Paré, Ambroise. *On Monsters and Marvels*. Translated and edited by Janis L. Pallister. Chicago: University of Chicago Press, 1982.

Parisi, Susan. "Acquiring Musicians and Instruments in the Early Baroque: Observations from Mantua." *Journal of Musicology* 14, no. 2 (1996): 117–50.

Park, Katharine. "Cadden, Laqueur, and the 'One-Sex Body.'" *Medieval Feminist Forum* 46, no. 1 (2010): 96–100.

———. "The Criminal and the Saintly Body: Autopsy and Dissection in Renaissance Italy." *Renaissance Quarterly* 47, no. 1 (1994): 1–33.

———. "The Rediscovery of the Clitoris." In *The Body in Parts: Fantasies of Corporeality in Early Modern Europe*, edited by David Hillman and Carolo Mazzio, 171–93. London: Routledge, 2013.

Park, Katharine, and Lorraine Daston. "Introduction: The Age of the New." In *The Cambridge History of Science*, vol. 3, *Early Modern Science*, edited by Katharine Park and Lorraine Daston, 1–18. Cambridge: Cambridge University Press, 2003.

Parker, Patricia. "Gender Ideology, Gender Change: The Case of Marie Germain." *Critical Inquiry* 19, no. 2 (1993): 337–64.

Pasqualigo, Zaccaria. *Decisiones morales iuxta principia theologica*. Verona, 1641.

Pastore, Alessandro, and Giovanni Rossi, eds. *Paolo Zacchia: Alle origini della medicina legale: 1584–1659*. Milan: Franco Angeli, 2008.

Pattenden, Miles. *Electing the Pope in Early Modern Italy, 1450–1700*. Oxford: Oxford University Press, 2017.

Paul of Aegina. *The Seven Books of Paulus Aegineta*. Translated by Francis Adams. 3 vols. London: Printed for the Sydenham Society, 1844–47.

Payne, Alina A. *The Architectural Treatise in the Italian Renaissance: Architectural Invention, Ornament, and Literary Culture*. Cambridge: Cambridge University Press, 1999.

Peraino, Judith A. "The Same, but Different: Sexuality and Musicology, Then and Now." *Journal of the American Musicological Society* 66, no. 3 (2013): 825–31.

Perella, Nicolas J. *The Critical Fortune of Battista Guarini's "Il Pastor Fido."* Florence: Olschki, 1973.

Peritz, Jessica Gabriel. *The Lyric Myth of Voice: Civilizing Song in Enlightenment Italy*. Oakland: University of California Press, 2022.

Peschel, Enid Rhodes, and Richard E. Peschel. "Medicine and Music: The Castrati in Opera." *Opera Quarterly* 4, no. 4 (1986): 21–38.

Phillips, John. "Sade." *French Studies* 68, no. 4 (2014): 526–33.

Pinch, Trevor, and Karin Bijsterveld. "New Keys in the World of Sound." In *The Oxford Handbook of Sound Studies*, edited by Trevor Pinch and Karin Bijsterveld, 3–35. New York: Oxford University Press, 2012.

Pirrotta, Nino. *Li due Orfei da Poliziano a Monteverdi*. Turin: Einaudi, 1975.

———. *Music and Culture in Italy from the Middle Ages to the Baroque: A Collection of Essays*. Cambridge, MA: Harvard University Press, 1984.

———. "Theater, Sets, and Music in Monteverdi's Operas." In *Music and Culture in Italy from the Middle Ages to the Baroque*, 254–71. Cambridge, MA: Harvard University Press, 2013.

Plato. *Phaedrus*. Translated by Robin Waterfield. New York: Oxford University Press, 2002.

———. *The Republic of Plato*. Translated by Francis MacDonald Cornford. London: Oxford University Press, 1945.

———. *Timaeus*. Translated by Donald J. Zeyl. Indianapolis, IN: Hackett, 2000.

Poizat, Michel. *The Angel's Cry: Beyond the Pleasure Principle in Opera*. Ithaca, NY: Cornell University Press, 1992.

Pollens, Stewart. "Michele Todini's Golden Harpsichord: Changing Perspectives." *Music in Art: International Journal for Music Iconography* 32, no. 1/2 (Spring–Fall 2007): 142–51.

Pomata, Gianna. *Contracting a Cure: Patients, Healers, and the Law in Early Modern Bologna*. Baltimore: Johns Hopkins University Press, 1998.

Powers, Wendy. "The Golden Harpsichord of Michele Todini (1616–1690)." Heilbrunn Timeline of Art History, New York, The Metropolitan Museum of Art, October 2003. https://www.metmuseum.org/toah/hd/todi/hd_todi.htm.

Pratt, Mary Louise. "Arts of the Contact Zone." *Profession* (1991): 33–40.

Prest, Julia. "In Chapel, on Stage, and in the Bedroom: French Responses to the Italian Castrato." *Seventeenth-Century French Studies* 32, no. 2 (2010): 152–64.

Priorato, Gualdo. *Historia della Sacra Real Maestà di Christina Alessandra Regina di Svetia*. Rome: Stamperia della Rev. Camera Apostolica, 1656.

———. *The History of the Sacred and Royal Majesty of Christina Alessandra Queen of Swedland* [ . . . ]. Translated by John Burbury. London, 1658.

Prodi, Paolo. *The Papal Prince: One Body and Two Souls; The Papal Monarchy in Early Modern Europe*. Translated by Susan Haskins. Cambridge: Cambridge University Press, 1987.

Przybyszewska-Jarmińska, Barbara. "The Careers of Italian Musicians Employed by the Polish Vasa Kings (1587–1668)." *Musicology Today* 6 (2009): 26–43.

Quaintance, Courtney. *Textual Masculinity and the Exchange of Women in Renaissance Venice*. Toronto: University of Toronto Press, 2015.

Quijano, Aníbal. "Coloniality of Power, Eurocentrism, and Latin America." *Nepantla: Views from the South* 1, no. 3 (2000): 533–80.

Quintilian. *Institutio oratoria*. Translated by H. E. Butler. 4 vols. New York: G. P. Putnam's Sons, 1921–1922.

Radano, Ronald, and Tejumola Olaniyan. "Introduction: Hearing Empire—Imperial Listening." In *Audible Empire: Music, Global Politics, Critique*, edited by Ronald Radano and Tejumola Olaniyan, 1–22. Durham, NC: Duke University Press, 2016.

Radcliffe, Edmonds G. *Redefining Ancient Orphism: A Study in Greek Religion*. Cambridge: Cambridge University Press, 2013.

Raguenet, François. *Les Monumens de Rome, ou Description des plus beaux ouvrages de peinture, de sculpture et d'architecture qui se voyent à Rome aux Environs*. Amsterdam, 1701.

———. *A comparison between the French and Italian musick and operas. Translated from the French; with some remarks. To which is added a critical discourse upon operas in England, and a means proposed for their improvement*. London: William Lewis, 1709.

Rancière, Jacques. *The Names of History: On the Poetics of Knowledge*. Translated by Hassan Melehy. Minneapolis: University of Minnesota Press, 1994.

Rau, Carl August. *Loreto Vittori: Beiträge zur historisch-kritischen Würdigung seines Lebens, Wirkens und Schaffens*. Munich, 1916.

Reardon, Colleen. *A Sociable Moment: Opera and Festive Culture in Baroque Siena*. New York: Oxford University Press, 2016.

Redondi, Pietro. *Galileo Heretic*. Translated by Raymond Rosenthal. Princeton, NJ: Princeton University Press, 1987.

Reed, Cory A. "Ludic Revelations in the Enchanted Head Episode in *Don Quijote* (II, 62)." *Cervantes: Bulletin of the Cervantes Society of America* 24, no. 1 (2004): 189–219.

Reeser, Todd W. "Montaigne on Gender." In *The Oxford Handbook of Montaigne*, edited by Philippe Desan, 562–80. New York: Oxford University Press, 2016.

Reeves, Eileen. "Galileo, Oracle: On the History of Early Modern Science." *I Tatti Studies in the Italian Renaissance* 18, no. 1 (2015): 7–22.

———. "Hearing Things: Organ Pipes, Trumpets, and Telescopes." *The Starry Messenger, Venice 1610: From Doubt to Astonishment*. Washington, DC: Library of Congress, 2013.

Reilly, Kara. *Automata and Mimesis on the Stage of Theatre History*. New York: Palgrave Macmillan, 2011.

Reiner, Stuart. "Preparations at Parma 1618, 1627–28." *Music Review* 25 (1964): 273–301.

*The Remarkable Trial of the Queen of Quavers and Her Associates, for Sorcery, Witchcraft, and Enchantments at the Assizes Held in the Moon, for the County of Gelding before the Rt. Hon. Sir Francis Lash, Lord Chief Baron of the Lunar Exchequer*. [London]: Printed for J. Bew, [1777?].

Richards, Annette. "Automatic Genius: Mozart and the Mechanical Sublime." *Music & Letters* 80, no. 3 (August 1999): 366–89.

Richter, Irma A., and Thereza Wells, eds. *Leonardo da Vinci: Notebooks*. New York: Oxford University Press, 2008.

Richter, Jean Paul, ed. *The Notebooks of Leonardo da Vinci*. New York: Dover, 1970.

Rietbergen, Peter. *Power and Religion in Baroque Rome: Barberini Cultural Policies*. Boston: Brill, 2006.

Ringrose, Kathryn M. *The Perfect Servant: Eunuchs and the Social Construction of Gender in Byzantium*. Chicago: University of Chicago Press, 2007.

Rinne, Katherine Wentworth. *The Waters of Rome: Aqueducts, Fountains, and the Birth of the Baroque City*. New Haven, CT: Yale University Press, 2010.

Riskin, Jessica. "The Defecating Duck, or, the Ambiguous Origins of Artificial Life." *Critical Inquiry* 29, no. 4 (2003): 599–633.

Roach, Joseph R. "Cavaliere Nicolini: London's First Opera Star." *Theatre Journal* 28, no. 2 (1976): 189–205.

Rognoni, Francesco. *Selva de varii passaggi secondo l'uso moderno, per cantare, & suonare con ogni sorte de stromenti*. Milan, 1620.

Romey, John. "Singing the Fronde: Placards, Street Songs, and Performed Politics." *Early Modern French Studies* 41, no. 1 (2019): 52–73.

Rosand, Ellen. "Barbara Strozzi, *virtuosissima cantatrice*: The Composer's Voice." *Journal of the American Musicological Society* 31, no. 2 (1978): 241–81.

———. "The Descending Tetrachord: An Emblem of Lament." *Musical Quarterly* 65, no. 3 (1979): 346–59. http://www.jstor.org/stable/741489.

———. *Opera in Seventeenth-Century Venice: The Creation of a Genre*. Berkeley: University of California Press, 1991.

Rose, Gloria. "Agazzari and the Improvising Orchestra." *Journal of the American Musicological Society* 18, no. 3 (Autumn 1965): 382–93.

Rospigliosi, Giulio. *Il palazzo incantato*. Edited by Danilo Romei. Revised July 23, 2012. http://www.nuovorinascimento.org/n-rinasc/testi/pdf/rospigliosi/palazzo.pdf.

Rosselli, John. "The Castrati as a Professional Group and a Social Phenomenon, 1550–1850." *Acta Musicologica* 60, no. 2 (1988):143–79.

Rossi, Gian Vittorio. *Iani Nicii Erythraei Dialogi Septendecim*. Cologne: Idocum Kalcovium, 1645.

———. *Iani Nicii Erithraei Pinacotheca Altera Imaginum, Illvstrium, doctrinae vel ingenii laude, Virorum, qui, auctore superstite, diem suum obierunt*. Amsterdam, 1645.

Rousseau, George. "Policing the Anus: Stuprum and Sodomy; According to Paolo Zacchia's Forensic Medicine." In *The Sciences of Homosexuality in Early Modern Europe*, edited by Kenneth Borris and Georges Rousseau, 75–91. London: Routledge, 2008.

Rousseau, Jean-Jacques. *A Dictionary of Music*. Translated by William Waring. London: Printed for J. French, 1775.

———. *Complete Dictionary of Music*. Translated by William Waring. London: Printed for J. French, 1779.

———. *Essay on the Origin of Languages and Writings Related to Music*. Translated and edited by John T. Scott. Hanover, NH: University Press of New England, 1998.

Rowland, Ingrid. "'Th' United Sense of th' Universe': Athanasius Kircher in Piazza Navona." *Memoirs of the American Academy in Rome* 46 (2001): 153–81.

Rowland, Ingrid D. "Athanasius Kircher's Guardian Angel." In *Conversations with Angels: Essays towards a History of Spiritual Communication, 1100–1700*, edited by Joad Raymond, 250–70. London: Palgrave Macmillan, 2011.

Russo, Piera. "L'Accademia degli Umoristi. Fondazione, strutture e leggi: Il primo decennio di attività." *Esperienze letterarie: Rivista trimestrale di critica e di cultura* 4, no. 4 (1979): 47–61.

Rustici, Craig M. *The Afterlife of Pope Joan: Deploying the Popess Legend in Early Modern England*. Ann Arbor: University of Michigan Press, 2006.

Rütten, Thomas. "Hippocrates and the Construction of 'Progress' in Sixteenth-and Seventeenth-Century Medicine." In *Reinventing Hippocrates: The History of Medicine in Context*, edited by David Cantor, 37–58. New York: Routledge, 2016.

Sabba da Castiglione. "Ricordo cerca gli ornamenti della casa." *Ricordi, overo ammaestramenti di Monsignor Sabba Castiglione Cavalier Gierosolimitano*. Bologna, 1546.

Sabbatini, Nicola. *Pratica di fabricar scene e machine ne' teatri*. Ravenna: Pietro de' Paoli and Gio. Battista Giovannelli, 1638.

Sakai, Tatsuo. "Historical Evolution of Anatomical Terminology from Ancient to Modern." *Anatomical Science International* 82, no. 2 (2007): 65–81.

Salatino, Kevin. *Incendiary Art: The Representation of Fireworks in Early Modern Europe*. Los Angeles: Getty Research Institute for the History of Art and the Humanities, 1998.

Sánchez, Tomás. *Disputationum de sancto matrimonii sacramento*. 3 vols. Antwerp: M. Nutii, 1620.

Sandys, George. *Ovid's Metamorphosis Englished, Mythologiz'd, and Represented in Figures: An Essay to the Translation of Virgil's Aeneis*. Oxford, 1632.

———. *A Relation of a Journey Begun An. Dom. 1610: 4 Bookes, Containing a Description of the Turkish Empire, of Aegypt, of the Holy Land, of the Remote Parts of Italy, and Ilands Adioyning*. London, 1615. Reprint. New York: Da Capo, 1973.

Sanger, Alice E. "Sensuality, Sacred Remains and Devotion in Baroque Rome." In *Sense and the Senses in Early Modern Art and Cultural Practice*, edited by Alice E. Sanger and Siv Tove Kulbrandstad Walker, 199–215. Burlington, VT: Ashgate, 2012.

San Juan, Rose Marie. *Rome: A City Out of Print*. St. Paul: University of Minnesota Press, 2001.

———. "The Transformation of the Río de la Plata and Bernini's Fountain of the Four Rivers in Rome." *Representations* 118, no. 1 (2012): 72–102.

Saslow, James M. *The Medici Wedding of 1589: Florentine Festival as Theatrum Mundi*. New Haven, CT: Yale University Press, 1996.

Savage, Roger, and Matteo Sansone. "'Il Corago' and the Staging of Early Opera: Four Chapters from an Anonymous Treatise circa 1630." *Early Music* 17, no. 4 (1989): 495–512.

Sawday, Jonathan. *Engines of the Imagination: Renaissance Culture and the Rise of the Machine*. London: Routledge, 2007.

Schaffer, Talia. *The Forgotten Female Aesthetes: Literary Culture in Late Victorian England*. Charlottesville: University Press of Virginia, 2000.

Schiltz, Katelijne. "Gioseffo Zarlino and the *Miserere* Tradition: A Ferrarese Connection?" *Early Music History* 27 (2008): 181–216.

Schlag, Pierre. "Cannibal Moves: An Essay on the Metamorphoses of the Legal Distinction." *Stanford Law Review* 40, no. 4 (1988): 929–72.

Schott, Gaspar. *Magiae universalis naturae et artis pars secunda: Acoustica in VII libros*. Bamberg, 1677.

———. *Mechanica hydraulico-pneumatica*. Würzburg: Pigrin, 1657.

———. *Mechanica hydraulico-pneumatica cum figuris Aneis, et privilegio sacrae Cesare Majestatis*. Frankfurt: Herbipoli, 1657.

Schrade, Leo. *Monteverdi, Creator of Modern Music*. New York: W. W. Norton, 1950.

Schraven, Minou. "Roma Theatrum Mundi: Festivals and Processions in the Ritual City." In *A Companion to Early Modern Rome, 1492–1692*, edited by Pamela M. Jones, Barbara Wisch, and Simon Ditchfield, 247–65. Boston: Brill, 2019.

Scott, John, and John Taylor, eds. "The Opera." *London Magazine and Review* 2 (August 1825): 516–20.

Seifert, Lewis C. "Masculinity and Satires of 'Sodomites' in France, 1660–1715." *Journal of Homosexuality* 41, no. 3/4 (2002): 37–52.

Selwyn, Jennifer D. *A Paradise Inhabited by Devils: The Jesuits' Civilizing Mission in Early Modern Naples*. London: Routledge, 2017.

Shakespeare, William. *Henry the Eighth*. The Norton Shakespeare, edited by Stephen Greenblatt et al. New York: W. W. Norton, 2016.

Sherr, Richard. "*Ex Concordia Discors*: Popes, Cardinals, Nationalities, Conflict, Deviance, and Irrational Behavior; A Crisis in the Papal Chapel in the Pontificate of Paul IV." *Journal of Musicology* 32, no. 4 (2015): 494–523.

———. "Gugliemo Gonzaga and the Castrati." *Renaissance Quarterly* 33, no. 1 (1980): 33–56.

———. "The 'Spanish Nation' in the Papal Chapel, 1492–1521." *Early Music* 20, no. 4 (1992): 601–10.

Sikes, Alan. "'Snip Snip Here, Snip Snip There, and a Couple of Tra La Las': The Castrato and the Nature of Sexual Difference." *Studies in Eighteenth-Century Culture* 34, no. 1 (2005): 197–229.

Simeoni, Gabriello. *La Vita et Metamorfoseo d'Ovidio*. Lyon: Per Giovanni di Tornes nella via Resina, 1559.

Simons, Patricia. "Manliness and the Visual Semiotics of Bodily Fluids in Early Modern Culture." *Journal of Medieval and Early Modern Studies* 39, no. 2 (2009): 331–73.

———. *The Sex of Men in Premodern Europe: A Cultural History*. New York: Cambridge University Press, 2011.

Siraisi, Nancy G. *Avicenna in Renaissance Italy: The Canon and Medical Teaching in Italian Universities after 1500*. Princeton, NJ: Princeton University Press, 2014.

———. *Medieval and Early Renaissance Medicine: An Introduction to Knowledge and Practice*. Chicago: University of Chicago Press, 1990.

Smith, Pamela H. *The Body of the Artisan: Art and Experience in the Scientific Revolution*. Chicago: University of Chicago Press, 2004.

Snorton, C. Riley. *Black on Both Sides: A Racial History of Trans Identity*. Minneapolis: University of Minnesota Press, 2017.

Sole, Giovanni. *Castrati e cicisbei: Ideologia e moda nel Settecento italiano*. Soveria Mannelli: Rubbettino, 2008.

Solerti, Angelo, ed. *Le origini del melodramma: Testimonianze dei contemporanei*. Turin: Fratelli Bocca, 1903.

Solie, Ruth A., ed. *Musicology and Difference: Gender and Sexuality in Music Scholarship*. Berkeley: University of California Press, 1993.

Sonnino, Eugenio. "Strutture familiari a Roma alla metà del Seicento." In *Popolazione e società a Roma dal medioevo all'età contemporanea*, edited by Eugenio Sonnino, 247–59. Rome: Il calamo, 1998.

Stapleton, Michael L. "Nashe and the Poetics of Obscenity: *The Choise of Valentines*." *Classical and Modern Literature* 12 (1991): 29–48.

St. Augustine. *City of God*. Translated by Gerald G. Walsh et al. New York: Doubleday, 1958.

———. *Expositions on the Book of Psalms*. Vol. 6, *Psalm CXXVI–CL*. Oxford: John Henry Parker, 1857.

Stendhal, *Life of Rossini*. Translated by Richard N. Coe. New York: Criterion Books, 1957.

Sterne, Jonathan. "Sonic Imaginations." In *The Sound Studies Reader*, edited by Jonathan Sterne, 1–17. New York: Routledge, 2012.

Sternfeld, Frederick W. "Orpheus, Ovid and Opera." *Journal of the Royal Musical Association* 113, no. 2 (1988): 172–202.

Stolzenberg, Daniel, ed. *The Great Art of Knowing: The Baroque Encyclopedia of Athanasius Kircher*. Stanford, CA: Stanford University Libraries, 2001.

Strunk, Oliver, ed. and trans. *Source Readings in Music History*. New York: W. W. Norton, 1950.

Strunk, Oliver, ed. *Source Readings in Music History: The Renaissance*. New York: W. W. Norton, 1965.

Strunk, Oliver, and Leo Treitler, eds. *Source Readings in Music History*. Rev. ed. New York: W. W. Norton, 1998.

Swafford, Jan. "Nature's Rejects: The Music of the Castrato." *Slate*. November 9, 2009. http://www.slate.com/articles/arts/music_box/2009/11/natures_rejects.html.

Sweet, Rosemary. *Cities and the Grand Tour: The British in Italy, c. 1690–1820*. New York: Cambridge University Press, 2012.

Taegio, Bartolomeo. *La villa*. Milan: Francesco Moscheni, 1559.

Tamburini, Elena. "Dietro la scena: Comici, cantanti e letterati nell'Accademia romana degli Umoristi." *Studi secenteschi* 50 (2009): 89–112.

Taruskin, Richard, and Piero Weiss, eds. *Music in the Western World*. New York: Schirmer, 1984.

Tchikine, Anatole. "Giochi d'acqua: Water Effects in Renaissance and Baroque Italy." *Studies in the History of Gardens & Designed Landscapes* 30, no. 1 (2010): 57–76.

Teets, Anthony. "Singing Things: The Castrato in Vernon Lee's Biography of a 'Culture-Ghost.'" *The Sibyl: A Journal of Vernon Lee Studies*. http://thesibylblog.com/singing-things-the-castrato-in-vernon-lees-biography-of-a-culture-ghost-by-anthony-teets/.

Teresa of Avila. *The Collected Works of St. Teresa of Avila*. Translated by Kieran Kavanaugh and Otilio Rodriguez. 2nd ed. 3 vols. Washington, DC: ICS, 1987.

Testi, Fulvio. *Lettere*. Edited by Maria Luisa Doglio. 3 vols. Bari: Giuseppe Lateranza e Figli, 1967.

Theweleit, Klaus. "Monteverdi's *L'Orfeo*: The Technology of Reconstruction." In *Opera through Other Eyes*, edited by David J. Levin, 146–76. Stanford, CA: Stanford University Press, 1994.

Tick, Judith. "Reflections on the Twenty-Fifth Anniversary of Women Making Music." *Women and Music: A Journal of Gender and Culture* 16, no. 1 (2012): 113–32.

Todini, Michele. *Dichiaratione della galleria armonica*. Rome: F. Tizzoni, 1676. Edited by Patrizio Barbieri. Lucca: Libreria musicale italiana, 1988.

Tognotti, Eugenia. "The Rise and Fall of Syphilis in Renaissance Europe." *Journal of Medical Humanities* 30, no. 2 (2009): 99–113.

Tomlinson, Gary. "Ideologies of Aztec Song." *Journal of the American Musicological Society* 48, no. 3 (1995): 343–79.

———. "Montaigne's Cannibals' Songs." *Repercussions* 7/8 (1999/2000): 209–35.

———. *Monteverdi and the End of the Renaissance*. Berkeley: University of California Press, 1987.

———. *Music in Renaissance Magic: Toward a Historiography of Others*. Chicago: University of Chicago Press, 1993.

———. "Musicology, Anthropology, History." *Il saggiatore musicale* 8, no. 1 (2001): 21–37.

———. *The Singing of the New World: Indigenous Voice in the Era of European Contact*. Cambridge: Cambridge University Press, 2007.

———. "Vico's Songs: Detours at the Origins of (Ethno)Musicology." *Musical Quarterly* 83, no. 3 (1999): 344–77.

Tougher, Shaun Fitzroy, ed. *Eunuchs in Antiquity and Beyond*. London: Classical Press of Wales; Duckworth, 2002.

———. "Eunuchs in the East, Men in the West? Dis/unity, Gender and Orientalism in the Fourth Century." In *East and West in the Roman Empire of the Fourth Century: An End to Unity?*, edited by Roald Dijkstra, Sanne van Poppel, and Daniëlle Slootjes, 147–63. Boston: Brill, 2015.

———. "Two Views on the Gender Identity of Byzantine Eunuchs." In *Changing Sex and Bending Gender*, edited by Alison Shaw and Shirley Ardener, 60–73. New York: Berghahn, 2005.

Tracy, Larissa, ed. *Castration and Culture in the Middle Ages*. London: D. S. Brewer, 2013.

Traub, Valerie. *Thinking Sex with the Early Moderns*. Philadelphia: University of Pennsylvania Press, 2016.

Treadwell, Nina. *Music and Wonder at the Medici Court: The 1589 Interludes for* La pellegrina. Bloomington: Indiana University Press, 2008.

Tresch, John. "The Machine Awakens: The Science and Politics of the Fantastic Automaton." *French Historical Studies* 34, no. 1 (2011): 87–123.

Tresch, John, and Emily I. Dolan. "Toward a New Organology: Instruments of Music and Science." *Osiris* 28, no. 1 (2013): 278–98.

Tronchin, Lamberto. "The 'Phonurgia Nova' of Athanasius Kircher: The Marvellous Sound World of 17th Century." *Proceedings of Meetings on Acoustics* 4 (2008): 1–9. https://doi.org/10.1121/1.2992053.

Truitt, Elly Rachel. *Medieval Robots: Mechanism, Magic, Nature, and Art*. Philadelphia: University of Pennsylvania Press, 2015.

Uberti, Grazioso. *Contrasto musico: Opera dilettevole*. Rome: Lodovico Grignani, 1630. Facsimile edition. Lucca: LIM, 1991.

Valleriani, Matteo. "Galileo's Abandoned Project on Acoustic Instruments at the Medici Court." *History of Science* 50, no. 1 (2012): 1–31.

van Orden, Kate. *Music, Authorship, and the Book in the First Century of Print*. Berkeley: University of California Press, 2013.

*Vera Relatione del viaggio fatto dalla Maestà della regina di Svetia per tutto lo Stato Ecclesiastico, del suo ricevimento et ingresso nell'alma città di Roma il dì 20 di dicembre 1655*. Rome: Francesco Cavalli, 1655.

Vesalius, Andreas. "Andreas Vesalius on the Larynx and Hyoid Bone: An Annotated Translation from the 1543 and 1555 editions of *De humani corporis fabrica*." Translated by Daniel H. Garrison and Malcolm H. Hast. *Medical History* 37, no. 1 (1993): 3–36.

———. *Observatonum anatomicarum Gabrielis Falloppi examen*. Venice, 1564.

———. *On the Fabric of the Human Body: Book VI, the Heart and Associated Organs. Book VII, the Brain: A Translation of* De humani corporis fabrica libri septem. Translated by William Richardson and John Carman. Novato, CA: Norman, 2009.

Vespucci, Amerigo. *The Letters of Amerigo Vespucci and Other Documents Illustrative of His Career*. Translated by Clements R. Markham. London: Printed for the Haylukt Society, 1894.

Vignoli, Maria Portia. *L'Obelisco di Piazza Navona*. Rome: Francesco Moneta, 1651.

Villoslada, R. Garcia. *Storia del Collegio Romano dal suo inizio alla soppressione della compagnia di Gesù*. Rome: Typis Pontificiae Universitatis Gregorianae, 1954.

Virgil. *The Aeneid*. Translated by Robert Fitzgerald. New York: Vintage Books, 1983.

———. *Eclogues, Georgics, Aeneid*. Translated by H. R. Fairclough. Loeb Classical Library 63, 64. Cambridge, MA. Harvard University Press. 1916. http://www.theoi.com/Text/VirgilEclogues.html.

Vittori, Loreto. *La Galatea*. Edited by Thomas D. Dunn. Middleton, WI: A-R Editions, 2002.

Voskuhl, Adelheid. *Androids in the Enlightenment: Mechanics, Artisans, and Cultures of the Self*. Chicago: University of Chicago Press, 2013.

Waddell, Mark A. "Magic and Artifice in the Collection of Athanasius Kircher." *Endeavour* 34, no. 1 (2010): 30–34.

Waquet, Françoise. *Le modèle français et l'Italie savante: Conscience de soi et perception de l'autre dans la République des Lettres (1660–1750)*. Rome: École Française de Rome, 1989.

Warwick, Genevieve. *Bernini: Art as Theatre*. New Haven, CT: Yale University Press, 2012.

———. "Ritual Form and Urban Space in Early Modern Rome." In *Late Medieval and Early Modern Ritual: Studies in Italian Urban Culture*, edited by Samuel Cohn Jr., Marcello Fantoni, Franco Franceschi, and Fabrizio Ricciardelli, 297–328. Turnhout: Brepols, 2013.

Weaver, Andrew H. *Sacred Music as Public Image for Holy Roman Emperor Ferdinand III: Representing the Counter-Reformation Monarch at the End of the Thirty Years' War*. London: Routledge, 2012.

Weber, Alison. *Teresa of Avila and the Rhetoric of Femininity*. Princeton, NJ: Princeton University Press, 1990.

Weitz, Eric D. *Weimar Germany: Promise and Tragedy*. Princeton, NJ: Princeton University Press, 2007.

Werett, Simon. *Fireworks: Pyrotechnic Arts and Sciences in European History*. Chicago: University of Chicago Press, 2010.

———. "Watching the Fireworks: Early Modern Observation of Natural and Artificial Spectacles." *Science in Context* 24, no. 2 (2011): 167–82.

Whenham, John. *Claudio Monteverdi: Orfeo*. Cambridge: Cambridge University Press, 1986.

———. "The Gonzagas Visit Venice." *Early Music* 21, no. 4 (1993): 525–38.

Wilbourne, Emily. "The Queer History of the Castrato." In *The Oxford Handbook of Music and Queerness*, edited by Fred Everett Maus and Sheila Whiteley. New York: Oxford University Press, 2018.

———. "A Question of Character: Artemisia Gentileschi and Virginia Ramponi Andreini." *Italian Studies* 71, no. 3 (2016): 335–55.

Wilkins, Ann Thomas. "Bernini and Ovid: Expanding the Concept of Metamorphosis." *International Journal of the Classical Tradition* 6, no. 3 (1999): 383–408.

Wilson, Blake. *Singing to the Lyre in Renaissance Italy: Memory, Performance, and Oral Poetry*. Cambridge: Cambridge University Press, 2019.

Wolfe, Cary. *Animal Rites: American Culture, the Discourse of Species, and Posthumanist Theory*. Chicago: University of Chicago Press, 2003.

———. *What Is Posthumanism?* Minneapolis: University of Minnesota Press, 2010.

Wolfe, Jessica. *Humanism, Machinery, and Renaissance Literature*. Cambridge: Cambridge University Press, 2004.

Wynter, Sylvia. "1492: A New World View." In *Race, Discourse, and the Origin of the Americas: A New World View*, edited by Vera Lawrence Hyatt and Rex Nettleford, 5–57. Washington, DC: Smithsonian Institution Press, 1995.

Yates, Peter. *Twentieth Century Music: Its Evolution from the End of the Harmonic Era into the Present Era of Sound.* [New York?]: Minerva Press, 1967.

Zacchia, Paolo. *Quaestiones medico-legales.* 4th ed. Edited by Johannes Daniel Horstius. 3 vols. Venice, 1789.

Zacconi, Lodovico. *Prattica di musica.* Bologna: Forni, 1592.

———. "Prattica di Musica (1596): Part 1, Book 1, Chapter 66." Translated by Sion M. Honea. Brisch Center for Historical Performance, Historical Translation Series, University of Central Oklahoma. https://www.uco.edu/cfad/files/music/zacconi-prattica.pdf.

Zanatta, Alberto, Fabio Zampieri, Giuliano Scattolin, and Maurizio Rippa Bonati. "Occupational Markers and Pathology of the Castrato Singer Gaspare Pacchierotti (1740–1821)." *Scientific Reports* 6, no. 28463 (2016): 1–9. https://doi.org/10.1038/srep28463.

Zappi, Giovanni Maria. *Annali e memorie di Tivoli.* Edited by Vincenzo Pacifici. Tivoli: n.p., 1920.

Zarlino, Gioseffo. *Sopplimenti musicali.* Venice: Francesco de Franceschi, 1588.

Zorach, Rebecca E. "The Matter of Italy: Sodomy and the Scandal of Style in Sixteenth-Century France." *Journal of Medieval and Early Modern Studies* 28, no. 3 (1998): 581–609.

# INDEX

Page numbers in italics refer to illustrations.